Coastal and Ocean Engineering Practice

Series on Coastal and Ocean Engineering Practice

Series Editor: Young C Kim
(California State University, USA)

Vol. 1: Coastal and Ocean Engineering Practice
edited by Young C Kim

Series on Coastal and Ocean Engineering Practice – Vol. 1

Coastal and Ocean Engineering Practice

Edited by

Young C Kim

California State University, Los Angeles, USA

World Scientific

NEW JERSEY · LONDON · SINGAPORE · BEIJING · SHANGHAI · HONG KONG · TAIPEI · CHENNAI

Published by

World Scientific Publishing Co. Pte. Ltd.

5 Toh Tuck Link, Singapore 596224

USA office: 27 Warren Street, Suite 401-402, Hackensack, NJ 07601

UK office: 57 Shelton Street, Covent Garden, London WC2H 9HE

British Library Cataloguing-in-Publication Data
A catalogue record for this book is available from the British Library.

Series on Coastal and Ocean Engineering Practice — Vol. 1
COASTAL AND OCEAN ENGINEERING PRACTICE

Copyright © 2012 by World Scientific Publishing Co. Pte. Ltd.

Desk Editor: Tjan Kwang Wei

ISBN-13 978-981-4360-56-2
ISBN-10 981-4360-56-2

Typeset by Stallion Press
Email: enquiries@stallionpress.com

Printed in Singapore.

Contents

Preface   vii

The Editor   ix

Contributors   xi

1. Impact of the Delta Works on the Recent Developments
    in Coastal Engineering   1
    Krystian W. Pilarczyk

2. Coastal Structures in International Perspective   39
    Krystian W. Pilarczyk

3. Coastal Structures: Action from Waves and Ice   95
    Alf Tørum

4. Kaumālapa'u Harbor: Design and Construction
    Challenges of an Exposed Deepwater Breakwater   253
    Scott P. Sullivan

5. Waterfront Developments in Harmony with Nature   287
    Karsten Mangor, Ida Brøker, Peter Rand, and Dan Hasløv

6. Risk-Based Channel Depth Design Using Cadet   319
    Michael J. Briggs, Andrew L. Silver and Paul J. Kopp

Preface

The first Specialty Conference on Coastal Engineering Practice was held in Long Beach, California, in May 1992 under the leadership of Steven A. Hughes of the U.S. Army Engineer Research and Development Center. In his preface, Dr. Hughes stated, "Successful engineering of coastal projects relies upon practical experience in combination with the tools and technical knowledge available to the engineer. In many situations, the physical processes taking place at a project site are too complex for theoretical description, and this puts a greater burden on the coastal engineer to exercise sound judgment when working on coastal projects." Thus the practical experience and sound judgment are the keys to the successful completion of coastal and ocean engineering projects.

The second Specialty Conference on Coastal Engineering practice is scheduled to be held at San Diego, California, in 2011 under the leadership of eminent coastal engineer, Orville T. Magoon of Coastal Zone Foundation.

Because coastal engineering experience constitutes an important cornerstone of the profession, both conferences were intended to promote a forum for the exchange of views and information between practicing coastal engineers.

As with the conferences, this book series will focus on the latest technology applied in design and construction, effective engineering methodology, unique projects and problems, design and construction challenges, and other lessons learned. In addition, unique practice in planning, design, construction, maintenance, and performance of coastal and ocean engineering projects will be explored.

This book series will focus on the practical aspects of coastal and ocean engineering practice. This series will consist of books, monographs, and review volumes, and is expected to produce one title a year.

Finally, the author wishes to express his deep appreciation to Ms. Kimberly Chua of World Scientific Publishing Company who gave him invaluable support and encouragement from the inception of this series to its realization.

Young C. Kim
Los Angeles, California
May 2010

The Editor

Young C. Kim, Ph.D., is currently a Professor Emeritus of Civil Engineering at California State University, Los Angeles. The other academic positions he held include a Visiting Scholar of Coastal Engineering at the University of California, Berkeley (1971); a NATO Senior Fellow in Science at the Delft University of Technology in the Netherlands (1975); and a Visiting Scientist at the Osaka City University for the National Science Foundations' U.S.–Japan Cooperative Science Program (1976). For more than a decade, he served as Chair of the Department of Civil Engineering (1993–2005) and was Associate Dean of Engineering in 1978. For his dedicated teaching and outstanding professional activities, he was awarded the university-wide Outstanding Professor Award in 1994.

Dr. Kim was a consultant to the U.S. Naval Civil Engineering Laboratory in Port Hueneme and became a resident consultant to the Science Engineering Associates where he investigated wave forces on the Howard-Doris platform structure, now being placed in Ninian Field, North Sea.

Dr. Kim is the past Chair of the Executive Committee of the Waterway, Port, Coastal and Ocean Division of the American Society of Civil Engineering (ASCE). Recently, he served as Chair of the Nominating Committee of the International Association of Hydraulic Engineering and Research (IAHR). Since 1998, he served on the International Board of Directors of the Pacific Congress on Marine Science and Technology (PACON). He is the past President of PACON. Dr. Kim has been involved in organizing 10 national and international conferences, has authored three books, and has published 52 technical papers in various engineering journals. Recently, he served as an editor for the *Handbook of Coastal and Ocean Engineering*, which was published by the World Scientific Publishing Company in 2009.

Contributors

Michael J. Briggs
Research Hydraulic Engineer
Coastal and Hydraulic Laboratory
U.S. Army Engineer Research and Development Center
Vicksburg, Mississippi
Michael.J.Briggs@usace.army.mil

Ida Broker
Head, Coastal and Estuarine Dynamics
DHI
Horsholm, Denmark
ibh@dhigroup.com

Dan Haslov
Architect and Director
Haslov & Kjaarsgaard I/S. Architects and Planners M.A.A.
Copenhagen, Denmark
dbh@hogk.dk

Paul J. Kopp
Naval Architect
Seakeeping Division
Naval Surface Warfare Center
West Bethesda, Maryland
Paul.Kopp@navy.mil

Karsten Mangor
Chief Engineer Shoreline Management
DHI
Horsholm, Denmark
km@dhigroup.com

Krystian W. Pilarczyk
(Formerly) Rijkswaterstaat
Hydraulic Engineering Institute
Delft, The Netherlands
Hydropil
Zoetermeer, The Netherlands
krystian.pilarczyk@gmail.com

Peter Rand
Senior Biologist
Ecology and Aquaculture
DHI
Horsholm, Denmark
prd@dhigroup.com

Andrew L. Silver
Senior Seakeeping Specialist
Seakeeping Division
Naval Surface Warfare Center
West Bethesda, Maryland
andrew.silver@navy.mil

Scott P. Sullivan
Vice President
Sea Engineering, Inc.
Waimanalo, Hawaii
ssullivan@seaengineering.com

Alf Torum
Professor Emeritus
Department of Civil and Transport Engineering
Norwegian University of Science and Technology
Trondheim, Norway
alf.torum@ntnu.no

Chapter 1

Impact of the Delta Works on the Recent Developments in Coastal Engineering

Krystian W. Pilarczyk

Former Rijkswaterstaat, Hydraulic Engineering Institute,
Delft, The Netherlands
krystian.pilarczyk@gmail.com

A disaster in 1953 was the turning point in the Dutch policy on flood protection. The Delta Works, which followed this disaster, contributed significantly to the recent worldwide developments in hydraulic and coastal engineering. A brief overview is presented of some important items specifically related to closing barriers, closure techniques, erosion, scour, and protection. More detailed information can be found in the references.

1.1. Introduction

The Netherlands is situated on the delta of three of Europe's main rivers: the Rhine, the Meuse, and the Scheldt. As a result of this, the country has been able to develop into an important, densely populated nation. But living in the Netherlands is not without risks. Large parts of the Netherlands are below mean sea level and water levels, which may rise on the rivers Rhine and Meuse (Fig. 1.1). High water levels due to storm surges on the North Sea or due to high discharges from the rivers are a serious threat to the low-lying part of the Netherlands. A total of about 3,000 km of primary flood protection structures protect areas that are vital for the existence of the Dutch nation. Flood protection measures have to provide sufficient safety to the large number of inhabitants and the ever-increasing investments in the Netherlands. Construction, management, and maintenance of flood protection structures are essential conditions for the population and further development of the country.

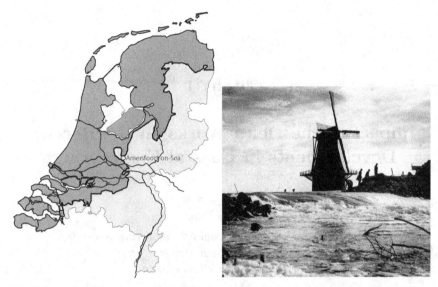

The Netherlands without flood protection (black area flooded by sea) and flood 1953 (Zeeland)

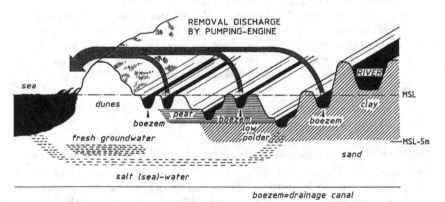

Principles of drainage system in the Netherlands

Fig. 1.1.   Characteristics features of the netherlands.

History shows that flooding disasters nearly always resulted in actions to improve the situation by raising dikes or increasing the discharge capacity of the rivers. The disastrous flood of 1953 in the Netherlands marks the start of a national reinforcement of the flood protection structures. The recent river floods of 1993 and 1995 did accelerate the final stages of this reinforcement program. History also shows that neglect is the overture for the next flooding disaster. In an attempt to improve on this historical experience, the safety of the flood protection structures in the Netherlands will be assessed regularly. Maintaining the strength of the dikes at levels according to the legally prescribed safety standards is the main goal of this safety assessment.

1.2. History

To understand the historical development of the protection by dikes in the Netherlands, it is essential to know the aspect of the gradual land subsidence in combination with rise of the sea level with respect to the land and also the decreasing deposits of soil in the North Sea and the rivers (Fig. 1.2).[1]

Mechanization and industrialization have led to improved drainage. Making use of these techniques, even land under water could be recovered (i.e., some lakes and Zuyder Sea). At the same time, however, the lowering of the ground level accelerated because of which the effects of possible flooding only increased. Strengthening flood defences was seen in the

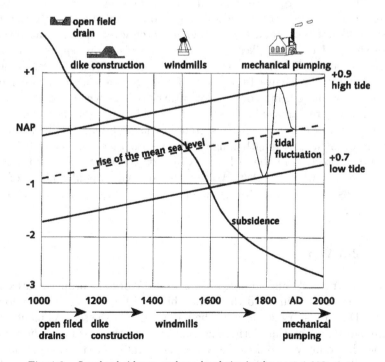

Fig. 1.2.   Land subsidence and sea-level rise in the past 1,000 years.

past as a means of effectively addressing this threat. In this way, from the middle of the 13th century until today, about 550,000 ha of land in total was reclaimed. However, during the course of centuries, much of the previously reclaimed land was lost by attack of the sea mainly due to storm surges, which many times caused destruction of the dikes. Another phenomenon was the occurrence of landslides along tidal channels (most dikes were situated on loose soils), thus causing disappearance of dikes. During high storm surges, the sea also eroded this land. Nevertheless, every time the spirit of the people to push back the sea succeeded. Most of the lost land was reclaimed again, despite the ever-occurring storm surges.

However, the continuous land subsidence and increased sea level rise have serious implications for the safety of the land protected by dikes, dunes, and other defence structures along the coast and the lower parts of the main rivers. Sea level rise increases the risks of overtopping and the ultimate collapse of these structures during storms.

The history of the Netherlands is marked by storm-surge disasters. The most recent disaster took place on February 1, 1953 when a north-westerly storm struck the south-western part of the Netherlands (Delta area). The storm-surge level reached 3 to 3.5 m above normal high water and exceeded design storm-surge levels by about 0.5 m at some places. Some dikes could not withstand these surging water levels, so that at several hundreds of places, the dikes were damaged and/or broken, over a total length of 190 km. Through nearly 90 breaches, 150,000 ha of polder land was inundated. This caused the death of 1,835 people and 100,000 persons had to be evacuated; moreover, a lot of livestock drowned and thousands of buildings were damaged or destroyed. This disaster gave a new impulse to improve the whole sea defence system in the Netherlands. The resulting Delta Plan included strengthening of the existing dikes and shortening the length of protection in the Delta area by closing off estuaries and a tidal river. Together with the Delta Works, the low-lying "polder land" of the Netherlands has reached already a relative high degree of safety against storm surges.[1]

1.3. Delta Works

To indicate the role and importance of Delta Works in proper context, it is necessary to keep in mind the vulnerability of the Netherlands to floods and the Dutch history concerning the battle against the sea. As mentioned before, over half of the Netherlands lies below sea level. Just how vulnerable the country is to flooding was demonstrated by storm-surge on the night of February 1, 1953. The results were catastrophic as many dikes were breached. Flooding caused by storm surges was nothing new to the Netherlands, but this time the nation was stunned by the extent of a disaster unparalleled for centuries. It was the Netherlands' worst disaster in 300 years. The hardest hit areas were the province of Zeeland, the southern part of South Holland and the western part of North Brabant. The bewilderment and shock felt by people in the rest of the Netherlands when they learnt of the extent of the flooding soon gave way to determination, and great efforts were made to reseal the breached dikes. The last breach, near Ouwerkerk on Schouwen-Duiveland, was resealed at the beginning of November 1953.

Rarely have the people of the Netherlands been as united as when they decided that such a catastrophe should never happen again. We may say that Disaster 1953 was a turning point in Dutch approach to flood protection and flood management. The outcome of this determination was the Delta Project.

1.3.1. *The Delta Project*

On February 21, 1953, the Delta Commission was created, headed by the director-general of the Department of Waterways and Public Works. Its aim was to draw up a plan to achieve the following two goals:

1. Drain the areas that are flooded regularly because of high water levels and protect them from water.
2. Protect the land from getting brackish.

The body of measures proposed by this committee forms the Delta Plan. The aim of the Plan was to enhance safety by radically reducing the length of and reinforcing the coastline. These measures are laid down in the Delta Act of 1958.

The Delta Plan Project's principal goal was to improve the safety of the southwest Netherlands by considerably shortening and reinforcing the coastline. It was decided that dams should be constructed across inlets and estuaries, considerably reducing the possibility of the sea surging into the land again. Freshwater lakes would be formed behind them. Roads along the dams would improve access to the islands of Zeeland and South Holland.

Dams could not be constructed across the New Waterway or the Western Scheldt, as these important shipping routes to the seaports of Rotterdam and Antwerp had to be kept open. The safety of these areas was to be guaranteed by substantially reinforcing the dikes.

The Delta Project is one of the largest hydraulic engineering projects that has ever been carried out anywhere in the world. New hydraulic engineering techniques were gradually developed for the construction of 11 dams and barriers of various sizes, which were built over a period of 30 years.

In the early 1970s, preserving as much of the natural environment as possible was increasingly realized by people and this point of view has left its mark on the Delta Project. As a result, the original plans were changed (see Sec. 1.3.2).

The Delta Project was likely carried out in a random order. Work began on the relatively simpler parts, so that the experience gained could be used during the construction of larger, more difficult dams across inlets and estuaries with strong tidal currents. That was how the Delta Works progressed: new hydraulic engineering techniques were first applied on a small scale and then used in the larger, more complicated, projects. In this way, learning as much as possible during the construction process was possible. Also, when determining the order of execution of projects, the materials and manpower available were taken into account. On the basis of these considerations, it was decided to carry out the works in the following order (date of completion) (Fig. 1.3).

1.3.2. *The Eastern Scheldt (Oosterschelde)*

While the Haringvliet dam and the Brouwers dam were nearing completion (1971), preparations had already begun for the construction of the dam across the mouth of the Eastern Scheldt, the last, largest and also most complex part of the Delta Project. Three islands were constructed: Roggenplaat, Neeltje Jans, and Noordland. A pumped sand dam was built between the latter two. In the remaining channels, the first steel towers were built for the cableway, as it was planned to dam the Eastern Scheldt using this well-tried method. Its completion date was set for 1978.

At the end of the 1960s, however, protests were voiced about the project. Scientists became aware of the special significance of the flora and fauna in and around the Eastern Scheldt. The sandbars and mud flats exposed at low tide are important feeding grounds for birds, and the estuary is a nursery for fish from the North Sea. Fishermen and action groups made sure that the scientific findings were heard by the Dutch government and Parliament. A heated debate flared up. Opponents of the dam believed that the safety of the region could be guaranteed by raising the height of the dikes along the Eastern Scheldt. The inlet would then remain open and saline. The supporters of the solid dam, for example agricultural and water boards, appealed to the emotions of the Zeelanders, asking whether the consequences of the flood disaster of 1953 had already been forgotten.

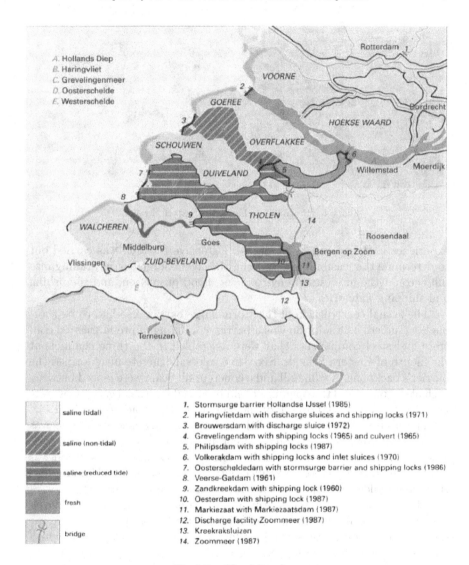

A. Hollands Diep
B. Haringvliet
C. Grevelingenmeer
D. Oosterschelde
E. Westerschelde

saline (tidal)

saline (non-tidal)

saline (reduced tide)

fresh

bridge

1. Stormsurge barrier Hollandse IJssel (1985)
2. Haringvlietdam with discharge sluices and shipping locks (1971)
3. Brouwersdam with discharge sluice (1972)
4. Grevelingendam with shipping locks (1965) and culvert (1965)
5. Philipsdam with shipping locks (1987)
6. Volkerakdam with shipping locks and inlet sluices (1970)
7. Oosterscheldedam with stormsurge barrier and shipping locks (1986)
8. Veerse-Gatdam (1961)
9. Zandkreekdam with shipping lock (1960)
10. Oesterdam with shipping lock (1987)
11. Markiezaat with Markiezaatsdam (1987)
12. Discharge facility Zoommeer (1987)
13. Kreekraksluizen
14. Zoommeer (1987)

Fig. 1.3.   The delta plan.

1.3.3.   *The Storm-Surge Barrier*

Finally, a compromise was reached in 1976: a storm-surge barrier, which would stay open under normal conditions but which could be closed in case of very high tides (Fig. 1.4), was to be constructed. However, building a storm-surge barrier needed expertise that was yet to be developed and

Fig. 1.4.   Storm-surge barrier in Eastern Scheldt.

experience that had yet to be gained. Extensive research was carried out to determine the feasibility of building the storm-surge barrier, taking into full account the interests of environment, flood protection, and the fishing and shipping industries.

The actual construction of the storm-surge barrier also had to be thoroughly studied. The solution was a barrier consisting of pre-fabricated concrete and steel components that were assembled in the three channels at the mouth of Eastern Scheldt. Sixty-five colossal concrete piers formed the barrier's backbone. A stone sill and a concrete sill beam were placed between each of the piers, and the openings could be closed with steel gates. Concrete box girders were placed on top of the piers to form a road deck (Fig. 1.5).[2]

The seabed also needed special consideration. A new technique was required to prevent the strong current in the mouth of the river from washing away the sand on which the piers were to stand. The solution was to place the piers on mattresses filled with graded layers of sand and gravel that would allow the water to flow through but trap the sand.[2,3]

The construction of the storm-surge barrier also required the development of special equipment. The *Mytilus* made its appearance in the estuary to compact the seabed, followed by the *Jan Heijmans*, which laid asphalt and dumped stones, the *Cardium* to position the mattresses, the *Ostrea* to lift, transport and position the piers and the mooring and cleaning pontoon *Macoma*. These are very special ships designed for just one purpose: to construct the storm-surge barrier. New measuring instruments and computer programs were also developed so that engineers working 30 to 40 m below the surface could position components with such precision that the maximum error would be just a few centimeters.

The *Cardium* laid the first mattress in November 1982 and the *Ostrea* placed the first pier in August 1983. Work progressed quickly. There were

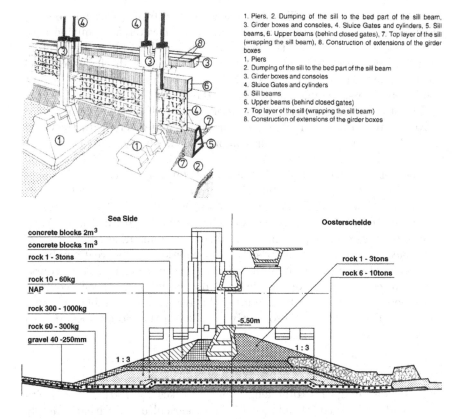

1. Piers, 2. Dumping of the sill to the bed part of the sill beam, 3. Girder boxes and consoles, 4. Sluice Gates and cylinders, 5. Sill beams, 6. Upper beams (behind closed gates), 7. Top layer of the sill (wrapping the sill beam), 8. Construction of extensions of the girder boxes

1. Piers
2. Dumping of the sill to the bed part of the sill beam
3. Girder boxes and consoles
4. Sluice Gates and cylinders
5. Sill beams
6. Upper beams (behind closed gates)
7. Top layer of the sill (wrapping the sill beam)
8. Construction of extensions of the girder boxes

Fig. 1.5. Structural elements and typical cross-section of the storm-surge barrier in Eastern Scheldt.

virtually no technical setbacks; only the cost turned out to be higher than expected. The storm-surge barrier was 30% more expensive than estimated. On October 4, 1986, Her Majesty Queen Beatrix officially opened the storm-surge barrier. The Eastern Scheldt has remained open and also flood protection has been achieved. On average, the barrier has to be closed once a year because of storms.

1.3.3.1. *Eastern scheldt project*

The Eastern Scheldt Project comprises more than the construction of a storm-surge barrier in the mouth of the estuary. In the eastern part of Markiezaatskade, Oesterdam and the Philipsdam have been constructed along with shipping locks to create a tide-free navigation way between

Rotterdam and Antwerp. The Zoommeer forms a fresh water lake behind these dams. To prevent the salty Eastern Scheldt water from mixing with the fresh Zoommeer water, the locks are fitted with a special freshwater–saltwater separation system. From the Zoommeer, a discharge channel carries freshwater into the Western Scheldt. The Eastern Scheldt Project also involved far-reaching adaptations of the South Beveland canal.

The storm-surge barrier in the Eastern Scheldt has been commissioned by Rijkswaterstaat (Public Works Department), and was built by the Oosterschelde Stormvloedkering Bouwcombinatie (Dosbouw/Ostem), a joint venture by a number of contractors.

The detailed description of Eastern Scheldt Barrier can be found in *Design Plan Oosterschelde* (1994).[2] The total design documentation consists of five books. This book is a translation of Book 1: *Total design and design philosophy*. The other books are only available in Dutch. More general findings obtained from Delta Works are summarized in *Closure of Tidal Basins*.[4] Both books are based on the experiences of a great number of engineers and other professionals involved in this large project. In the course of the project, a wealth of knowledge and experience has been acquired. These books have been written to make this knowledge available for future use. These experiences are also partly included in the recent *Rock Manual* (2007).[5]

1.3.3.2. *Significance of the delta works*

Besides shortening the total length of the dikes by 700 km, the Delta Works had many other advantages. First, the agricultural freshwater supply was improved. Because the border between freshwater and saltwater was moved further west, less freshwater was required to balance the freshwater–saltwater division. The excess water could be transported to the north of the Netherlands, in the direction of the Ijsselmeer (Ijssel lake), where extra freshwater was welcomed to improve the water conditions. Second, the complete water balance of the Delta area was improved. Thanks to the construction of the major and auxiliary dams, the streams in this area were able to be manipulated more easily. Different types of sluices made it possible to allow freshwater in, or polluted or excess water out. Third, the construction of the Delta Works encouraged traffic between the many islands and peninsulas. Large parts of the province of Zeeland had literally been isolated for centuries. The building of the Zeeland Bridge together with a tunnel under the Westerschelde (2003) also helped increase mobility. Fourth, the inland waterways shipping was supported by the Delta Works. In 1976, Belgium and the Netherlands signed a contract to regulate shipping between the ports of Antwerp and Rotterdam. Obviously, this agreement had to be taken into account when building the Delta Works. It was realized by construction of compartment dams and navigation locks. Lastly, the Delta Works have influenced new developments in the areas of nature and recreation. Understandably, a number of nature reserves were irreparably damaged, but as a compensation, new nature reserves have emerged at different sites. Nowadays, dry shores are sometimes used as recreational areas. Whether or not nature has benefited from the Delta Works will remain an unsolved debate. However, there is no doubt on the need for durable water management, in which safety, prosperity and nature are taken into account. The past decisions were supported by policy studies.[6]

Other developments: In addition to the construction of new dams and barriers, at several places, existing dams had to be heightened. This was especially required in the western parts of the islands (Walcheren, Schouwen, and Goerree) and along the waterway of Rotterdam and the Western Scheldt. The dikes needed reinforcement because they were not directly protected by large works. It is a common misconception that the Delta Works were only built to replace dikes. In most of the cases, building a delta work was much quicker and cheaper than reinforcing existing dikes. Since the building and strengthening of dikes are time-consuming and expensive, two other Delta Works were built to the west of Rotterdam at the end of the 20th century. The movable barrier, called the "Maeslant Barrier," can close off the New Waterway when water levels are threatening the dikes in the environment. The second, smaller one, "Hartel Barrier," was constructed in the Europort area.

Due to the recent climate change and the rise in sea level, high water levels are more likely to occur near the coasts of Zeeland and Holland. The number of people who live in the polders, several meters below sea level, has actually increased since the flood of 1953. The general consensus among scientists is that the reinforcement of dikes and the construction of dams and barriers are in no way the final answer in the battle against the sea.

1.3.3.3. *Maeslant barrier: New storm-surge barrier at Rotterdam*

The barrier in the New Waterway to Rotterdam, situated close to the sea entrance at Hoek van Holland, was one of the last construction parts of the Delta Plan. To provide a better financial control of the project, it was decided to construct the storm-surge barrier following the basic principles of the market mechanism philosophy. As a result, a "design-and-construct contract" was issued as a tender according to European rules.[7] Initially six contractor consortia applied for the tender. The Rijkswaterstaat (Public Works Department) prepared the technical requirements for selecting the winner. The most important requirement for the design was that the barrier should not hinder shipping. The barrier should only be closed under exceptional circumstances — no more than once or twice every 10 years.

Based on these criteria, the sector gate and the segment door were chosen for further competition. In the final phase of the competition, the two consortia performed hydraulic model testing on their designs at the Delft Hydraulics Laboratory to eliminate all the remaining uncertainties. After a technical evaluation, the Rijkswaterstaat finally selected the Bouwkombinatie Maeslant Kering (BMK) sector gate on the basis of costs. The main technical reason for selecting the BMK design was its simplicity of the technical concept. Moreover, the structure is easy to maintain, mainly in dry conditions with only limited parts remaining under water.

The BMK Barrier consists of two hollow semi-circular gates attached by means of steel arms to a pivotal point on both banks (Fig. 1.6). The sector gates (wings), each 237 m in length and 20 m high, are a prominent visual part of the storm-surge barrier. Their function is to transmit directly to the ball-joint the loads exerted on the retaining walls in the closure process. During closure, the gates will ride up and by approximately 40 cm into a chamber, levelling out again when the load eases. The construction was completed in the period 1991–1997. Some details and simulations can be found in the website http://www.keringhuis.nl/engels/maeslantkering/index.html.

The decision to construct a storm-surge barrier in the New Waterway also necessitated a barrier in the Europort area. Otherwise, when the New Waterway Storm Surge Barrier is closed, too much seawater can flow in through the Europort area threatening the safety of South Holland. Therefore, the second, smaller barrier, called "Hartel Barrier," was constructed in the arm of the Old Maas in the Europort area. The movable barrier consists of large elliptical gates suspended between oval towers (see also http://www.keringhuis.nl/engels/maeslantkering/index.html).

The New Waterway Storm Surge Barrier (Maeslant Barrier, Fig. 1.6), a limited dike-strengthening program and the Hartel Barrier completed the Delta Project in 1997, thus protecting South Holland against high water.

1.4. Contribution Delta Works to Developments in Hydraulic and Coastal Engineering

The Netherlands has long and varied experience of hydraulic engineering, and particularly of constructing dikes, digging canals, draining polders, and building locks, bridges, tunnels, and ports. That experience is also put to use not only in the offshore industries — in the construction of production

1. Steel sector gate
2. Sill and bottom protection
3. Parking dry dock
4. Abutment
5. Control centre
6. Ball joint foundation block
7. Locomobile and guidance tower

Fig. 1.6.   View of the maeslant barrier.

platforms, but also in foreign projects in South Korea, Bangladesh, Bahrain, and others, for example. Working in and with water has given the Dutch a worldwide reputation, and the Zuyder Zee project, which not only protected large areas of the country from flooding but also provided about 160,000 ha of new land, and the Delta project, which is also to protect the Netherlands from the ravages of the sea, are outstanding examples of their expertise in this field. The impact of the Delta Works is manifested in a number of transitions in hydraulic engineering approaches, for example, from deterministic into probabilistic approach, from traditional to innovative techniques and materials, etc.

At present, it is difficult to recognize the origin of present-day working methods. Nevertheless an attempt has been made to list at least some of the achievements including[8]

— Quantitative insight in scour
— Extensive study of wave impact forces
— Large-scale application of geotextile and asphalt mats for scour protection
— Closure of tidal channels by dumping material from cable cars
— Application of geocontainers
— Use of discharge caissons
— Large-scale use of asphalt and sand asphalt for slope protection instead of labor-intensive stone revetments
— High capacity dredging of sand
— Extensive use of pre-stressed concrete in the marine environment
— Compaction of sand by vibration to a depth of NAP −60 m
— Placement of foundation mattresses consisting of three layers of granular material

— Lifting and accurate positioning of extremely heavy elements in water depth of 35 m and velocities up to 4 m/s
— Closing tidal channels with sand only
— Development of probabilistic design methods
— Technology of granular filters
— Accuracy of dredging and providing foundation layers for caissons.

Some elements of these developments and expertise will be briefly highlighted below. More information can be found in the references.

1.4.1. *Design Methodology and Innovative Execution*

The Delta Project was a challenge to Dutch hydraulic engineers. It was evident that past experience and existing techniques would not be sufficient to construct dams across the wide and deep tidal channels. The tidal range in the Delta is approximately 3 m and the water flows in and out twice a day with powerful currents shifting enormous quantities of sand. Weather conditions in the estuaries were often unfavorable and North Sea storms produced powerful waves. New techniques had to be developed quickly so that the Delta Project could be carried out.

The towering caissons were adapted and improved. Man-made fibers (geosynthetics) were used for the first time to protect the seabed and to clad the dikes. The traditional seabed protection method, which involved covering the seabed with large mats made of willow wood and weighted with stone, was gradually replaced.

Changes took place step by step. It was decided to implement the Delta Project by working from small to large so that technological progress would keep pace with the growth in experience.

Pre-fabrication became a common technique and in addition to new materials, new equipment was also very valuable. Sluice caissons were developed. A cableway with gondolas was developed to tip stone into the channels.

Hydrodynamic study techniques were refined by developing laboratory tests. The computer gradually made its entry into the process. Measuring techniques and weather forecasts became more accurate. The Delta Project took about 25 years to complete. This marked the beginning of a new age for hydraulic engineering also in an international context.

Design, management, and maintenance of hydraulic structures and flood protection systems are of great importance. The probabilistic design technique is suitable for this task. This technique gives a clear view of the weak points of a structure and the various ways in which it can be optimised. Because of complexity of Eastern Scheldt project, the probabilistic approach has been applied on a large scale.[3]

Probabilistic approach: Probabilistic methods were introduced in the design of the storm-surge barrier mainly for two reasons.[3] After the Disaster 1953, the Delta Committee stipulated that primary sea-retaining structures had to provide full protection against storm-surge levels with an excess frequency of 2.5×10^{-4} times per year. In case of conventional defences, such as dikes, an extreme water level (combined with a maximum extrapolated single wave) may be used as a design criterion, because overtopping is considered to be the most important threat to dikes. However, this approach was unsuitable for a storm-surge barrier.

The structure consists of various components (concrete piers, steel gates, a sill, a bed protection, and a foundation), which have to be designed on different load combinations providing most dangerous threat to the structural stability. The probability density function of the load was derived by integrating the multidimensional probability density function of wave spectra, storm-surge levels and basin levels using the transfer function of the structure. To ensure consistent safety throughout the structure, probabilistic analyses taking into account the stochastic character of the loads and the structural resistance are performed for the main components. To assess the safety of the barrier as a sea defence system, a risk analysis was performed using the fault tree technique (Fig. 1.7). More information on this approach can be found in Refs. 2 and 3.

The application of probabilistic approach in the design of storm-surge barrier in Eastern Scheldt was a starting point of introducing this technique in other fields of hydraulic and coastal engineering in the Netherlands and elsewhere. Probabilistic approach and risk analysis are actually standard items included in civil engineering education (refer to http://www.hydraulicengineering.tudelft.nl).

Failure modes: The probabilistic approach gives the probability of failure of a (elements of a) structure and takes into account the stochastic character of the input variables. This is in contrast to a deterministic design method, which is based on fixed values, for example, mean or extreme values. Studies have been performed using a failure mechanism based on the stability and displacement of structural elements and protection materials.

A focal point in the feasibility study was whether or not the subsoil would be able to deliver the necessary counteracting forces against storm surge. Together with specialists from several countries, the Delft Soil Mechanics Laboratory completed a series of laboratory tests to assess

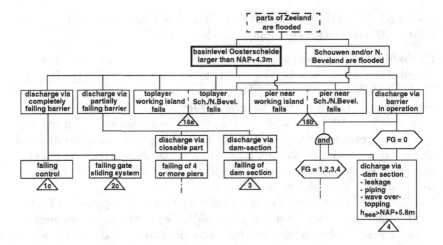

Fig. 1.7.   Example of fault tree for storm-surge barrier in Eastern Scheldt.

subsoil characteristics. In addition, a full-scale test was conducted outdoors to study the behavior of a caisson subjected to cyclic loading. New computation techniques were developed to predict the behavior of the subsoil under design conditions. Delft Hydraulics Laboratory was given the task of predicting design loads on the barrier due to tides, storm surges, and waves. Model experiments were performed in the wind-wave facilities of the laboratory using small-scale models of the proposed structures, which were subjected to random seas.

Verification of design: Not all hydraulic or coastal structures or their components are understood completely; moreover, the existing design techniques represent only a certain schematization of reality. Therefore for a number of structures and/or applications, the verification of design by more sophisticated techniques are still be needed.

Delta Works were an excellent example for the application of large-scale verification. A large number of prototype tests were executed for verification of small-scale results and validation some design assumptions. Examples of this approach were tests on caissons, prototype scour tests, tests of bank protection systems, etc., but also analyses of failures from prototype.

While certain aspects, particularly in the hydraulic field, can be relatively accurately predicted, the effect of the subsequent forces on the structure (including transfer functions into sub-layers and subsoil) cannot be represented with confidence in a mathematical form for all possible configurations and systems. Essentially this means that the designer must make provisions for perceived failure mechanisms either by empirical rules or from past experience. However, using this approach, it is likely that the design will be conservative. In general, coastal structures (i.e., dikes, revetments, sea walls, etc.) are extended linear structures representing a high level of investment. The financial constraints on a project can be so severe that they may restrict the factors of safety arising from an empirical design. It is therefore essential from both the aspects of economy and structural integrity that the overall design of a structure should be subject to verification. Verification can take several forms: physical modeling, full-scale prototype testing, lessons from past failures, etc.

Engineers are continually required to demonstrate value for money. Verification of a design is often expensive. However, taken as a percentage of the total costs, the cost is in fact very small and can lead to considerable long-term savings in view of the uncertainties that exist in the design of hydraulic and coastal structures. The client should therefore always be informed about the limitations of the design process and the need for verification in order to achieve the optimum design.

1.4.2. *Closure Techniques: Sand Closures*

A large number of closure techniques were applied in the scope of Delta Works (Fig. 1.8).

More information hereabout can be found in "Closure Tidal Basins"[4] and in Refs. 4, 5, 9. Only sand closures will be discussed more in detail.[10]

Sand closures: One of the latest developments in the closure techniques for tidal gaps is using sand as the only building material for closure. The sand closure is carried out by supplying sand either by dumping from the boat or by a pipeline that takes a water-sediment suspension to the head of the dam. The suspension runs off over the fill, where the sand deposits due to decelerating water velocities and the dam is built this way. Due to narrowing of the closure gap, which causes increasing velocities, and due to the fact that a part of the sediment supplied by the pipe will reach the closure gap in suspension, loss of some amount of sand is inevitable. The material (sand) should have either sufficient weight (diameter) to resist erosion or be supplied in such large quantities that the main portion is not carried away by the current.

In 1960s, a number of tidal channels in the Netherlands, and some in Germany and Venezuela, were closed successfully by pumping sand in the gap. The experience was restricted to the gaps with the maximum current velocities in the final closure stage not higher than $3\,\mathrm{m/s}$.

Until 1982 in all cases the dredger (= pumping) capacities that were needed were estimated by the semi-empirical method as described in Ref. 4.

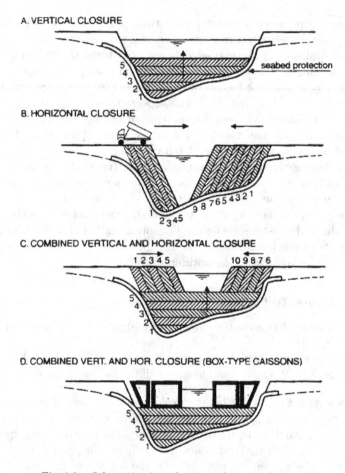

Fig. 1.8.   Schematization of various closure techniques.

The new developments with larger gaps, higher velocities, and new calculation and execution techniques are described in the *Manual on Sand Closures*.[10]

Before calculations of the dredger capacity and sand loss start, some assumptions have to be made on the dimension of the closure dam (Fig. 1.9).

The crest elevation of the closure dam (temporary function) depends on astronomical tide and wind set-up (usually a recurrence interval of 5–10 years is taken) and the run-up. The crest width depends on the number of pipes and the working space needed for bulldozers on the fill. The pipes will be used in pairs: one for production while the other can be lengthened by connecting a new pipe section ($B \sim 20\,\mathrm{m}$ for one pair of pipes). The

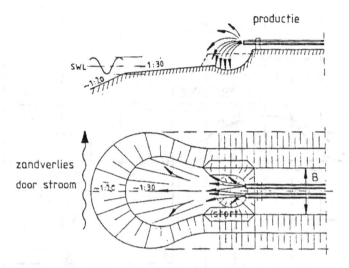

Fig. 1.9. Situation sketch (production and sand losses).

slopes of the dam in the tidal range vary from 1:30 to 1:50, depending on the wave action. The slopes under low-water are in the order of 1:10 to 1:20. The side-slopes of the dam are mostly created by bulldozers and can be taken to be equal to 1:5. In order to reduce the sand loss in the final stage of the closure, the remaining gap should be situated in a shallow part of the channel. This is due to the fact that less sand should be brought in to make the same progress as in the case of a deeper gap.

A distinction is made between gross and net loss of sand. The gross loss is the sand that deposits outside of the closure dam profile but inside the final dam profile and is of importance for the time involved in the closure operation. The net loss is the sand that is taken beyond the profile of the final dam and will determine the actual loss. However, in the standard calculation method, no distinction is made between these two kinds of loss and only the total loss is calculated (Fig. 1.10).

The duration (and cost) of the operation depends further on the location and the available diameter of sand where the sand for the dam is dredged. Travel distance should be kept as short as possible. The increasing grain size will diminish the losses and thus will influence the success of the operation in a positive way. Success also depends on the working method, the layout of the pipelines, the sand–water ratio of the spoil and the effective working time of the dredgers. In the final gap, bulldozers and draglines operate on the fill to control the sand deposition above the water level (Fig. 1.11). However, this very important influence of the working method on the total

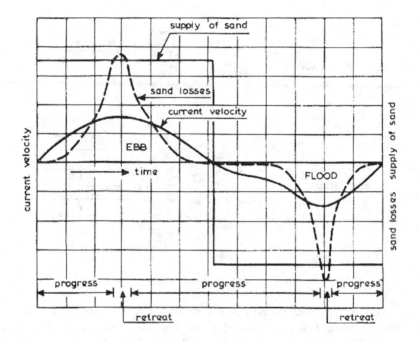

Fig. 1.10.   Sand losses as a function of time and tidal motion.

Fig. 1.11.   Execution of sand closure.

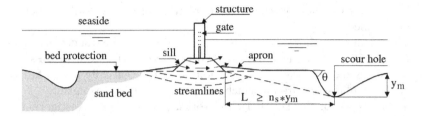

Fig. 1.12.   Schematization of a storm-surge barrier.

sand losses is not taken into account in the calculating procedure. This positive effect can be treated as an extra safety in calculating the results, especially in the light of many limitations of this method. The detailed overview of past sand closures and design information can be found in *Manual on Sand Closures*.[10]

1.4.3. *Scour and Bottom Protection*

Scour is a natural phenomenon caused by the flow of water in rivers and streams. Scour occurs naturally as part of the morphological changes of rivers and as result of man-made structures. Scour prediction and bed protection during closure operations are essential aspects to be considered during design and execution (Fig. 1.12).

In case of Eastern Scheldt Barrier, to protect the seabed from erosion caused by the increased speed of the currents, a 500–600-m wide area on either side of the barrier had to be covered. Had this not been done, the channels might have eroded to such an extent that the barrier would be endangered. The protection consists of concrete weighted erosion mats, aprons of stone-filled asphalt and stone-weighted mastic asphalt slabs. Highly advanced techniques were developed especially for this operation. The concrete-weighted erosion mats, manufactured in the factory at the Sophia harbor later used to make the block mattresses, were laid by a special vessel. The asphalting operations were carried out by the *Jan Heijmans*, which has later been converted into a gravel/stone dumper to be used for filling in the space between the foundation filter mattresses.

Scour and flow slides: Generally the scour process can be split up into different time phases. In the beginning, the development of scour is very fast, and eventually an equilibrium will be reached. During closing operations of dams in estuaries or rivers, which take place within a limited time, equilibrium will not be attained. Not only during the construction, but also in the equilibrium phase, it is vital to know both the development of scour as function of time and whether the foundation of the hydraulic structure is undermined progressively.

Experience has shown that due to shear failures or flow slides or small-scale shear failures at the end of the bed protection, the scour process can progressively damage the bed protection, leading eventually to the failure of the hydraulic structure for which the bed protection was meant.

In the scope of the Dutch Delta Works, a systematic investigation of time scale for two- and three-dimensional local scour in loose sediments was conducted by Delft Hydraulics and "Rijkswaterstaat" (Department of Transport and Public Works).[4] From model experiments on different scale and bed materials, relations were derived between time scale and scales for velocity, flow depth, and material density.[11,12] In addition, empirical relations were found in order to predict the steepness of the upstream scour slope.[13]

Besides the systematic investigation, design criteria for the length of the bed protection were deduced, which were based on many hundreds of shear failures and flow slides that had occurred along the coastline of estuaries in the south-western part of the Netherlands.[4]

Verification with prototype data: Within the scope of research activities with respect to scour behind the storm-surge barrier and compartment dams in Eastern Scheldt, some field experiments were carried out.[13] For this purpose, the sluice in the Brouwersdam was chosen, which was built to refresh the brackish water in the Grevelingen Lake for environmental reasons. A 5.4-m high sill was constructed at the lake side of the sluice with two side constrictions equal to 2.5 m on the left side and 1.5 m on the right side. The flow depth was about 10 m and the length of the bed protection from the toe of the sill measured about 60 m. The effective roughness of the bed protection is estimated to be 0.4 m. The experiments were executed to study the influence of clay layers to scour and to verify scour relations obtained from scale models.[13−15] The agreement between the model and the experiment was very satisfactory except for the stability of the upstream scour surface. In general, the prediction of upstream slope of scour hole is still a weak point and needs further investigation (see also http://www.wldelft.nl/rnd/publ/docs/Pub242.pdf).

The Dutch efforts to gain further knowledge of scour were completed in a PhD thesis by G. Hoffmans[14] and finally compiled in the *Scour Manual*.[16]

1.4.4. *Stability of Cover Layers*

A large number of various sloping structures were realized in the scope of Delta Works. Such structures as abutments of barriers, jetties, compartment dams, navigation channels, and some dikes needed a proper slope protection. However, the inventory studies at the end of 1970s indicated a lack of proper design criteria for revetments. Therefore, a number of basic studies were initiated for this purpose. The results of these studies were later transformed and translated into more general design standards, which often became the international standard.

1.4.4.1. *Rubble structures and riprap*

All the older designs were mostly based on (simplified) formula of Hudson dating back to 1950s, which gained its popularity due to its simplicity and the stature of US Army Corps. However, the problems with using this formula started in 1970s with introduction of random waves and the necessity of transformation of regular waves into irregular waves. In 1980s, the number of testing facilities and test results with random waves became so large that the necessity of new design formulas became evident. The new research in 1980s provided an enhanced understanding of failure mechanisms and development of more sophisticated new formulae on stability and rocking of rubble mound structures and artificial armour units.[17−20]

In the early 1980s, Rijkswaterstaat had commissioned to the Delft Hydraulics a systematic research on static and dynamic stability of granular materials (rock and gravel) under wave attack. This program was successfully completed under direct guidance by Van der Meer in 1988.[17] Formulae developed by Van der Meer by fitting to model test data, with some later modifications, became the standard international design formulations.[5,21]

The work of Van der Meer is now generally applied by designers and it has considerably reduced (but not eliminated) the need to perform model experiments during the design process. We always have to remember that each formula represents only a certain schematization of reality. Moreover, as far as these formulas are based on experiments and not based on a complete physical understanding and mathematical formulations of processes involved, each geometrical change in the design may lead to deviation in the design results and a need for model investigation. Another advantage of the Van der Meer formulae over the formula of Hudson is the fact that the statistical reliability of the expression is given, which enables the designer to make a probabilistic analysis of the behavior of the design.

Following the same philosophy, Van der Meer and others have modified and extended the formulae for the stability of rock to many other aspects of breakwater design such as stability of some artificial units, toe stability, overtopping and wave transmission. What has been proposed for slopes under wave attack is also largely valid for slopes and bottom protection under currents.

The designer has a number of black box design tools available,[5,21] but the understanding of the contents of the black box is far from complete.[22] Specifically when these black box design formulae are used in expert systems, one may in the end be confronted with serious mistakes. The designer still must be properly trained in the shortcomings and limitations of the black box formula.

1.4.4.2. *Block revetments*

Placed block revetments are very popular in the Netherlands because of no one source of rock. However, till early 1970s, there were no proper design methods for these revetments; the design was mainly based on experience. That was the reason for new research with the aim to provide proper design criteria for dams and dikes within the Delta Project and, on a long-term, create more physical understanding and quantification of physical processes in block revetments.[23,24]

Wave attack on revetments will lead to a complex flow over and through the revetment structure (filter and cover layer), which are quantified in analytical and numerical models.[25–27] The stability of revetments with a granular and/or geotextile filter (pitched stones/blocks, block mats, and concrete mattresses) is highly influenced by the permeability of the entire revetment system. The high uplift pressures, induced by wave action, can only be relieved through the joints in the revetment (Fig. 1.13). The permeability of the revetment system is a decisive factor determining its stability, especially under wave attack, and also it has an important influence on the stability of the subsoil.

The usual requirement that the permeability of the cover layer should be larger than that of the under layers cannot usually be met in the case of a closed block revetment. The low permeable cover layer introduces uplift pressures during a wave attack. In this case, the permeability ratio of the cover layer and the filter, represented in the so-called leakage length, is found to be the most important structural parameter, determining the uplift pressure. The effect of the leakage length on the stability of semi-permeable revetments is described in *The Design Manual*.[26]

1.4.5. *Filters*

A filter, one of the most common elements in civil engineering practice, plays an important role in the total stability of hydraulic structures.[4,28,29] That was also the case in Delta Works. However, in case of breakwaters, offshore platforms, and closing barriers, the designer must take into account the cyclic hydraulic lading due to waves. In the design of the foundation

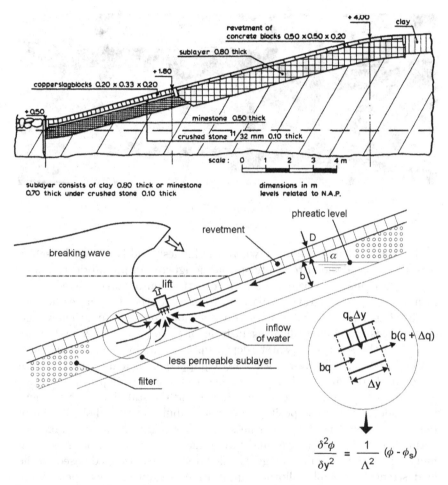

Fig. 1.13. Example of slope protection (Oester dam) and physical processes in revetment structure.

of the Eastern Scheldt Barrier, the combined effect of high static pressure head (when closed) and high waves (causing a strong cyclic flow in the foundation) was defined as a representative loading for filters (Figs. 1.12 and 1.13). The classical filter rules were not sufficient for this design. Because the design rules for semi-stationary and cyclic loading were not available at that time, it was decided to perform a systematic research at Delft Hydraulics and Delft Geotechnics on filter design under various loading conditions, and with hydraulic gradients acting parallel and vertically to the bed[4,28] (see also http://www.wldelft.nl/rnd/publ/docs/Pub287.pdf).

Fig. 1.14. Cross-section of the filter mat.

For foundation of the piers, where the highest load is acting, for safety reasons, it was decided to apply special designed and pre-fabricated filter mats. These mats consisted of a filter construction of three graded layers (sand + fine gravel + gravel) separated from each other by permeable geotextile (Fig. 1.14). The filter mats were assembled on shore and then wound on to a gigantic cylinder floating in front of a plant. The filter mats were placed on the sea bed in water (depth up to 35 m) by the specially built vessel *Cardium*. This vessel was also equipped with a dust pan suction nozzle to dredge and level off the sea bed to the correct depth.[2]

The research on revetments has proved that the stability of a revetment is dependent on the composition and permeability of the whole system of the cover layer (Fig. 1.13). Formule have been derived to determine the permeability of a cover layer and filters, including a geotextile. Also, stability criteria for granular and geotextile filters were developed based on the load–strength principle, allowing application of geometrically open filters, and thus allowing optimization of composition and permeability of revetments. It is obvious that only a certain force exceeding a critical value can initiate the movement of a certain grain in a structure. This also means that applying geometrically closed rules for filters often may lead to unnecessary conservatism in the design and/or limitation in optimization freedom. Also, it often results in execution problems especially when strict closed filter (with many layers) has to be executed under water under unstable weather conditions.[4,29,30]

In the scope of these studies, the internal strength of subsoil has also been studied in terms of critical hydraulic gradients. It was recognized that to reduce the acting gradients below the critical ones, a certain thickness of the total revetment is needed. This has resulted in additional design criteria on the required total thickness of revetment to avoid the instability of the subsoil. This also means that granular filter cannot always be replaced by geotextile only. For high wave attack (usually, wave height larger than 0.5 m or high turbulence of flow), the geotextile functioning as a filter must be often accompanied by a certain thickness of the granular cushion layer for damping hydraulic gradients. All these design criteria can be found in Refs. 26, 29, 30.

The main problem in extension of these achievements to other applications (other revetments, filter structures, bottom protection, breakwaters, etc.) is the lack of calculation methods on internal loads (i.e., hydraulic gradients) for different structural geometries and compositions. Also the geometrically open filter criteria need further development. Research in these fields is still needed and will result in more reliable and cost-effective designs.

1.4.6. *Navigation Channels and Bank Protection*

In 1970s, it was decided to build a navigation channel (known as Schelde–Rhein connection) as a connection between Rotterdam and Antwerp. Such a large project required proper information for design.[31]

The water motion induced by ships can often lead to the erosion of the bank protection of ship canals. The problem becomes more serious nowadays because of recent developments in inland shipping involving the increasing engine power of the vessels and the introduction of push-tow units on smaller canals. A systematic fundamental research program on ship-induced water motion and the related bank protection design has resulted in design guidelines based on small-scale hydraulic models. Small-scale tests however only yield limited information. Therefore, it was decided to carry out two series of full-scale measurements in Hartel Canal to verify the results of the model tests (1981, 1983) (Fig. 1.15). In addition, the influence of the subsoil on the stability of bank protection, which generally cannot be studied in small-scale models, was studied at full scale.[31] The results are also implemented in PIANC report.[32]

1.4.7. *Freshwater–Saltwater Separation Systems*

During the execution of the Delta project, the awareness of the environmental values of estuaries was growing strongly. This awareness has led to a reconsideration of the original proposal to close off the Eastern Scheldt (Oosterschelde) and to an alternative view of the management of the estuarine areas and of the civil engineering structures. Now, the completed storm-surge barrier and two compartmentization dams, together with the

Fig. 1.15.   Prototype tests Hartel canal.

reinforced dikes, guarantee sufficient safety against flooding. Also various
civil engineering structures have been built for water management purposes,
which make it possible to influence the quantity and quality of water in the
compartments.

Freshwater–saltwater exchange occurs at all ship locks in tidal areas.
This might be for various unwanted reasons; it reduces the economic value
of water for irrigation and industrial purposes and ecologic damage can
also occur. To prevent or to reduce this exchange, several options can
be considered.[33] Some of these were also applied in the Eastern Scheldt
Project.

The Eastern Scheldt (Oosterschelde) Project comprises more than the
construction of a storm-surge barrier in the mouth of the estuary. In
the eastern part, Oester and the Philips dams have been constructed
(in 1987) along with shipping locks to provide tidal-free inland navigation
way between Rotterdam and Antwerp. The enclosed area behind the compartment dams is called Volkerak or more general Zoommeer. As mentioned
earlier, Oester and Philips dams were constructed using sand and are the
largest sand dams in the world to have been constructed in flowing water.

To prevent the salty Oosterschelde water from mixing with the fresh
Zoommeer water, locks are fitted with a special freshwater–saltwater

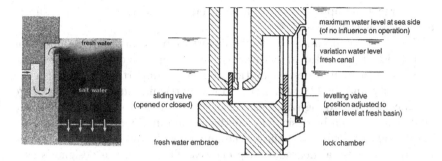

Fig. 1.16. Principle of saltwater–freshwater separation.

separation system. From the Zoommeer, a discharge channel (Bath canal) carries freshwater into the Western Scheldt. The Oosterschelde Project also involved far-reaching adaptations of the South Beveland canal.

A sophisticated saltwater–freshwater separation system has been built into the Krammer lock complex (in Philips dam) to prevent the freshwater of the lake from mixing with the salt water of the Eastern Scheldt, and conversely to prevent dilution of the Eastern Scheldt with freshwater.[33,34] Two locks were built for commercial shipping and are suitable for four-barge push-tows. Separate locks were built for pleasure craft. The system works on the principle that saltwater is denser than freshwater (Fig. 1.16). During the lockage of ships to the Zoommeer, the saltwater is replaced with freshwater, and the procedure is reversed for ships travelling in the opposite direction. The saltwater is drained through perforations in the chambers while the freshwater is allowed through slots in the wall located near the water surface. As mentioned, the system can also be operated in the reverse order. The design of this system was supported by very extensive and systematic research using detailed overview of hydraulic models and numerical calculations at Delft Hydraulics.

A similar system has also been used in the Kreekrak locks to prevent the brackish and polluted water from the industrial area around Antwerp from entering into Zoommeer. The originally planned lift lock in the Oesterdam was finally replaced by a conventional lock. The saltwater–freshwater exchange quantities at this location are rather small. More information can be found in Refs. 33–35.

1.4.8. *Materials and Systems*

In the Netherlands, due to the lack of natural rock resources, the application of waste materials and geotextiles in civil engineering has a long tradition.

The large-scale application of these alternative materials started during execution of Delta Works.[36]

The cost of production and transportation of materials required for hydraulic and coastal structures is an important consideration when selecting a particular design solution. Thus it is important to establish the availability and quality of materials for a particular site at an early stage when considering design options. Using the available tools and models, the civil structure can be designed to perform the functional requirements. An additional problem is that these functions may change with time in service because of material degradation processes. Therefore, the designer's skill must also encompass consideration of durability and degradation processes. Degradation models for materials and structures should be developed so that the whole-life consequences may be considered at the design stage itself.

1.4.8.1. *Waste and industrial by-products as alternative materials*

Domestic and industrial wastes and industrial by-products form a still growing problem especially in highly industrialized countries or highly populated regions. A careful policy on application of these materials in civil engineering may (partly) help to reduce this problem. Current European policies aim to increase the use of waste materials of all kinds and to find economic, satisfactory, and safe means of their disposal. The use of waste materials in hydraulic and coastal structures is limited by their particle size distribution, mechanical, and chemical stabilities and the need to avoid materials that present an actual or potential toxic hazard.[5,36]

The extensive research on properties of waste materials helps us in proper selection of disposal methods depending on environmental requirements. Waste materials such as silex, quarry wastes, dredging sludge (depending on the source/location), and many minestone wastes have little or no hazardous contamination. These materials can be used as possible core, embankment fill, or filter material. The engineering properties of many waste materials are often comparable to or better than traditional materials. Slags have good friction properties due to their angularity and roughness and typically have high density. Mine wastes sometimes have poor weathering characteristics, but are usually inert and have satisfactory grading for deep fills. The fine materials such as fly ashes and ground slags are already in general use as cement replacement and fillers. Good quality control, not only for limiting the potential for toxic hazard, but also of the mechanical properties of waste materials can considerably increase the use of such low-cost materials in appropriately designed coastal and bank protection structures.[36]

1.4.8.2. *Geosynthetics and geosystems*

Geosynthetics are relatively a new type of construction material, which has gained huge popularity especially in geotechnical engineering and as component for filter structures. There are a large number of types and properties of geosynthetics, which can be tailored to the project requirements.[30] Geosynthetics have already transformed geotechnical engineering to the point that it is no longer possible to do geotechnical engineering without geosynthetics; they are used for drainage, reinforcement of embankments, reduction of settlement, temporary erosion control, and hazardous waste containment facilities. Of late, they are very often used in land reclamation along the shores.

When geosynthetic materials or products are applied in civil engineering, they are intended to perform particular functions for a minimum expected time, called the design life. Therefore, the most common (and reasonable) question when applying geosynthetics is "what is the expected/guaranteed lifespan of these materials and products." There is no straight answer to this question. Actually, it is still a matter of "to believe or not to believe." Both the experimental theory and practice cannot answer this question yet. However, the Dutch evaluation of the long-term performance of the older applications of geotextiles (dating back to 1968) has proved that the hydraulic functioning was still satisfactory. A similar conclusion has been drawn from evaluation of the long-term performance of nonwoven geotextiles from five coastal and bank-protection projects in USA.[37]

The technology of geosynthetics has improved considerably over the years. Therefore, one may expect that with all the modern additives and UV stabilizers, the quality of geosynthetics is (or can be, on request) much higher than in the 1960s. Therefore, for the "unbelievers" among us, the answer about the guaranteed design life of geosynthetics can be at least 50 years. For "believers," one may assume about 100 years or more for buried or underwater applications. These intriguing questions on the lifespan of geosynthetics are the subject of various studies and the development of various test methods all over the world. Also, the international agencies related to normalization and standardization (CEN, ISO, ASTM) are very active in this field.

In recent years, traditional forms of river and coastal works and structures have become very expensive to build and maintain. Various structures and systems can be of use in hydraulic and coastal engineering, from traditional rubble and/or concrete systems to more novel materials and systems such as geotextiles/geosynthetics, natural (geo)textiles, gabions, waste materials, etc. The shortage of natural rock in certain geographical regions can also be a reason for turning to other materials and systems. This all has prompted a demand for cheaper, less massive, and more environmentally acceptable engineering. However, besides the standard application in filter constructions, the application of geosynthetics and geosystems in hydraulic and coastal engineering still has a very incidental character, and it is usually not treated as a serious alternative to the conventional solutions. That was the main reason for this author to write about the state-of-the-art application of geosynthetics and geosystems in hydraulic and coastal engineering.[30]

These new (geo)systems (geomattresses, geobags, geotubes, seaweed, geocurtains, and screens) were applied successfully in a number of countries and they deserve to be applied on a larger scale. Recently, geocontainers filled with dredged material have been used in dikes and breakwaters in a number of projects around the world, and their use in this field is growing very fast.

The main obstacle in their application, however, is the lack of proper design criteria (in comparison with rock, concrete units, etc.). In the past, the design of these systems was mostly based on rather vague experience than on the general valid calculation methods. More research, especially concerning the large-scale tests and the evaluation of the performance of projects already completed, is still needed. In Pilarczyk,[30] an overview is given of the existing geotextile systems, their design methods (if available), and their applications. Where possible, some comparison with traditional materials and/or systems is presented. The recent research on some of these systems has provided better insight into the design and applications.

1.5. Conclusions

Large projects (likes Delta Works) usually need some specific solutions. It stimulates new research and innovation, which contribute strongly to new developments in hydraulic and coastal engineering.

The emphasis on research and innovation during the Delta Project was not restricted to only government agencies. Private enterprises, including the offshore and dredging industry, was challenged and sometimes urged to participate in this process. As such, the Delta Project has contributed significantly to the evolution of the hydraulic and coastal engineering profession in general and the offshore and dredging industry in particular, from a vocational profession into a modern science-based industry. With so much emphasis on innovation, there must have been a tremendous spin-off. At present, it is often difficult to recognize the origin of present-day working methods but many of them originate from this project.[8]

The Delta Project was originally designed in a period when awareness of the environment and the ecological effects of civil engineering works scarcely existed. Moreover, the decision to carry out the project was taken in an emotional context immediately after a major disaster that took 1,800 lives. It is therefore not surprising that during the execution of the project priorities changed. The growing level of prosperity and the growing attention for the quality of life strengthened the concern for the environment.[6,8]

Knowhow and technology transfer is an important ingredient in the sustainable development of nations. Technology transfer is sustainable when it is able to deliver an appropriate level of benefits for an extended period of time, after major financial, managerial, and technical assistance from an external partner is terminated. Apart from clearly identified objectives for technology transfer projects, proper project design and well-managed project execution, essential factors conditioning the survival of projects include: policy environment in recipient institution/country, appropriateness of technology, and management organizational capacity.[38]

Dutch coastal and hydraulic engineers believe that the knowledge and expertise they gained in the 30 years it took to complete the Delta Project should also benefit other countries in addition to the Netherlands. Therefore this new knowledge is relatively well documented (see references) and implemented in the scope of various international projects. Projects in South Korea, Bangladesh, Bahrain, and more recently New Orleans are examples and evidence of that effort.

There are a large number of hydraulic and coastal structures. For some of them, workable design criteria have been developed in recent years within the scope of Delta Works and further improved in the scope of international research and projects (closure dams/techniques, rubble-mound breakwaters, riprap, block revetments, filter structures, scour and bottom protection, geosystems, etc.). However, many of these criteria, formulae, and techniques are still not quite satisfactory, mainly because they lack physical background, which makes their extrapolation beyond the present range of experience rather risky. To solve this problem, it will be necessary to continue physical model experiments (on scale and in prototype) to develop, validate, and calibrate new theories.

Research on hydraulic and coastal structures should benefit from more cooperation among researchers and the associated institutions. Publishing basic information and systematic (international) monitoring of completed projects (including failure cases) and evaluation of the prototype and laboratory data may provide useful information for verification purposes and further improvement of design methods. Inventory, evaluation, and dissemination of existing knowledge and future needs, and creating a worldwide accessible data bank are urgent future needs and some actions in this direction should be undertaken by international organizations involved. Adjustment of the present education system as a part of capacity building for solving future problems should also be recognized as one of the new challenges in hydraulic and coastal engineering. Finally, there is a continuous development in the field of hydraulic and coastal engineering, and there is always a certain time gap between new developments (products and design criteria) and publishing them in manuals or professional books. Therefore, it is recommended to follow professional literature on this subject for updating the present knowledge and/or exchanging new ideas.

In the light of above mentioned, the Delta Works can be seen as a good example of integration of research and practice in solving practical problems in an integrated way, but also in documentation and translation of this new knowledge into the Design Manuals, and dissemination of this knowledge to the international hydraulic and coastal society.[38]

We may finally conclude with the closing words by Professor Kees d'Angremond from 2003[8]:

> Looking back 50 years, it is clear the Delta Project has certainly been very effective in reducing the risk of inundation. With the growing concern about the rise in sea level, this aspect is gaining emphasis. Also, the side effects of the project have been positive for the economic and social development of the region, including the opportunities for tourism and recreation.
>
> It is perhaps some small solace to those who lived through the disaster and grief of the storm flood of 1953, that without it and the Delta Project as its logical consequence, hydraulic engineering and dredging would not have advanced as dramatically as they have, preventing similar disasters since and providing us with solutions for the impending problems of the 21st century.

References

1. K. W. Pilarczyk (ed.), *Dikes and Revetments* (A. A. Balkema, Rotterdam, The Netherlands, 1998); http://books.google.nl/books?ct=title &q=Coastal+Protection+,+Pilarczyk&lr=&sa=N&start (see also http:// safecoast.nl/editor/databank/File/NL%20The%20Delta%20Project.pdf).
2. RWS, *Design Plan Oosterschelde Storm-Surge Barrier: Overall Design and Design Philosophy*, Ministry of Transport, Public Works and Water Management, Rijkswaterstaat (RWS) (A. A. Balkema, Rotterdam, The Netherlands, 1994).

3. RWS (Rijkswaterstaat), *The Closure of Tidal Basins: Closing of Estuaries, Tidal Inlets and Dike Breaches* (Delft University Press, The Netherland, 1984, 1987), http://repository.tudelft.nl/file/600734/373576.

4. RWS, *Hydraulic Aspects of Coastal Structures: Developments in Hydraulic Engineering Related to the Design of the Oosterschelde Storm Surge Barrier in The Netherlands* (Delft University Press, The Netherlands, 1980); also, Eastern Scheldt Storm Surge Barrier, *Proc. Delta Barrier Symposium*, Rotterdam, October 13–15, 1982; Magazine CEMENT, Den Bosch.

5. Rock Manual, The use of rock in hydraulic and coastal engineering, CIRIA-CUR-CETMEF (2007); also CUR/RWS, 1995, Manual on the use of rock in hydraulic engineering, CUR report 169, Centre for Civil Engineering Research and Codes (CUR), P.O. Box 420, 2800 AK Gouda, A. A. Balkema, Rotterdam; http://www.kennisbank-waterbouw.nl/rockmanual/.

6. C.-J. van Westen, H. Janssen and H. Brouwer, A Well-Considered Choice; The Usefulness of Policy Analysis Instruments in Dam Construction, Rijkswaterstaat (2006); http://www.nethcold.org/nethcold/index.php?c=publications.

7. J. P. F. M. Janssen, A. van Ieperen, B. J. Kouwenhoven, J. M. Nederend, A. F. Pruijssers and H. A. J. de Ridder, The design and construction of the new waterway storm surge barrier in The Netherlands: Technical and contractual implications, *28th International Navigation Congress (PIANC)*, Seville, Spain (1994) (see also Maeslant Barrier, New Waterway Storm Surge Barrier, http://www.keringhuis.nl/engels/home_flash.html).

8. Kees D'Angremond, From disaster to Delta Project: The storm flood of 1953, *Terra et Aqua* **90** (March 2003).

9. K. D'Angremond and F. C. Van Roode, *Breakwaters and Closure Dams* (Delft University Press, The Netherlands, 2001); http://mail.vssd.nl/hlf/f011contents.pdf.

10. CUR, Sand Closures, Centre for Civil Engineering Research and Codes, Report No. 157, P.O. Box 420, Gouda, The Netherlands (1992).

11. H. N. C. Breusers, Conformity and time scale in two-dimensional local scour, *Proc. Symp. model and prototype conformity*, Hydraulic Research Laboratory, Pune, India (1966).

12. H. N. C. Breusers and A. J. Raudkivi, *Scouring, Hydraulic Structures Design Manual*, IAHR (A. A. Balkema, Rotterdam, The Netherlands, 1991).

13. A. F. F. De Graauw and K. W. Pilarczyk, Model-prototype conformity of local scour in non-cohesive sediments beneath overflow-dam, *19th IAHR Congress, New Delhi* (1981); also Delft Hydraulics, Publication No. 242; http://www.wldelft.nl/rnd/publ/docs/Pub242.pdf.

14. G. J. C. M. Hoffmans, Two-dimensional mathematical modelling of local-scour holes, PhD. thesis, Faculty of Civil Engineering, Hydraulic and Geotechnical Engineering Division, Delft University of Technology, Delft, The Netherlands (1992).

15. G. J. C. M. Hoffmans and K. W. Pilarczyk, Local scour downstream of hydraulic structures, *J. Hydraulic Engineering, ASCE* **121**(4), 326–340 (1995).

16. G. J. C. M. Hoffmans and H. J. Verheij, *Scour Manual* (A. A. Balkema, Rotterdam, The Netherlands, 1997); also, http://www.wldelft.nl/rnd/publ/docs/Ho_Ve_2002.pdf.

17. J. P. Van der Meer, Rock slopes and gravel beaches under wave attack, PhD. thesis, Delft University of Technology (April 1988); also available as Delft Hydraulics Communication 396); http://www.vandermeerconsulting.nl/.

18. J. P. Ahrens, Large Wave Tank Tests of Riprap Stability, C. E. R. C. Technical Memorandum No. 51, (May 1975).

19. K. W. Pilarczyk, *Interaction Water Motion and Closing Elements, The Closure of Tidal Basins* (Delft University Press, Delft, 1984), pp. 387–405.

20. E. Van Hijum and K. W. Pilarczyk, Gravel beaches; equilibrium profile and longshore transport of coarse material under regular and irregular wave attack, Delft Hydraulics Laboratory, Publication No. 274 (Delft, The Netherlands, July 1982).

21. CEM, *Coastal Engineering Manual*, US Army Corps of Engineers, Vicksburg (2006).

22. M. R. A. Van Gent, Wave interaction with permeable coastal structures, PhD. thesis, Delft University of Technology, ISBN 90-407-1182-8, Delft University Press (1995); http://www.library.tudelft.nl/ws/search/publications/theses / index.htm?to =2008&de =Hydraulic+Engineering&n=10&fr=2008&s=1&p=2.

23. M. B. De Groot, A. Bezuijen, A. M. Burger and J. L. M. Konter, The interaction between soil, water and bed or slope protection. *Int. Symp. Modelling Soil-Water-Structure Interactions*, A. A. Balkema, Delft (1988).

24. A. M. Burger, M. Klein Breteler, L. Banach, A. Bezuijen and K. W. Pilarczyk, Analytical design formulas for relatively closed block revetments; *J. Waterway, Port, Coastal and Ocean Eng., ASCE* **116**(5) (September/October 1990).

25. ASCE, *Wave Forces on Inclined and Vertical Wall Structures, Task Committee on Forces on Inclined and Vertical Wall Structures* (ASCE, New York, 1995).

26. CUR, Design manual for pitched slope protection, CUR Report No. 155, ISBN 90 5410 606 9, Gouda, The Netherlands (1995).

27. Krystian W. Pilarczyk, Design of alternative revetments, Rijkswaterstaat, Delft (2002); www.tawinfo.nl (select: English, downloads).

28. A. F. F. de Graauw, T. van der Meulen and M. van der Does de Bye, Design criteria for granular filters, Delft Hydraulics Publication No. 287, January 28, 1983, pp.; http://www.wldelft.nl/rnd/publ/docs/Pub287.pdf.

29. G. J. Schiereck, *Introduction to Bed, Bank and Shore Protection* (Delft University Press, Delft, 2001); http://www.vssd.nl/hlf/f007.htm.

30. K. W. Pilarczyk, *Geosynthetics and Geosystems in Hydraulic and Coastal Engineering* (A. A. Balkema, Rotterdam, The Netherlands, 2000); http://

books.google.nl/books?id=iq0JCcSw38QC&pg=PA369&dq=Geosynthetics+
Geosystems + , + Pilarczyk&lr=&ei=4uefSYSBEIeyyQTXpfHDDQ.

31. H. M. Blaauw, M. T. de Groot, F. C. M. van der Knaap and K. W. Pilarczyk,
Design of bank protection of inland navigation fairways, in *Flexible Armoured
Revetments Incorporating Geotextiles* (Thomas Telford Ltd, London, 1984).

32. PIANC, Guidelines for the design and construction of flexible revetments
incorporating geotextiles for inland waterways, *PIANC, Supplement to Bulletin* No. 57, Brussels, Belgium (1987).

33. J. Kerstma, P. A. Kolkman, H. J. Regeling and W. A. Venis, *Water Quality
Control at Ships Locks; Prevention of Salt- and Fresh Water Exchange*
(A. A. Balkema, Rotterdam, The Netherlands, 1994).

34. J. G. Hillen, C. P. Ockhuysen and P. C. Kuur, The technical and economic
aspects of the water separation system of the Krammer locks, *Symp. Proc.
Integration of Ecological Aspects in Coastal Engineering Projects*, Rotterdam,
The Netherlands (1983). Also published in *Water Science and Technology*
16, 131–140 (1984).

35. P. A. Kolkman and J. C. Slagter, The Kreekrak Locks on the Scheldt-Rhine
Connection, Rijkswaterstaat Communications No. 24, The Hague (1976).

36. K. W. Pilarczyk, G. J. Laan and H. Den Adel, Application of some waste
materials in hydraulic engineering, *2nd European Conf. Environmental Technology*, Amsterdam (1987).

37. G. Mannsbart and B. R. Christopher, Long-term performance of nonwoven
geotextile filters in five coastal and bank protection projects, *Geotextiles and
Geomembranes* **15**(4–6) (1997).

38. K. W. Pilarczyk, Coastal engineering design codes and technology transfer
in the Netherlands, *Proc. Coastal Structures'99*, ed. Inigo J. Losada, Vol. 2,
pp. 1077–1089 (1999) (see also www.tawinfo.nl and http://safecoast.nl/
public_download/).

39. PIANC, (a) Analysis of rubble mound breakwaters, (b) Guidelines for the
design and construction of flexible revetments incorporating geotextiles in
marine environment, *PIANC, Supplement to Bulletin No. 78/79*, Brussels,
Belgium (1992).

Chapter 2

Coastal Structures in International Perspective

Krystian W. Pilarczyk

Former Rijkswaterstaat, Hydraulic Engineering Institute,
Delft, The Netherlands
krystian.pilarczyk@gmail.com

Selected international developments and perspectives on the policy, design, construction, and management aspects of coastal structures are discussed. This overview ranges from initial problem identification, boundary condition definition, and functional analysis, to design concept generation, selection, detailing, and includes an examination of construction and maintenance considerations and quality assurance/quality control aspects. It also indicates the principles and methods that are applied in the design procedure, and appropriate references to other documents are provided where necessary. However, it should be recognized that the design process is a complex iterative process and may be described in more than one way. A prediction on the possible future needs and/or trends in coastal engineering and structures in the larger international perspective is also presented briefly.

2.1. Introduction

Coastal structures (soft and hard) act as a solution to coastal damage and erosion problems on a heavily populated coast, especially where the population or facilities are significantly threatened by storms.[1]

To place hydraulic and coastal structures in international perspective of users, one has at first to define what we understand by this term. In general, each manmade structure in contact with a marine environment can be treated as a coastal structure, and when in contact with fresh water (river, reservoir, estuary) as an inland or hydraulic structure. Applying this definition, some traditional/standard civil structures (e.g., a sheet pile, a bulkhead, a concrete wall, etc.) are defined as coastal structures when placed in contact with marine environment, or as hydraulic structures when placed in contact with fresh water. Sometimes, the term hydraulic structure(s) is used as an umbrella term covering both inland and coastal structures. Usually, by coastal structures, we imply a number of typical structures such as breakwaters, jetties, groins, seawalls, sea dikes, sea revetments, etc. Usually the heavy hydraulic loading associated with the marine environment distinguishes these structures from more conventional land and inland applications (hydraulic structures). Therefore, the discussion around coastal structures in this chapter holds good for hydraulic structures also. The discussion, however, cannot remain limited to the technical subjects only. Most engineering structures have a large impact on the environment, and the compulsions of the society's security and sustainability force the engineering world to mitigate the negative effects of works (if any) on the environment.

To start the discussion of coastal structures, it is useful to first briefly indicate the type of structures and their terminology, and where coastal structures play a role in the marine technology (see Fig. 2.1(a)–(c)).[2-4]

As mentioned above, there are many ways of classification and presentation of coastal structures. For a quick orientation within the scope of functional selection, the classification proposed by Van der Weide can be applied (Fig. 2.1(b)).[3] Rock structures can be classified using the ratio between the hydraulic load (e.g., wave height, H_s) and the strength (e.g., ΔD, where Δ = relative mass density of material and D = representative size, i.e., stone diameter); $H_s/\Delta D$. This classification is shown in Fig. 2.1(c).[4]

Whenever possible, geographical differentiation and international comparison (various safety standards, use of local materials, equipment and labor, etc.) will be taken into account when discussing these structures. In general, it is worth noting that each region/country has its own problems for which solutions are designed in terms of technical feasibility and economic affordability.

2.2. Problem Identification and Design Process

In general, a coastal (or hydraulic) structure is planned as a practical measure to solve an identified problem. Examples are seawalls and dikes, planned to reduce the occurrence of inundation due to storm surges and/or flooding, or a shore or bank protection structures to reduce erosion.

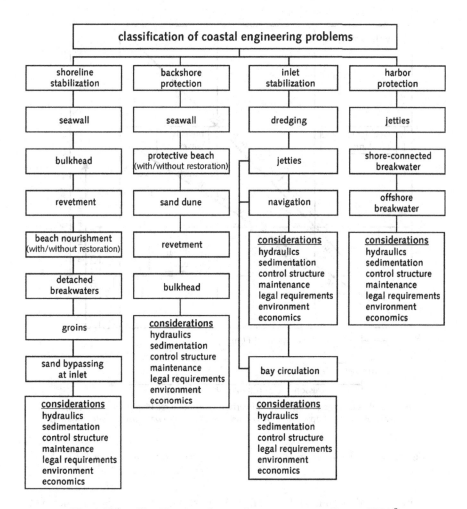

Fig. 2.1(a).   Classification of coastal structures according to SPM.[2]

Coasts and banks appear in many landforms, yet all coasts and banks have one element in common: they form the transition between land and water. With the water being a dynamic element, it is clear that coasts or banks also dynamic in nature. The design of erosion control structures is one of the most challenging task for an engineer because of its multifunctional character and multidisciplinary interactions and responses, namely, interaction between complex hydraulic loading, morphology, foundation (geotechnical aspects) and structural elements (stability).[5] It should be stressed that the integrated, multifunctional and multidisciplinary approach

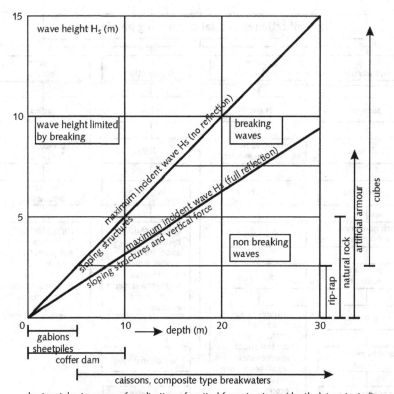

Fig. 2.1(b). Classification of coastal structures according to Van der Weide.[3]

to planning and design of hydraulic or coastal structures is still not yet
a common approach. Even in the education process of engineers, this
approach is not always followed.

As indicated earlier, hydraulic and coastal structures are one of the
means to solve a water management or a coastal problem. Coastal erosion
is one of the most frequent coastal problems. Erosion of a part of the coast,
which is often considered to be the most valuable part, viz., beach and dunes
(or mainland), is an example of such a problem. To understand the problem
and to find a proper control measure, one must understand the hydraulic
and morphological processes involved. Morphological processes cover those
physical processes that eventually result in the modification of the shape of
a coast. The hydraulic and morphological processes in the coastal zone are

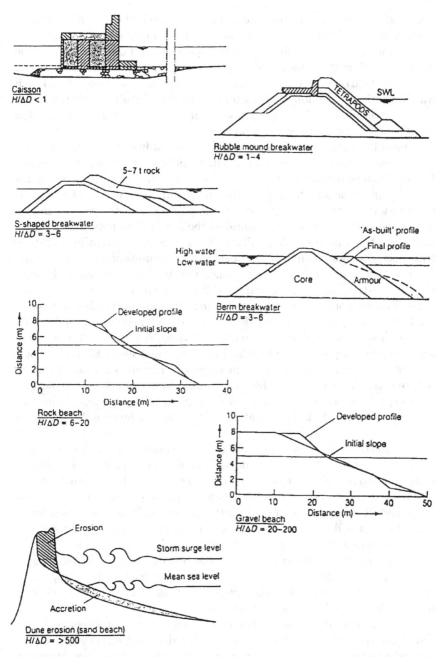

Fig. 2.1(c). Classification of rock structures using thee $H_s/\Delta D$ parameter proposed by Van der Meer.[4]

governed by two primary phenomena, namely, wind and tide. The winds are directly responsible for the generation of waves, currents, and water-level fluctuations and as a result for the transport of sand onshore and on the dry beach, while the tides influence periodic rising and falling of the water and in tidal currents. Strong winds result in extreme storm-surges and high waves, which are the dominant factor for structural stability of structures, flood protection, etc. They are "short" duration phenomena although the duration can be even in order of days.

Coastal erosion can be due to two fundamentally different processes[6,7]: (I) erosion during a severe storm surge and/or (II) structural erosion. Process I can be considered as a typical (often heavily, but temporary) redistribution phenomenon. Sand from the dunes and upper part of the beach is transported during the storm surge to deeper water in the sea/ocean and settles there. Under ordinary conditions, the sand will usually return partly to its pre-storm position. Assuming that there is no gradient in longshore transport, the total volume of sand between some limits of a dynamic cross-shore profile practically does not change due to the storm surge. Process (II), called structural erosion, is quite different from erosion due to a storm surge. This erosion is, in most cases, due to morphological gradients (mainly, gradients in longshore currents) along the coast. The volume of sand within a cross-shore profile reduces gradually with time. Without additional measures, the upper part of the profile (i.e., dune area) may also be lost permanently. Changing the gradient in longshore current provides a way to reduce this form of erosion. When an erosion control scheme to a structural erosion problem is designed, one always needs to take into account the consequences of the selected alternative for the erosion process during storm surges.

The methods of interference differ from each other in a way they interfere with the coast. Whereas the hard methods aim at reducing the sediment transport along the coast or at attempting to contain the sand on the beach or in the dunes, beach nourishment merely supplies sand that will be eroded again consequently. The latter implies that in most cases beach nourishments will have to be repeated regularly in order to protect an eroding coastline in the long term. When these beach nourishments can be reduced, for instance by applying offshore breakwaters, the investment of constructing the offshore breakwaters may be paid back by the (long term) reduction of the beach nourishments.[7] All these factors must already be included in the design of project scheme.

There is still much misunderstanding on the use of dikes and seawalls and their possible disadvantages related to the disturbance of the natural coastal processes and even acceleration of beach erosion. However, it should be said that in many cases when the upland becomes endangered by inundation (as in The Netherlands, Bangladesh, Vietnam, and other countries) or by high-rate erosion (possible increase of sea-level rise), leading to high economical or ecological losses, whether one likes it or not, the dike or seawall can even be a "must" for survival. The proper coastal strategy to be followed should always be based on the total balance of the possible effects of the countermeasures for the coast considered, including the economical effects or possibilities. It is an "engineering art" to minimize the negative effects of the solution chosen.[8]

In general, the designer has always to remember that an effective application of (hard) measures to stop or reduce the gradual erosion in the area under consideration result in a reduced input of sediments to the lee-side area. Often this reduced input leads to (increased) erosion in the lee-side area compared to the previous situation. Whether this is acceptable or not depends on the particular case. The lee-side consequences should always be taken properly into account in studying solutions for erosion problems.[9,10] In conclusion, before making a final choice of a specific measure, the effectiveness and consequences of applying such a measure should be investigated with all available means. Some of these means (models) can give probably only a qualitative answer (show tendencies), but still can be a very useful tool in helping to take a right decision.

Substantial developments have taken place in hydraulic and coastal engineering design in recent years. These have been principally due to an improved scientific understanding of the river and coastal environment and to the development of better analytical and predictive techniques, particularly through mathematical modeling. Although a number of calculation methods have been developed and are applied, the mathematical description of the hydromorphological processes and the consequent quantitative assessment of the influence of structures on the behavior of the coastline still are in a first stage. Further developing the description of these processes and incorporating them in computer programs will be required in order to have the tools available to design coastal structures and predict their impact more reliably. A number of promising attempts in this direction are already done in various research centers in Europe, the United States, and Japan.[11–13] A closer international cooperation in this field is needed to accelerate these developments, including a proper validation using different geographical site data. However, from the viewpoint of coastal structure design (groins, offshore breakwaters, sea walls, etc.), further developments in a considerable measure will still be necessary before the design can be carried out on a fully analytical basis. Experience and engineering judgment still therefore form the major elements of good design practice for hydraulic and coastal structures.

Despite these new developments, a large number of new designs all over the world have not always been successful and the investments on existing systems have not always yielded the anticipated benefits. As an example, inventory of functioning of groins applied along Dutch coast has indicated that about 50% of them do not fulfill the functional requirements, and sometimes even have an adverse effect. However, due to the relatively high investment involved, the designers and local authorities (to avoid to be blamed for a wrong choice) often defend a certain choice of insufficient or even nonfunctioning structures. But this wrong choice reflects often only a real state of our knowledge and not an inadequacy of a designer. There is an obvious need for guidelines on the use and effectiveness of coastal structure systems to assist engineers working in coastal engineering planning and management and to give them a basic understanding of design practice.

At the problem identification stage (see example in Fig. 2.2), the presence of an existing or future problem is acknowledged and defined. The acknowledgment of an existing or future problem is generally accompanied by a determination to find an appropriate solution to that problem. In the context of this book, this solution will probably consist of a coastal or shoreline structure, bed protection, or any kind of maritime works. Future problems may be foreseeable as a result of predictable changes or may be generated by proposed engineering works. A simple example of the latter would be the need of protecting the down drift part of a planned shoreline protection (flank protection). Where several options exist, the preferred solution should always be determined as a result of cost–benefit analysis and consideration of environmental impact.

Starting with the identification of a problem (e.g., inundation or shoreline erosion), a number of stages can be distinguished in the design process for (and life cycle of) a structure, the subsequent stages of which are determined by a series of decisions and actions cumulating in the creation of a structure to resolve the problem: functional requirements–natural environment–design–construction–operation & maintenance. Post-design stages (to be considered during design) are the construction and maintenance (monitoring and repair) of the structure and, finally, its removal or replacement (see also Fig. 2.3). From the design process/life cycle of a structure, one must be aware that the design of a structure may easily develop into a multidisciplinary process, including social conditions, economics, environmental impact, safety requirements, etc. In conjunction with identification of the problem, all of the boundary conditions, which influence the problem and its potential solutions, must also be identified. These boundary conditions are of various types and include aspects of

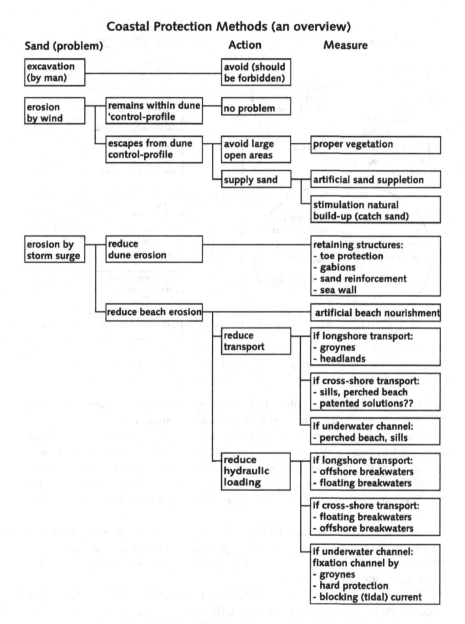

Fig. 2.2.   Identification of a coastal problem.

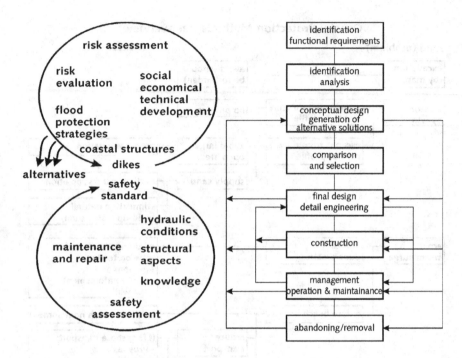

Fig. 2.3.   General integrated approach and principles of design process.

the following: planning policy (including environmental impact aspects), physical site conditions, and construction and maintenance.

Planning policy aspects involve political, legislative, and social conditions and include a definition of acceptable risk of failure/damage/loss of life and acceptable/desirable environmental impacts. Therefore, the functional design of hydraulic and/or coastal structures, which aims to combine all these functions and requirements, can be often a very heavy task for the designer.

Note: Prof. J.W. Kamphuis has recently published a number of papers on the modern approach to coastal engineering design process and coastal education. See also his website: http://civil.queensu.ca/people/faculty/kamphuis/publications/.[14]

Kamphuis identifies three limits to coastal engineering. First, the traditional approval processes often result in daring, experimental designs, which test the limit of existing coastal science and engineering knowledge. Second, contemporary project approvals require assessments of impacts with greater accuracy than we can provide. This limit of present knowledge in science and engineering is reached because of uncertainties in our data, models, and computations. Because of substantial uncertainties assessment of a project becomes necessarily a combination of objective input (hard numbers) and subjective input (experience). Since the public and legal processes have great difficulty understanding uncertainties and trusting subjective decisions, projects are easily delayed, postponed, or cancelled, either because of the uncertainties in the results or often because the contemporary approval process provides too many opportunities for opponents to block the project. This is a third limit to engineering in Kamphuis' point of view.

Kamphuis' views are mainly related to the traditional approach; however, some aspects of contemporary approach are also included.

2.3. Developments and Future Needs in Polices and Design Philosophies

Design and construction of hydraulic and/or coastal structures were for many centuries (actually, up to mid-20th century) based mainly on the system of trial and error, with little scientific assessment. However, there has been an increasing need in recent years for reliable information on design methodology and stability criteria of revetments exposed to wave and current action. This need arises partly from an increase in the number and size of applications that have to be realized accordingly to the higher safety standards, and partly from constructing structures at specific locations where they are exposed to more severe wave and current attack (artificial islands, offshore breakwaters, river and sea dikes, waterways and entrance channels with increased intensity and loading due to navigation, etc.).

In the past, we have seen only too often that local experience determined the selection of type and dimensions of the protection system. A satisfying structure of the neighbors was copied, although hydraulic loads and subsoil properties were different. This led to designs, which were unnecessarily conservative and consequently too costly, or were inadequate and thus leading to high maintenance costs. Actually, the technical feasibility and the dimensions of protective structures can easily be determined on a sounder basis and supported by better experience than in the past. Often, the solution being considered should still be tested on a scale model since no generally accepted design rules exist for all possible solutions and circumstances.

Applied design methods, usually site- and material-specific, require often different design parameters, and vary considerably in reliability. As a result, engineers experience particular difficulties when comparing alternative options for new structures and are very restricted in calculations of failure risk and residual life. Bringing more worldwide uniformity in design approaches is a very important factor for overall improvement of reliability of coastal structures. However, proper functioning of hydraulic and coastal structures as an instrument in solving water management and coastal problems is even more an important aspect. Both of these components include risks. Managing these risks, equally when there is a strong manmade (e.g., structure) or a natural component (e.g., flood protection), basically means assessing alternative options under uncertainty. The possibility of multiple fatalities is one of the factors that can vary between options. There are a number of publications that help to increase the awareness of societal risk, and show how to aggregate risks from major hazards, disseminate available knowledge of existing approaches, and exchange information of applications from various domains.[15] The basic components of this integrated approach are shown in Fig. 2.3.

This is very important when structures have to function for flood protection, especially for low-lying areas. The higher, stronger, and more reliable the flood defenses are, the lower the chance that they will collapse. Reducing the possibility of consequent damage is the essential benefit of the level of safety inherent in the flood defenses. To provide these benefits of strengthening the flood defenses demands major investment from society. This covers not only investment for construction of flood prevention structures and their maintenance. In many cases, such construction or improvement of the flood defense systems means damage to the countryside, natural life, or local culture. The demands that are made on the level of protection against high waters also have to be based on balancing of social costs against the benefits of improved flood defenses.[16] However, the balance between costs and benefits can also change as a result of changing social insights, at last but not at least, the actual occurrence of floods and flood damage, or the future climate change. To include all these aspects in the design, it is necessary to have the new design techniques centered on risk-based approach.

In future practice, the results of (much) improved calculations should give rise to the discussion whether the local standards have to be increased (that means also further strengthening of flood defenses or other risk reduction measures) in order to comply with existing standards for Group Risk or that the present situation is to be accepted as a good practice. However, this discussion can only take place based on an extensive policy analysis. At present, such policy analysis cannot be fully drawn up yet. A lot of technical and nontechnical data have to be collected and models need to be developed, but there are more than technical problems. This also implies the new requirements concerning the education of engineers and/or the need to work in multidisciplinary design teams.

2.3.1. *Level of Protection*

Most design manuals are based on a deterministic design philosophy assuming a design water level and a design wave height of predicted return period (say 1 in 50 or 100 years), and the structure is designed to resist that event with an acceptable degree of safety. Probabilistic design methods, applied firstly on a large scale (in coastal engineering) during design and construction of the Eastern Scheldt Barrier in the Netherlands in 1980s, are still not yet a common design philosophy in coastal engineering.[17] However, their use is highly increasing in recent years in Western countries. In probabilistic approach, the reliability of the structure is defined as the probability that the resistance of the structure exceeds the imposed loads. Extensive environmental (statistical) data is necessary if realistic answers are to be expected from a probabilistic analysis, and it is mainly for this reason that the procedures have not been frequently used in the past. However, the more uncertainty one has on environmental data and on structure response calculations, the more important it is to use a probabilistic approach. By using this approach one can estimate the uncertainties and their influence on the final result.

For the return period of environmental events used in the design of hydraulic and coastal structures, the actual value or values selected are generally considered both in relation to the level of protection required and the design life of the structure. Both have an important bearing on the subsequent cost–benefit study. Where high risk is involved and/or where the scheme has a disproportionately high capital cost (i.e., flood barrier scheme, dikes protecting low-lying high density housing/population) extreme return periods of up to 1 in 1,000 (or even 1 in 10,000) years are chosen to ensure an adequate factor of safety. In case of projects of national importance (i.e., flood protection scheme), usually very costly, grant aid is sough from central government (or international aid agency), in which case agreement is reached early in the design through consultation with the appropriate authority.[18]

2.3.2. *Design Life*

Existing Codes of Practice or Design Guidelines often provide some information on the minimum requirements for the design life of hydraulic and coastal structures (usually as 20 years for temporary or short term measures, 50 to 100 years for shore protection structures and 100 to 1,000 years or more for flood prevention structures). However, the proper choice of return period should be carefully investigated based on type and required function of the structure. Also, the probabilistic approach allows to carry out the calculation with respect to cost optimization, which can be a reasonable base for the proper choice of the design return period. Whatever the level of protection, there is always a risk of damage by storms more extreme than that anticipated during the design. Unless the structure is maintained in a good state of repair, the risk of damage is increased in time.

Similarly, the limitations in the serviceable life of some materials used in the construction (i.e., concrete subject to abrasion, steel subject to corrosion, timber subject to deterioration, etc.) means that they cannot be expected to last the overall life of the structure and, repairs and replacement must be allowed for (must be planned for already in the design stage).

It is unrealistic to expect to design any hydraulic or coastal structure such that it will be free of maintenance or repair during its lifetime. Nevertheless, the ever-increasing requirement to minimize maintenance costs in line with some (national) economic restrictions has a considerable influence on the type of solution ultimately accepted. Cost optimization often shows that it is beneficial to use heavier rock (often only with a little increase of cost) than normally used in a rubble structure, in order to reduce the risk of damage suffered and so reduce the maintenance requirement (especially in case when mobilizing of material and equipment can be a problem). Conversely, where access and maintenance are relatively easy and where the result of failure is less serious, low capital-cost works are often an economical and acceptable solution.

To continue the functioning of the hydraulic or coastal structures during the prescribed lifetime, their renovation/rehabilitation will be usually needed. In general, in designing rehabilitation or upgrading works, the engineer is restricted to a much more greater extent than in new works by the existing conditions. In some cases, complete demolition and reconstruction of the structure (or its part) can be considered as an optimum solution. The design of this type of works is primarily conditioned by the inadequacy of the previous structure to fulfill its original purpose or to meet the requirements of a new and more demanding standard or new boundary conditions (i.e., due to the climate change). In some cases, the wrong functioning of the structure can be proved (i.e., a loss of beach in front of the sea wall). In such a case, a radically different type of structural solution may be evolved, which is more compatible with coastal (or river) processes and the needs of conservation and amenity, (for example, introduction of a shingle beach, sometimes in combination with a groin system, instead of a sea wall, as it was often applied in United Kingdom).

2.3.3. *Failure Modes and Partial Safety Factors*

For the majority of coastal (or hydraulic) structures, however, like break-waters, groins, and revetments, there seems to be no generally accepted safety or risk levels, and very few design standards comprise such structures. Prof. Burcharth, a driving force in Europe for reliability standards for breakwaters, made an interesting attempt in 1999 at discussing the safety levels and ways of implementing them in the design procedure of break-waters, at least in the conceptual design stage.[19] In this stage, we basically are evaluating alternative designs and it is of course important that we compare designs with equal functional performance and equal safety.[19–23]

A (coastal) structure can fail or be damaged in several ways (Fig. 2.4). Consequently, it is very important that the designer considers all the relevant failure modes and assures a certain safety level for each of them. The safety of the whole structure can then be calculated by a fault tree analysis (see Fig. 2.5).

It is generally accepted that for each failure mode, implementation of safety in design should be done by the use of partial safety factors linked to the stochastic variables in the design equation, rather than by an overall safety factor on the design equation. An increasing number of national codes (for concrete structures, soil foundation, etc.) and the Euro Code are based on partial safety factors, because this allows a more precise consideration for the differences in parameter uncertainties than an overall safety factor.

The principle of partial safety factors and fault tree analysis related to flood defenses is explained in CUR/TAW document,[24] and for coastal structures is explained by Burcharth.[20,21] The formats of partial safety factor systems in existing codes and standards differ, but have the same

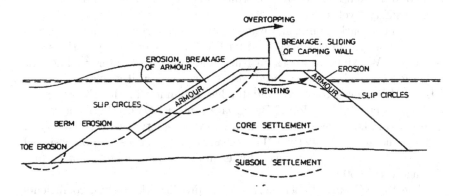

Fig. 2.4. Potential failure modes for breakwaters.[20]

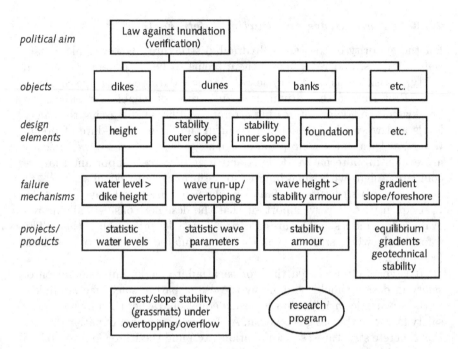

Fig. 2.5. Example of an event tree for dikes and correlation between aims and products.

shortcoming in that the specified coefficients are not related to a specific safety level. They are tuned to reflect the historically accepted safety of conventional designs, and are organized in broad safety classes for which the actual safety level is unknown. Such a format is not suitable for breakwaters and coastal defense structures for which no generally accepted designs and related safety levels exist.

Given this background, and using probabilistic techniques, a new partial safety factor system was developed for breakwaters, originally in the scope of PIANC Working Group 12 on Rubble Mound Breakwaters.[25] The new approach was subsequently used also for vertical wall structures in the PIANC Working Group 28, and further expanded in the EU-MAST 2 and 3 projects Rubble Mound Breakwaters and PROVERBS, respectively. The system allows design to any wanted target safety level (probability of failure) and structure lifetime. This means that structures can be designed to meet any target safety level, for example, 20% probability of certain damage within 50 years.

The PIANC partial coefficient system provides the partial coefficients for any safety level, but no recommendations about safety levels are given, as it is left to the designer to decide on this. Decision can and should be

Table 2.1.   Example of format for safety classes for permanent breakwater structures (numbers only illustrative).

Safety classes	Failure implies

Very low: *no* risk of human injury and *small* environmental and economic consequences

Low: *no* risk of human injury and *some* environmental and/or economic consequences

Normal: *risk* of human injury and *significant* environmental pollution or *high* economic or political consequences

High: *risk* of human injury and *significant* environmental pollution or *very high* economic or political consequences

Limit states: Ultimate Limit State (ULS) and Serviceability Limit State (SLS)

Failure probability within 50 years structure lifetime

			Safety class	
Limit state	Very low	Low	Normal	High
SLS	0.4	0.2	0.1	0.005
ULS	0.2	0.1	0.05	0.01

made on the basis of cost benefit analysis (risk analysis). However, because such analysis can be rather complicated and uncertain, there should be defined in the national or international codes and standards some design target safety classes for the most common types of coastal structures. This will certainly accelerate the use of partial safety factors. The safety classes and related failure probabilities could, for example, be formulated as given in Table 2.1.[19] The useful background information can be found in Ref. 20.

2.3.4. *Flood Protection and Management: Comparative Study for the North Sea Coast*

Increasing population and development have left coastal areas more vulnerable to a variety of hazards, including coastal storms, chronic erosion, and potential sea level rise. Development of coastal areas not only can create increased risk for human life, it also can create a substantial financial risk for individuals and the involved governmental agencies Absolute flood prevention will never be possible although its impact upon human activities can be mitigated in areas of flood hazard. The challenge of flood prevention therefore is to provide an acceptable degree of protection by physical infrastructure combined with alternative means of risk reduction against the most severe floods. A broad range of coastal management functions require good understanding of flooding in order to determine effective policy and response.

In most countries, the provision of flood defenses is undertaken by public authorities (national, regional, or local). Thus, the funding for flood defense infrastructure forms a part of public expenditure and competes with other services and budgets for a share of national and local revenue both for expenditure on new works and on maintenance of existing defenses. Public expenditure on flood defense may be judged on economic return at a national or regional level and is often constrained by political judgments on the raising and distribution of public finances. The time scales for such political judgments are driven by many factors including public opinion, national and international economic cycles, etc. It may be argued that the provision of effective flood defense can become a "victim" of its own success, with increased pressure to reduce expenditure on flood defense when the defenses appear to remove the flood hazard and the impacts of the previous flooding recede in the public and institutional memories. Thus we may hypothesize a cyclic variation of flood hazard determined by responses to major flood events.[26] Superimposed upon this cycle will be increases in vulnerability from economic and social development within (coastal) flood plains and changes in the climatic forcing and hydrological response.

The problem of flooding is too complex for a complete review. However, there are a large number of excellent publications where useful information on these problems and associated techniques can be found in Refs. 15, 18, 24, 27–29. Also, as an example, the results of comparative studies on coastal flooding for some countries along the North Sea are presented below.[30]

The countries along the North Sea coast enjoy both the advantages and disadvantages of a shared neighbor. All countries face the threat of coastal floods to some extent, although the potential consequences of a flooding disaster vary significantly. Each country has developed a system of flood protection measures according to the nature of the threat, potential damages, and its historical, social, political, and cultural background. These measures may range from coastal zone planning to evacuation in emergency situations. In all cases, however, construction and maintenance of flood defense structures is the core of these measures.

Recently, the North Sea Coastal Management Group (NSCMG) agreed to conduct a joint study on the different approaches to safeguarding against coastal flooding. The primary goal of research is to improve communication between the various countries on this subject. The study is limited to coastal defense structures in the five participating countries, namely, Belgium, the United Kingdom, Germany, Denmark, and the Netherlands.[30]

The safety offered by flood defense structures, generally expressed as return periods of extreme water levels, seems to vary quite a lot in the different countries. In the United Kingdom, no safety levels are prescribed.

Indicative safety levels range from less than 200 years to 1,000 years. In the Netherlands, on the other hand, the legally prescribed safety standards range from 2,000 to 10,000 years. The return period of an extreme water level however is only one indication of the actual safety provided by the flood defense structures. In practice, the applied data, design procedures, criteria, and safety margins determine the actual safety. In addition to all this, significant historical, social, cultural, and political differences contribute to the variety of flood protection policies, especially with regard to the authorities involved and responsibilities. The detailed results of this study can be found in the website www.tawinfo.nl (European studies).

Referring to this NSCMG project and the actual developments, it is important to mention that these activities are actually continued and broadened in the scope of the European projects: SafeCoast, ComCoast, and FLOODsite.

Project SafeCoast (http://safecoast.nl/) is about gaining and sharing knowledge and information on coastal flood and erosion risk management between coastal management authorities in five North Sea countries: Denmark, Germany, Netherlands, Belgium, and United Kingdom. What have we learned looking at each other's approaches? And how do we manage our coast in 2050? Project SafeCoast released in 2008 its final Synthesis Report in which all project results are reflected. See also http://safecoast.nl/links/index.php and http://safecoast.nl/national/.

ComCoast, Combining functions in Coastal Defence Zones (http://www.comcoast.org/), contains results of a four-year cooperation between five neighbors countries in the North Sea basin: Belgium, Denmark, Germany, the Netherlands, and the United Kingdom. This project provides the answers to the most common questions when creating Multifunctional Coastal Defense Zones.

FLOODsite project (http://www.floodsite.net/html/project_overview.htm) builds on existing knowledge and best practice for flood risk analysis and management at National, European, and International levels. FLOODsite aims to deliver tools and methodologies to support integrated flood risk analysis and management. The FLOODsite Project comprises an extensive program of research and development work addressing flood risk analysis and management. With a research team drawn from 37 different organizations in 13 different countries of the EU and addressing some 35 distinct tasks of work, there is a need to be methodical in ensuring effective, team-wide communication, and information dissemination. Effective communication and information dissemination are essential if the true potential of the research work is to be realized. Therefore, all publicly available material may be accessed on the website. In general, this project will provide generic guidance and tools regarding these latter activities based on lessons learnt from the pilot studies or collated from other sources, which can be interpreted locally for application as appropriate.

2.4. Manuals and Codes

Unlike the majority of engineering designs, the design of hydraulic and/or coastal works is not always regulated or formalized by codes of practice or centralized design and construction. In some countries (e.g., Japan and China) the design of hydraulic and coastal structures is based upon a national code of practice, but usually it is carried out following a design manual or standardized design guidelines. Such publications may have mandatory effect or be simply advisory. Design Codes of Practice are more useful for less developed countries (or countries with less maritime engineering tradition) where, due to a certain shortcomings in technological development, too much freedom can lead to the unreliable designs. However, such design codes must be prepared by experts (or at least verified by experts) and periodically upgraded. In general, it should be recommended to upgrade the codes every five years.

In Europe, most countries are using design guidelines instead of design code. A formal code of practice (or a strict formalized design manual) are usually considered to be inappropriate for coastal engineering in view of the somewhat empirical nature of the present design process, the diversity of factors bearing on the design solution (often site dependent), and still the major role that engineering judgment and experience plays in the design process. The term "guidelines" implies that guidance is given to the engineer responsible for planning and designing coastal structures in that steps in the design/planning process are described, and the considerations involved are discussed, alternative methodologies are set out and their present limitations explained. Usually these guidelines are officially formalized, but still they include a certain freedom in their use; the designer may deviate from these guidelines when reasonable arguments are provided or when better (more recent) approved design techniques are used. In this way, designers can follow the actual worldwide developments.

The Netherlands probably provides the best example of construction and use of design guidelines.[31] As a low-lying country, dependent on reliable water defense system, it requires high level of safety and thus also reliable design and construction techniques. The responsible departments of the Dutch government, under supervision of the Technical Advisory Committee for Water Defences (TAW), have supported the drawing up of a number of overall guidelines and technical reports that include guidelines on general strategies and design philosophies, and on specific technical subjects.[15] These guidelines cover not only the general design philosophy and methodology, but also technical details on failure modes and calculation methods (often developed in own research programs when not available, or not reliable enough on the market). These guidelines were often used as a reference by other countries (especially, countries around the North Sea) for establishment of their own guidelines. The usual period of upgrading these guidelines is about five years. There are national standards in the Netherlands on specification of materials (rock, concrete, timber, steel, geotextile, etc., which are gradually replaced by European Standards (EuroCodes).

In Germany, the Committee for Waterfront Structures has produced the design recommendations (EAU).[32] These are not mandatory regulations, and so can be simply updated annually if required. However, Germany is known as a country with a long standardization tradition in civil engineering applications (German DINs); it concerns specifications and design methods in a wide range of various materials (concrete, steel, timber, geosynthetics, earthworks, etc.). Also, structural safety is treated by one of these codes. Most of these codes will be actually replaced by Euro codes. When designing coastal structures or their components, reference is usually made to these DIN standards. Useful information can be found in the websites http://kfki.baw.de/ and http://safecoast.nl/national/.

In the United Kingdom, there is little centralized design for construction of coastal structures. In 1984, the British Standards Institution (BSI) issued a Code of Practice for maritime structures. This Code of Practice is not, however, intended to be of direct use in the design of coastal structures. While it considers some subject areas in detail, some other aspects of importance in the design of coastal structures receive very little attention and/or need updating. More information on organizational aspects (policy responsibility) and standards and technical guidelines in United Kingdom are provided by Fowler and Allsop.[33]

In Spain, design of coastal structures is regulated by a recent document: Recommendations for Maritime Structures, ROM 0.2–99.[34] The ROM documents gather the leading state-of-the-art knowledge, as well as the extensive experience in maritime engineering in Spain. The objective of the ROM is to define a set of rules and technical criteria that must be followed in the project design, operation, maintenance, and dismantling of maritime structures, irrespective of the materials and methods used in each of the project stages. Concerning the structural safety, the ROM proposes different levels of reliability analysis for each of the mutually exclusive and collectively exhaustive modes of failure, depending on the general and the operational nature of the maritime structure. Structures with small values of the "nature" (definition of the importance of structure and consequences of failure) can be verified with a "partial safety coefficient level," while those with high nature values are enforced to be verified with the application of a Probabilistic Level II method. An overall procedure is set up in order to guide the designer to fulfill the recommendations prescribed in the program ROM. A software program has been written in order to help designers to follow the ROM. A new revised version of ROM is prepared in 2002.

Japan is known as a country working with rather strict design standards. The history and recent developments on design standards for maritime structures in Japan are extensively outlined in the three papers at Coastal Structures'99. Originally, depending on the designation and usage of a particular region, coasts were managed by four governmental agencies (construction, transport, fishery, and agriculture), each with its own standards and regulations. This situation was very confusing for everybody, especially because of different design approaches and criteria. Recently, the Ministry of Construction and the Ministry of Transport were combined to form a new Ministry of Land, Infrastructure and Transport (MLIT), which clarifies the present coastal management situation in Japan. Actually, extensive revision of technical standards on coastal facilities is underway. Already in 1999, it was decided to revise the old "Technical Standards for Port and Harbour Facilities" and "Design Standard of Fishing Port Facilities."[35] Following that, the Japanese Committee on Coastal Engineering is actually preparing Design Manual on Coastal Facilities, which will form the basis for a new Japanese standard (Mizuguchi and Iwata, 1999). Also, the Japanese Coastal Act (dated from 1956) is under revision. The purpose of these revision activities is to harmonize the different approaches and to reflect progresses in coastal engineering from the recent years. The neutral body like Japanese Committee on Coastal Engineering has been asked to guide all these revision activities, and in this way to help to resolve the differences within the ministries involved.

In United States, where the U.S. Army Corps of Engineers is responsible for many coastlines, the most frequently used guide is the Shore Protection Manual.[2] This is not an official formalized national Design Code, but in practice it is treated in that way, especially within the U.S. Army Corps organization. This guide was a very modern tool in the 1970s and was used worldwide for the design of coastal structures. The advantage of this guide was its completeness, clear style, and calculation examples. The disadvantage was its conservatism and limited upgrading in new editions; upgrading was only accepted when the faults become very evident or when much experience with some new techniques was gained, usually outside the United States. As a result, the new design techniques developed often in the United States were at first applied in Europe or other countries where European consultants were active. Currently, SPM is being replaced by the new and updated Coastal Engineering Manual,[36] which becomes a very modern guide reflecting latest developments and which, in combination with recent Manuals on Rock,[37] can be recommended for worldwide use as a reference.

It should be mentioned here that during the Coastal Structures'99 Conference, a special session on "Guidelines, Standards and Recommendations on Maritime Structures" was held.[38] Speakers invited from different countries around the world (Denmark, France, Germany, Holland, Italy, Japan, UK, United States, and Spain) were invited to present the state of the art and the level of development of guidelines in their respective countries. Also, the PIANC Safety Factor System for Breakwaters was discussed. From the presentations, it seems that a large variety of codes, manuals, and guidelines with different scopes and objectives are now available; moreover, it is apparent that each country is writing its own standards in isolation, having no interaction with other countries. However, there are many similarities between these documents.

Concerning the structural safety, most of the standards are using the Method of the Limit States as a standard method for the verification of the failure modes. Many countries are still using overall safety factors, while others are developing or using partial safety factors. In order to facilitate the design of breakwaters to any target safety level, the PIANC PTC II Working Groups on breakwaters developed a system of partial safety factors corresponding to any wanted safety level, which can be considered as a practical engineering way of using a probabilistic level II method. The PIANC "method" is independent as such of the level of environmental data quality. The partial safety factors given are related to the data quality (poor or good data sets). More details on this subject can be found in the Proceedings of this Coastal Structures'99 conference.[38]

2.4.1. *Future Design Requirements (Codes)*

Technical developments will always play a crucial role, resulting in one or other way in further upgrading of our knowledge and improvement of our design standards. However, each new period brings some new elements and problems, which should explicitly be taken into account and planned in more structural way. The type of design that will need to be undertaken in the 21st century will reflect on one side the type and magnitude of existing work including their aging and continuing deterioration, which is seen to be mainly in the field of upgrading, extension, rehabilitation, and maintenance.[39] On the other side, the new problems arise due to the long-term changes affecting the coastal regimes, such as the structural erosion in front of coastal structures and the trend toward rising sea levels, steeping of foreshores, land subsidence, and continued reduction of sediment supply from the rivers. Some new flood elevation schemes in low-lying areas will probably be required due to new safety standards.

The upgrading and extension of existing structures to meet new defense standards may require the engineer to determine the structural stability of the existing structures in order to determine whether the increased loading is capable of being accommodated. To support the engineer in his new task, new techniques for safety assessment and new criteria for dealing with upgrading of existing structures should be developed. In this respect, the ability to benefit from lessons learnt from the previous works is of paramount importance. The Dutch guide on safety assessment of dikes can be seen as an example of such development[15,40] (www.tawinfo.nl).

Taking a long-term view, the nature of the requirements will partly depend on the increasing demands that might be made on the coastline due to the continuing upward trend in leisure pursuits, or a greater emphasis on conservation.[39] Such a development may well necessitate a radical change of strategy in coastal defenses but it is impossible to predict the type of changes that may result. However from the viewpoint of design, it is already recognized that there is a need to consider coastal engineering strategy over much greater lengths of coastline and over a longer period than at present. Some examples in this direction can be found in the Netherlands[15,41] and in United Kingdom.[42]

Strategic planning on this scale would be helped, if a comprehensive and detailed database of all the existing hydraulic and coastal defenses all over the world existed. National authorities and international organizations should initiate some actions for development of such a database (including the lessons learnt from failures) and preparing new guidelines. The state-of-the-art review, which follows, should be based on (international) discussion with design engineers, contractors, research scientists, and administrators to present a balanced presentation on a wide range of views. These guidelines should set out the state of the art in each subject area and comment on limitations that exist in available knowledge.

The introduction of (internationally recognized) guidelines should bring about an increase of reliability (reduction of risk) and a (possible) reduction in the overall cost to the nation of works by[39]

— helping the designer to identify the most effective design solution;
— improving the overall level of design practice, so as to reduce the number and cost of overdesigned and underdesigned works; and

— promoting common standards of planning and design, thus improving the effectiveness and co-ordination of coastline control nationally.

The future guidelines should serve a valuable role in setting out a common framework for future planning and design, and for helping to identify the most effective number of alternative design approaches with reference to the differences in geographical conditions and economic developments (abilities). Production of the guidelines must not reduce the need to carry out further research into the key areas of design but on the other hand, it should stimulate a new research in areas where our knowledge is still limited. Moreover, not all situations can be covered by guidelines, which refer more to standard cases, and there will always be need for additional research (i.e., model investigation) for special problems and high-risk projects. In order to maintain an overall coherence, the design guidelines should be reviewed (internationally) periodically (say, every 5 years, to a maximum of 10 years) to introduce the advances in the state-of-the-art technology and incorporate new experience.

2.5. Design Techniques

2.5.1. *Design Methodology*

The design of hydraulic and coastal structures subjected to currents and wave attack is a complex problem. The design process and methodology are summarized in Figs. 2.3 and 2.6.

When designing these structures, the following aspects have to be considered:

— The function of the structure
— The physical environment
— The construction method
— Operation and maintenance

The main stages, which can be identified during the design process, are shown in Fig. 2.3. The designer should be aware of the possible constructional and maintenance constrains. Based on the main functional objectives of the structure, a set of technical requirements has to be assessed.

When designing a hydraulic or a coastal structure (dike, seawall), the following requirements to be met can be formulated:

1. The structure should offer the required extent of protection against flooding at an acceptable risk.

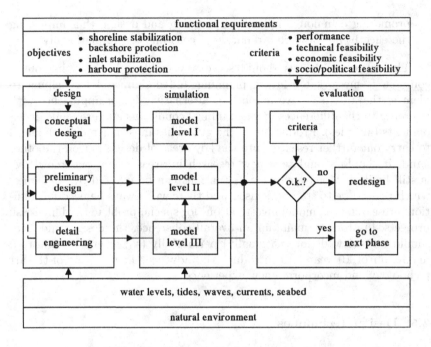

Fig. 2.6.   Design methodology and tools.

Note: The meaning of design levels I, II, and III in Fig. 2.6 is different from the terminology related to the design levels using probabilistic approach as discussed in previous sections (Level I: overall safety factors, Level II: only mean value and standard deviation for stochastic parameters, and Level III: actual distribution of stochastic parameters used).

2. Events at the dike/seawall should be interpreted with a regional perspective of the coast.
3. It must be possible to manage and maintain the structure.
4. Requirements resulting from landscape, recreational, and ecological viewpoints should also be met when possible.
5. The construction cost should be minimized to an acceptable/responsible level,
6. Legal restrictions.

A very important stage in the design process is that of making alternatives, both in case of a conceptual design (making choice of a solution) as well as in the final design (optimization of the structural design). These are the moments in the design where the cost of project/structure can be influenced. Elaboration of these points mentioned above depends on specific local circumstances as a type of upland (lowland or not) and its development (economical value), availability of equipment, manpower, and materials, and others. The high dikes/seawalls are needed for protection of lowlands against inundation while lower seawalls are often sufficient in other cases. The cost of construction and maintenance is generally a controlling factor in determining the type of structure to be used. The starting points for the design should be carefully examined in cooperation with the client or future manager of the project.

This design methodology is shown schematically in Fig. 2.6, including also various simulation models (design tools) required to evaluate the behavior of the structure in the various stages of design.[3,9] In general, it can be stated that in the course of the design process, more advanced methods are used. The actual choice, however, is dependent on the complexity of the problems, the size of the project, and an acceptable risk-level.

Depending on the objective, simulation can vary from crude approximations and rules of thumb (usually applicable at level I/conceptual design), through accepted empirical design formulae with their limitations (usually applicable at level II/preliminary design), to sophisticated reproductions of reality, using physical models, analogue techniques, or numerical models (usually applicable at level III/final detail engineering). This kind of methodology should be followed both in the case of functional design as well as in case of structural design. Examples of level I tools (rules of thumb) and level III tools (models) are presented below. Level II tools can be found in Design Manuals and on the websites:

http://ihe.nl/we/dicea/cress.htm or http://www.cress.nl (CRESS Program). Also PC-BREAKWAT (for breakwaters), ANAMOS (for block revetments), and Delft Hydraulics (http://www.wldelft.nl/soft/chess/index.html) belongs to this category.

2.5.2. *Level I Tools (Rules of Thumb)*

Some examples of rules of thumb (tools for first estimate) are given below:

(a) Stability of revetments under wave attack

$$\frac{H_s}{\Delta D} \leq F \frac{\cos \alpha}{\xi_p^{0.5}}; (\text{ctg}\alpha \geq 2) \tag{2.1}$$

with ξ_p = breaker similarity index on a slope;

$$\xi_p = \tan \alpha (H_s/L_op)^{-0.5} = 1.25\, T_p.H_s^{-0.5}.\tan \alpha,$$

H_s = significant wave height, Δ = relative mass density of units, D = thickness of cover layer (= D_{n50} for stone), α = angle of slope,

$F = 2$ to 2.5 for rock, $= 3$ for pitched stone, $= 4$ to 5 for place blocks, $= 5$ to 6 for interlocked blocks and cabled block mats.

(b) Maximum (depth-limited) wave height

$$H_{s,max} = (0.5\,\text{to}\,0.6)h \tag{2.2}$$

where h is the local depth.

More precisely, use Goda's graphs[44] or ENDEC: http://www.cress.nl

(c) Current attack:

$$\Delta D = \frac{U^2}{2g} \tag{2.3}$$

where U is the depth-average velocity and g is the gravity.

Multiply $(3/4)D$ for a uniform flow and $(3/2)D$ for a nonstationary, turbulent flow.

(d) Scour depth (h_{scour}) and length of toe protection ($L_{toe\ prot.}$)

$$h_{scour} \cong H_s \quad \text{and} \quad L_{toe\ prot} \geq 2h_{scour} \tag{2.4}$$

where H_s is the local wave height.

(e) Minimum length of protection of crest/splash area

$$L_s > 3D_{n50}(>0.5\,\text{m}) \tag{2.5}$$

(f) Granular filter (Fig. 2.7)

$$D_{50up}/D_{50down} < 6\ \text{to}\ 10 \tag{2.6}$$

For uniformly graded materials (for breakwaters ≤ 3 to 5), or more general:

$D_{15up}/D_{85down} \leq 5$; material/soil tightness (for breakwaters ≤ 3 to 4)

$D_{15up}/D_{15down} \geq 5$; permeability criterion

$D_{60}/D_{10} < 10$; internal stability (= uniform grading)

(up = upper layer, down = lower layer)

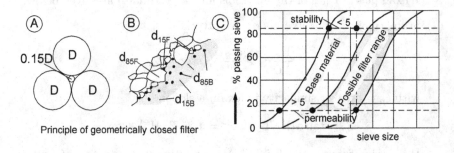

Fig. 2.7.   Geometrically closed granular filters.

(g) Geotextile filter (for uniform gradation of subsoil)

$$O_{90geot} < 2D_{90base} \text{ for low turbulence and } H_s < 0.5 \text{ m} \quad (2.7a)$$
$$O_{90geot} \leq D_{90base} \text{ for high turbulence and/}$$
$$\text{or dynamic loading, } H_s \geq 0.5 \text{ m} \quad (2.7b)$$

(h) Wave run-up, $R_{u2\%}$ (= 2% run-up exceeded by 2% of waves only)

$$R_{u2\%}/H_s = 8 \tan \alpha \text{ (for ctg}\alpha \geq 3 \text{ and } 0.03 < H_s/L_p < 0.05) \quad (2.8)$$

or more general

$$R_{u2\%}/H_s = 1.6\xi_p \quad \text{for } \xi_p < 2 \text{ to } 2.5$$

and

$$R_{u2\%}/H_s = 3.2 \quad \text{for } \xi_p > 2.5$$

where H_s = significant wave height, α = angle of slope and ξ_p = breaker index; for riprap, use $0.6 \, R_{u2\%}$.

(g) Stability of low-crested structures (LCS)

Usually for submerged structures, the stability at the water level close to the crest level will be most critical. Assuming depth limited conditions ($H_s = 0.5 \, h$, where h = local depth), the (rule of thumb) stability criterion becomes $H_s/\Delta D_{n50} = 2$ or, $D_{n50} = H_s/3$, or $D_{n50} = h/6$, where $D_{n50} = (M_{50}/\rho_s)^{1/3}$; $D_{n50} =$ nominal stone diameter and M_{50} and $\rho_s =$ average mass and density of stone. The upgraded stability formulas for LCS structures, including head effect and scour, can be found in Delos report (www.delos.unibo.it).[43]

2.5.3. *Level III Tools (Models) and Input Parameters*

2.5.3.1. *Models for input parameters*

Knowledge of the relevant wave climate is crucial to the design and construction of coastal structures (do remember: rubbish in rubbish out). Good reliable data measured over a long period is rarely available, and in many cases limitations of time and/or cost do not permit such data to be obtained. The alternative is to derivate long-term estimates of wave climate by hindcasting on the basis of wind data. Special attention still deserves the prediction of wave climate in areas exposed to hurricanes, typhoons, and tsunamis, especially for countries, which do not have own (proper equipped) forecasting services.

An essential parameter in the design of hydraulic or coastal structures is the probability of occurrence of severe events (high water levels, high waves). The design procedure based on the probabilities of water levels alone plus appropriate wave height is presently widely applied. However, this procedure does not reflect the full picture, as it does not allow for the possible correlation between the various parameter causing extreme events (tide, surge magnitude, wind direction, wave height, and period). For an optimum design, the joint probability of all these parameters should be taken into account. A number of recent manuals and guidelines have included this as a recommended approach; however, its application needs more detailed statistical data and correlations, and probabilistic calculation methods.[24,25,36,37] Further developments in this direction should be stimulated, especially concerning the more user-friendly programs for reworking of statistical data into the required full joint design probability.

The next important design activity is to transfer the "offshore" wave conditions into shallow water. Actually, the most frequently used methods are 1-D models as developed by Goda[44] and by Battjes & Janssen (ENDEC-model).[45] This item is currently the subject of the further improvement of a number of computer programs of various levels of sophistication. The most advanced method is at this moment probably the Simulating Waves Nearshore (SWAN) model, which was developed by the Technical University of Delft in the Netherlands and is a public domain model.[46] However, even this model still needs further validation under various conditions. Shallow foreshores considerably affect wave propagation and hence wave impact and runup on coastal structures. This concerns, for instance, the evolution of wave height distributions and wave energy spectra between deep water and the toe of coastal structures.

As it was already mentioned, there are a number of models available. As an example, two numerical models have been applied in recent studies in the Netherlands to model the wave propagation over the foreshore and one numerical model has been applied to model wave motion on the structure. The models applied for wave propagation over the shallow foreshore are a spectral wave model (SWAN) and a time-domain Boussinesq-type model (TRITON).[47] The model applied for modeling wave motion on the structure (a dike) is a time-domain model based on the nonlinear shallow-water wave equations (ODIFLOCS).[48]

The general impression from examining the wave energy spectra is that the time-domain model simulates both the spectral shapes and the energy levels rather accurately for the conditions with significant energy dissipation due to severe wave breaking. Also the energy shift to the lower frequencies is modeled surprisingly well. The accuracy of the predictions for the wave periods is higher than obtained with the spectral wave model, though at the cost of higher computational efforts. The model is considered as suitable to provide estimates of the relevant parameters for wave runup and wave overtopping on dikes with shallow foreshores. Based on the investigations the following conclusions can be drawn.

The spectral wave model, applied for wave propagation of short waves over the foreshore (SWAN), yields valuable insight in the evolution of wave energy spectra over the foreshore. It also shows that the computed energy levels in the short waves are rather accurately predicted, considering the rather extreme energy dissipation in the tests. The wave parameters H_{m0} and $T_{m-1,0}$ at the toe of the structure are both under predicted (13% and 21%, respectively), using the default settings of this numerical model. Modifications of the numerical model settings for this kind of applications might improve the results. Further improvements of this model could be dedicated to decrease wave energy transfer to higher frequencies and to increase wave energy transfer to lower frequencies.

The time-domain wave model applied for wave propagation over the foreshore (TRITON) shows accurate results for the wave parameters H_{m0} and $T_{m-1,0}$ at each position. The deviations at the toe of the structure remain below 10% and 5%, respectively (based on the energy in the short waves). The evolution of the wave energy spectra is rather accurately simulated despite the extreme energy dissipation. Also the energy transfer to lower frequencies is clearly present. The model to include wave breaking in this Boussinesq-type model appears to be effective in reducing the wave energy without significant loss of accuracy in the simulation of wave energy spectra. Further validations of this model include 2DH-situations with angular wave attack, directional spreading, and nonuniform depth contours.

The time-domain wave model applied for the simulation of wave interaction with the dike (ODIFLOCS) shows that accurate results on wave runup levels can be obtained if use is made of measured surface elevations of the incident waves (on average 10% under predictions of the wave runup levels). The use of incident waves based on numerical results from the spectral wave model (SWAN) doubles the mean differences (18%) because both numerical models lead to too much wave energy dissipation. The use of incident waves calculated by the time-domain wave model (TRITON) reduces the differences significantly (on average less than 3%) because the numerical models lead to counteracting errors. Applying the two models together led to relatively accurate predictions of the wave runup levels for the present data set.

Recently Battjes presented the actual review of advances in understanding and modeling of hydrodynamic and morphodynamic processes in the coastal zone as well as some challenges for further developments.[49]

Different processes are distinguished, such as sea level, tides, and storm surges, large-scale coastal currents, episodic events, wind waves and swell, wave–current interaction, surf beat, morphodynamic processes and modeling, and for each of these, essential characteristics of the state of the art are mentioned, with emphasis on recent developments, as well as open questions that are considered important. Structural and other aspects that are relevant in the broader domain of coastal engineering are not considered.

Some of these new approaches (i.e., spectral wave definitions, using SWAN for transformation of waves in shallow water, etc.) have already been incorporated in the recent manuals.[36,37,50] The influence of different definition of input parameter on calculation results is discussed (Verhagen et al.[51])

2.5.4. Stability of Cover Layers: Some Examples

Breakwaters. The sudden intensification of research on rubble mound breakwaters in the 1980s was triggered by several damage to a series of relatively new breakwaters (Sines in Portugal, Arzew in Algeria, and Tripoli in Libya). Also a substantial number of damage cases in Japan led to review of existing design formulae and the development of the famous Goda's formula for vertical breakwaters.[44] It was evident that there were fundamental problems with our methods of designing rubble mound breakwaters. The failure of Sines, Arzew, and Tripoli breakwaters are described in Burcharth.[52] The main causes of the failures were as follows:

— The relative decrease in armor unit strength with increasing size was not considered and/or taken into account. This was crucial for slender and complex types of armor units.
— The second reason for the major failures of rubble mound breakwaters was underestimation of the wave climate.
— The third reason was bad model testing with incorrect modeling of the structure and the seabed topography.

A lot of research on the strength of slender armor units followed these failures resulting in strength-design formulae for Dolosse and Tetrapods by which one can estimate the tensile stresses as function of incident waves, size of the units, and for (Dolosse) the waist ratio. The tensile strength can then be compared to the concrete tensile strength in order to estimate if breakage takes place or not. The formulae also provide the relative number of broken units given the wave climate, the size, and the tensile strength of the concrete.[53] The failures involving broken slender concrete units resulted in two trends: return to bulky units like cubes and development in stronger complex (multileg) units, still with hydraulic stability higher than, for example, cubes. The Accropode is a result of this development.

All the older designs were mostly based on (simplified) formula of Hudson dated from 1950s, which gained its popularity due to its simplicity and the status of U.S. Army Corps. However, the problems with using this formula started in 1970s with introduction of random waves and the necessity of transformation of regular waves into irregular waves. In 1980s, the number of testing facilities and test results with random waves became so large that the necessity of new design formulas became evident. The Hudson's formula is still preferred in some countries, e.g., United States, for shallow water conditions.[36]

The new research in 1980s provided more understanding of failure mechanisms and more new sophisticated formulae on stability and rocking of rubble mound structures and artificial armor units.

Formulae developed by Van der Meer by fitting to model test data, with some later modifications, became standard design formulations.[4] However, the reason of this development was quite different from those above mentioned. To explain this we have turn back to 1970s when the author was involved in Delta project, the largest project of damming tidal gaps in the Netherlands, following the necessary actions after flood disaster in 1953. The author has discovered at that time that there was little known on stability of cover layers under wave attack, and that the existing formulations (Hudson, Irribaren, Hedar) were not perfect. His doubts were strengthened by research of John Ahrens, a brilliant researcher from US. Corps of Engineers, who was probably too far ahead in time for a general acceptance. Even in his home organization, he never gained the recognition, as he deserved; his work was never mention in U.S. Shore Protection Manual. Ahrens performed extensive tests on riprap stability and the influence of the wave period in the mid-1970s; the tests were conducted in the CERC large wave tank (with regular waves).[54] Pilarczyk[55,56] continued to replot Ahrens's data and obtained surprising similarity with the later design graph by Van der Meer.[4] This work by Ahrens and his later research on dynamic stability of reefs and revetments, together with work by van Hijum on gravel beaches, were the reason for the author to prepare a set of proposals for a systematic research on static and dynamic stability of granular materials (rock and gravel) under wave attack.[55,56] This program, commissioned in early 1980s to the Delft Hydraulics, was successfully realized under direct guidance by Van der Meer in 1988.

The basic structure of the Van der Meer formula is such that the stability number $H/\Delta D$ is expressed in terms of natural or structural boundary conditions, for example (the sample formula is valid for rock under plunging waves):

$$\frac{Hs}{\Delta D_{n50}} = 6.2 P^{0.18} \left(\frac{S}{\sqrt{N}} \right)^{0.2} \frac{1}{\sqrt{\xi_m}} \qquad (2.9)$$

in which H_s = significant wave height, Δ = relative mass density, D_{n50} = nominal stone diameter, P = permeability coefficient representing composition of the structure, S = damage level, and ξ_m = surf similarity parameter (Iribarren number).

The work of Van der Meer is now generally applied by designers and it has considerably reduced (but not eliminated) the need to perform model experiments during design process.[36,37,43] We have always to remember that each formula represents only a certain schematization of reality. Moreover, as far as these formulas are based on experiments and not based on fully physical understanding and mathematical formulations of processes involved, each geometrical change in the design may lead to deviation in the design results, and to the need for performance of model investigation. Another advantage of the Van der Meer formulae over the formula of Hudson is the fact that the statistical reliability of the expression is given, which enables the designer to make a probabilistic analysis of the behavior of the design.[5] The statistical uncertainty analysis for almost all formulae (including Hudson) is given by Burcharth in CEM.[36]

Following the same philosophy, Van der Meer and others have modified and extended the formulae for the stability of rock to many other aspects of breakwater design such as stability of some artificial units, toe stability, and overtopping and wave transmission. However, these latest formulae (overtopping and transmission) are still in a very rudimentary stage and need further improvement and extension. It concerns especially such structures as submerged reefs with a wide crest where all design aspects (stability, transmission, and functional layout) are not understood properly yet. Some experience with these structures is obtained in Japan and United States, and summarized in the scope of DELOS project[43]; however, the generally valid design criteria are still absent.

What has been said for slopes under wave attack is largely valid for slopes and horizontal bottom protection under currents. The designer has a number of black box design tools available, but the understanding of the contents of the black box is far from complete. Specifically when these black box design formulae are used in expert systems, one may in the end be confronted with serious mistakes. If an experienced designer still realizes the shortcomings and limitations of the black box formula, the inexperienced user of the expert system can easily overlook the implication of it.

Although a reliable set of design formulae is available, the main challenge in the field of rubble mound structures is to establish a conceptual model that clarifies the physical background of it. This will require careful experimental work, measuring the hydrodynamic conditions in the vicinity of the slope and inside the breakwater. Burcharth *et al.*[22] studied the internal flow process in physical models at different scales and in prototype and developed a method for scaling of core material, thus minimizing scale effects on stability. Possibly, an intermediate step has to be taken by developing a 2D mathematical model (often called "numerical flume") that describes the pressures and flow field with sufficient accuracy, and examples of such development can be found in references.[48,57] Experimental work also is necessary to assess the influence of turbulence. A second challenge is further exploring the opportunity to use single instead of double armor layers and further modification of filter rules. This will lead to considerable savings.

The first results of these new developments are promising, but it will take quite some time before this research leads to engineering tools. Another development (and also challenge for future) is the multifunctional design of breakwaters (and may be other structures), for example, a combination of protective function with energy production (wave or tide power), aquaculture, public space for (restaurants, underwater aquarium/parks), etc. Many activities in this direction are already undertaken in Japan.

Influence of input parameters on calculation results. As it was already mentioned above, various definitions of wave parameters are currently used in various formulae on stability, runup, and overtopping. The effect of these input parameters is very evident when applied for shallow water, as it was discussed by Verhagen *et al.*[51] Their findings are summarized below. Recent research has shown that for wave structure interaction in case of shallow water, the spectral period based on the first negative moment of the energy spectrum (Tm-1.0) is a better descriptor than a mean period or the peak period of the spectrum. On this basis in the Rock Manual[37], several equations for runup, overtopping, and structural stability are presented. Also in the new EurOtop Overtopping Manual,[50] this parameter is used. The Rock Manual also indicates that for structural stability the parameter $H2\%$ is a better descriptor than the Hs or the $Hm0$. However, the determination of these parameters is not yet a standard procedure, and often conversion values are used (e.g., $H2\%/Hm0 = 1.4$ and $Tm\text{-}1,0/Tp = 1.1$; and therefore implicitly assuming a Rayleigh distribution and a Jonswap spectral shape with peak enhancement factor $\gamma = 3.3$ and an f-5 spectral tail). However, by using these standard conversion factors, the advantages of the new approach completely disappear, as for the conversion factors are different for nonstandard coasts because the near-shore spectral shape differs from deep water. Exactly in these cases, the new approach is valuable.[51] To compare the effect of the different parameters a calculation is made for the required rock size ($Dn50$, $W50$) and the expected runup (Ru) for a construction build in front of the coast. The structure will be on the plateau at MSL -8 m.

Five different cases are calculated:

1. Use $H2\%$ and $Tm\text{-}1,0$
2. Use $H2\%$ and $Tp/1.1$
3. Use 1.4 $Hm0$ and $Tm\text{-}1,0$
4. Use 1.4 $Hm0$ and $Tp/1.1$
5. Use local $Hm0$ and $Tm0$ and deep water formula for stability

This resulted in the following values summarized in Table 2.2.

This sample calculation shows that the required stone size may vary between 1,200 and 2,200 kg, depending on the choice of parameters. It stresses the importance of a correct choice of the parameters.

The $Tm\text{-}1,0$ parameter can be calculated using SWAN model and its software package can be downloaded from:

- http://www.kennisbank-waterbouw.nl/Software or
- http://www.hydraulicengineering.tudelft.nl
  For information on SWAN, the reader is referred to the official SWAN homepage:
- http://www.swan.tudelft.nl or
- http://fluidmechanics.tudelft.nl/swan

More information on new developments in design techniques and alternative design of maritime structures, and the future needs for research, can be found in Refs. 36,37,50,58,59. However, the new shallow water equations presented in the Rock Manual and the Overtopping Manual are only useful in case one is able to determine the shallow water boundary conditions with sufficiently high accuracy, and not only with conversion numbers. The tool

Table 2.2. Comparison of calculation results.[51]

	1	2	3	4	5
$Dn50$ (m)	0.78	0.85	0.85	0.93	0.87
$W50$ (kg)	1244	1606	1635	2127	1719
Ru (m)	8.4	10.0	9.2	10.7	—

SWAN or SwanOne is able to perform these computations in a user-friendly way.[51]

Revetments. It should be stressed that the proposed developments for breakwaters were partly stimulated/initiated by early developments in understanding and quantification of physical processes in block revetments. [40,58,60−63]

Wave attack on revetments will lead to a complex flow over and through the revetment structure (filter and cover layer), which are quantified in analytical and numerical models.[58,61] The stability of revetments with a granular and/or geotextile filter (pitched stones/blocks, block mats and concrete mattresses) is highly influenced by the permeability of the entire revetment system. The high uplift pressures, induced by wave action, can only be relieved through the joints or filter points in the revetment (Fig. 2.8). The permeability of the revetment system is a decisive factor determining its stability, especially under wave attack, and also has an important influence on the stability of the subsoil. The permeability of a layer of closely placed concrete blocks on a filter layer with and without a geotextile has been investigated in recent years in the Netherlands in the scope of the research program on stability of revetments.

The usual requirement that the permeability of the cover layer should be larger than that of the under layers cannot usually be met in the case

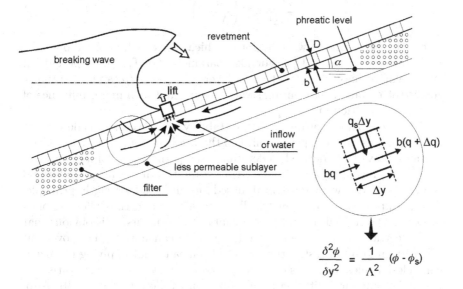

Fig. 2.8.   Physical processes in revetment structure.

of a closed block revetment and other systems with low permeable cover layer (i.e., concrete geomattresses). The low permeable cover layer introduces uplift pressures during a wave attack. In this case, the permeability ratio of the cover layer and the filter, represented in the leakage length, is found to be the most important structural parameter, determining the uplift pressure. The schematized situation can be quantified on the basis of the Laplace equation for linear flow (Fig. 2.8). In the analytical model, nearly all-physical parameters that are relevant to the stability have been incorporated in the "leakage length" factor. For systems on a filter layer, the leakage length Λ is given as:

$$\Lambda = \sqrt{\frac{bDk}{k'}} \tag{2.10}$$

where: Λ = leakage length (m), D = thickness of the revetment cover layer (m), b = thickness of the filter layer (m), k = permeability of the filter layer or subsoil (m/s), and k' = permeability of the top (cover) layer (m/s).

The pressure head difference, which develops on the cover layer, is larger with a large leakage length than with a small leakage length. This is mainly due to the relationship k/k' in the leakage length formula. The effect of the leakage length on the dimensions of the critical wave for semi-permeable revetments is apparent from the following equation:

$$\frac{H_{scr}}{\Delta D} = f\left(\frac{D}{\Lambda \xi_{op}}\right)^{0.67} \tag{2.11}$$

where H_{scr} = significant wave height at which blocks will be lifted out (m); $\xi_{op} = \tan\alpha/\sqrt{(H_s/(1.56T_p^2))}$ = breaker parameter (-); T_p = wave period (s); Δ = relative mass density of cover layer = $(\rho_s - \rho)/\rho$, and f = stability coefficient mainly dependent on structure type and with minor influence of Δ, $\tan\alpha$, and friction.

This research has proved that the stability of a revetment is dependent on the composition and permeability of the whole system of the cover layer. Formulas have been derived to determine the permeability of a cover layer and filters, including a geotextile. Also, stability criteria for granular and geotextile filters were developed based on the load–strength principle, allowing application of geometrically open filters, and thus allowing optimization of composition and permeability of revetments. It is obvious that only a certain force exceeding a critical value can initiate the movement of a certain grain in a structure. That also means that applying geometrically closed rules for filters often may lead to unnecessary conservatism in the design and/or limitation in optimization freedom (see Fig. 2.9). Also, it often results in execution problems especially when strict closed filter

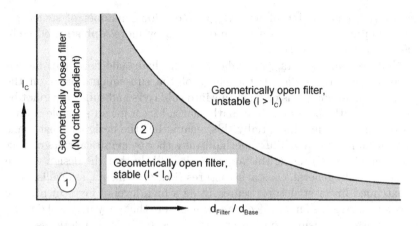

Fig. 2.9. Application of granular filters.[66]

(with many layers) has to be executed under water under unstable weather conditions.[55,61,64−66]

In the scope of these studies, also the internal strength of subsoil has been studied in terms of critical hydraulic gradients. It was recognized that to reduce the acting gradients below the critical ones a certain thickness of the total revetment is needed. This has resulted in additional design criteria on the required total thickness of revetment to avoid the instability of the subsoil. That also means that granular filter cannot always be replaced by geotextile only. For high wave attack (usually, wave height larger than 0.5 m or high turbulence of flow) the geotextile functioning as a filter must be often accompanied by a certain thickness of the granular cushion layer for damping hydraulic gradients. All these design criteria are summarized by Pilarczyk.[67]

The main problem in extension of these achievements to other applications (other revetments, filter structures, bottom protection, breakwaters, etc.) is the lack of calculation methods on internal loads (i.e., hydraulic gradients) for different structural geometries and compositions. Also the geometrically open filter criteria need further development. Research in these fields is still needed and will result in more reliable and cost effective designs.

2.5.5. *Verification of Design*

Not all hydraulic or coastal structures or their components are understood completely; moreover, the existing design techniques represent only a certain schematization of reality. Therefore for a number of structures and/or applications, the verification of design by more sophisticated techniques can still be needed.

While certain aspects, particularly in the hydraulic field, can be relatively accurately predicted, the effect of the subsequent forces on the structure (including transfer functions into sublayers and subsoil) cannot be represented with confidence in a mathematical form for all possible configurations and systems. Essentially this means that the designer must make provisions for perceived failure mechanisms either by empirical rules or past experience. However, using this approach it is likely that the design will be conservative. In general, coastal structures (i.e., revetments, sea walls, etc.) are extended linear structures representing a high level of investment. The financial constraints on a project can be so severe that they may restrict the factors of safety arising from an empirical design. It is therefore essential from both the aspects of economy and structural integrity that the overall design of a structure should be subject to verification. Verification can take several forms: physical modeling, full-scale prototype testing, lessons from past failures, etc.

Engineers are continually required to demonstrate value for money. Verification of a design is often expensive. However, taken as a percentage of the total costs, the cost is in fact very small and can lead to considerable long-term savings in view of the uncertainties that exist in the design of coastal structures. The client should therefore always be informed about the limitations of the design process and the need for verification in order to achieve the optimum design.

2.6. Developments in Materials and Systems: Some Examples

The cost of production and transportation of materials required for hydraulic and coastal structures is an important consideration when selecting a particular design solution. Thus it is important to establish the availability and quality of materials for a particular site at an early stage when considering design options. Using the available tools and models, the structure can be designed to perform the functional requirements. An additional problem is that these functions will change with time in service because of material degradation processes. Therefore the designer's skill must also encompass consideration of durability and degradation processes. A degradation model for materials and structures should be developed so that the whole-life consequences may be considered at the design stage.[37]

2.6.1. *Wastes and Industrial by-Products as Alternative Materials*

Domestic and industrial wastes and industrial by-products form a still growing problem, especially in high-industrialized countries or highly populated regions. A careful policy on application of these materials in civil engineering may (partly) help to reduce this problem. Current European policies aim to increase the use of waste materials of all kinds and to find economic, satisfactory, and safe means of their disposal. The use of waste materials in hydraulic and coastal structures is limited by their particle size distribution, mechanical and chemical stabilities, and the need to avoid materials that present an actual or potential toxic hazard.[37,68]

In the Netherlands, due to the lack of natural rock resources, the application of waste materials in civil engineering has already a long tradition. The extensive research on properties of waste materials allows making a proper selection depending on environmental requirements. Waste materials such as silex, quarry wastes, dredging sludge (depending on the source/location), and many minestone wastes have little or no hazardous contamination. These materials can be used as possible core, embankment fill, or filter material. The engineering properties of many waste materials are often comparable or better than traditional materials. Slags have good friction properties due to their angularity and roughness and typically have high density. Mine wastes sometimes have poor weathering characteristics, but are usually inert and have satisfactory grading for deep fills. The fine materials such as fly ashes and ground slags are already in general use as cement replacement and fillers. Good quality control, not only for limiting the potential for toxic hazard, but also of the mechanical properties of waste materials can considerably increase the use of such low-cost materials in appropriately designed coastal and bank protection structures.

2.6.2. *Geosynthetics and Durability*

Geosynthetics are relatively a new type of construction material and gained a large popularity especially in geotechnical engineering and as component for filter structures. There is a large number of types and properties of geosynthetics, which can be tailored to the project requirements.[67] Geosynthetics have already transformed geotechnical engineering to the point that it is no longer possible to do geotechnical engineering without geosynthetics; they are used for drainage, reinforcement of embankments, reduction of settlement, temporary erosion control, land reclamation, and hazardous waste containment facilities.

When geosynthetic materials or products are applied in civil engineering, they are intended to perform particular functions for a minimum expected time, called the design life. Therefore, the most common (and reasonable) question when applying geosynthetics is "what is the expected/guaranteed lifespan of these materials and products." There is no a straight answers to this question. Actually, it is still a matter of "to believe or not to believe." Both the experimental theory and practice cannot answer this question yet. However, the Dutch evaluation of the long-term performance of the older applications of geotextiles (back to 1968) has proved that the hydraulic functioning was still satisfactory. A similar conclusion has been drawn from the recent evaluation of the long-term performance of nonwoven geotextiles from five coastal and bank protection projects in the United States.[69]

The technology of geosynthetics has improved considerably in the years. Therefore, one may expect that with all the modern additives and UV stabilizers, the quality of geosynthetics is (or can be, on request) much higher than in the 1960s. Therefore, for the "unbelievers" among us, the answer about the guaranteed design life of geosynthetics can be at least 50 years. For "believers," one may assume about 100 years or more for buried or underwater applications. These intriguing questions are the subject of various studies and the development of various test methods over the world. Also, the international agencies related to normalization and standardization are very active in this field. The recent guide (European Standard) of the European Normalization Committee presents the actual "normalized knowledge" on this subject.[70] The object of this durability assessment is to provide the designing engineer with the necessary information (generally defined in terms of material reduction or partial safety factors) so that the expected design life can be achieved with confidence.

2.6.3. *Geosystems*

Various structures/systems can be of use in hydraulic and coastal engineering, from traditional rubble and/or concrete systems to more novel materials and systems such as geosynthetics, geosystems, gabions, waste materials, etc. Moreover, there is a growing interest in low-cost or novel engineering methods, particularly as the capital cost of defense works and their maintenance continue to rise. The shortage of natural rock in certain geographical regions can also be a reason for looking to other materials and systems. This all has prompted a demand for cheaper, less massive, and more environmentally acceptable engineering where geosystems are used.[67] Some of the recent applications are shown in Fig. 2.10.

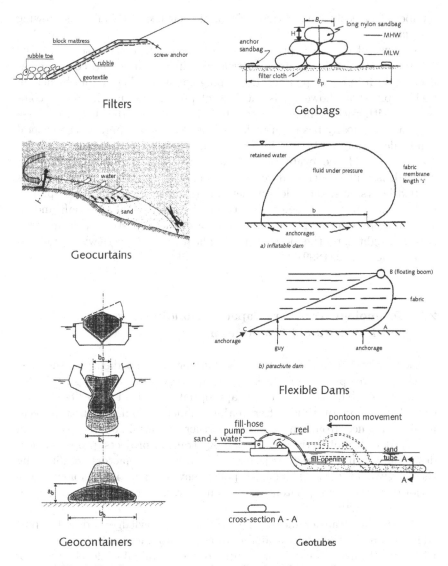

Fig. 2.10. Some concepts on the application of geotextile systems.

These new (geo)systems (geomattresses, geobags, geotubes, seaweed, geocurtains, and screens) were applied successfully in a number of countries and they deserve to be applied on a larger scale. Recently, geocontainers filled with dredged material have been used in dikes and breakwaters in a number of projects around the world, and their use in this field is growing very fast.

Because of the lower price and easier execution, these systems can be a good alternative for protective structures in hydraulic and coastal engineering both in developed and developing countries. The main obstacle in their application, however, is the lack of proper design criteria (in comparison with rock, concrete units, etc.). In the past, the design of these systems was mostly based on rather vague experience than on the general valid calculation methods. More research, especially concerning the large-scale tests and the evaluation of the performance of projects already realized, is still needed. Pilarczyk[67] presents an overview of the existing geotextile systems, their design methods (if available), and their applications. Where possible, some comparison with traditional materials and/or systems is presented. The recent research on these systems has provided better insight into the design and applications (http://www.digibib.tu-bs.de/?docid=00021899).

2.7. Technology Transfer, Capacity Building, and International Cooperation

Know-how/technology transfer is an important expedient in the sustainable development of nations. Technology transfer is sustainable when it is able to deliver an appropriate level of benefits for an extended period of time, after major financial, managerial, and technical assistance from an external partner is terminated. Apart from clearly identified objectives for technology transfer projects, proper project design and well-managed project execution, essential factors conditioning the survival of projects include: policy environment in recipient institution/country, appropriateness of technology, and management organizational capacity.[31,71]

Technology transfer means the transfer of knowledge and skills, possibly in combination with available tools, to institutions and individuals, with the ultimate aim to contribute to the sustainable development of the receiving institution (national) or country (international). Professional educational institutes are likely to restrict their work to specialized education and training of individuals. Institutions, however, generally aim at a broader approach. The objective of technology transfer should be to reinforce the capability of institutions and individuals to solve their problems independently. The required support can be indicated in a diagram differentiating between the analysis and solution phases of engineering problems.

Obviously, knowledge and experience is required on three levels to obtain optimum results:

— Knowledge of the processes
— Knowledge and experience in the use of process simulation techniques
— Experience in practical applications

Knowledge and experience can best be transferred in phases during a project that runs over several years. In many cases, however, budget restrictions call for another approach. The advisor may be called upon as a consultant, and the project includes only some of the phases mentioned above. Not only knowledge and skills must be transferred to the client's staff, but it must also be integrated into the client's organization for future, independent use. In order to guarantee an efficient interaction between the transfer and the integration activities, distinction in phases is required. One may distinguish the following realization phases:

1. Professional education (general scientific/technological level)
2. Professional training (new/specific technology/skills/tools)
3. Development phase (physical/logistic adjustments at recipient party)
4. Institute support (advisory services/exchange visits during project)

The necessary number of phases may vary depending of situation (country, type of project).

Capacity building is important precondition for the realization of future challenges and transfer of know-how, especially for developing countries.

Hydraulic and coastal engineering is a complex art. At this moment a limited number of phenomena can be understood with the help of the laws of physics and fluid mechanics. For the remaining ones, formulas have been developed with a limited accuracy. In addition, input data are limited availability and form another source of uncertainty. Consequently, a sound engineering approach is required, based on practical experience and supported by physical and numerical models, to increase the understanding of many phenomena and to come up with sustainable solutions. Especially, standard solutions do not exist in coastal engineering; solutions very much depend on the local circumstances as well as the social and political approach toward coastal engineering. Consequently, the transfer of coastal engineering knowledge is a complex art as well.

Sustainable transfer of hydraulic and coastal engineering technology at the postgraduate level should therefore aim at increasing the capacities and skills of the engineers such that they are able to analyze a problem correctly and identify possible directions of solutions. Simply learning formulas and learning standard solutions for standard problems are not fruitful and even dangerous: such training does not increase the engineer's understanding of the underlying processes and serious failures may be the result. Transfer of hydraulic and coastal engineering technology should therefore be problem oriented, practical in nature, and geared toward the specific needs of the engineers following the training programme.[72-74]

A general trend can be observed from applying rules to more conceptual thinking. Rules will change fast, so it is more important for an engineer to know the design philosophy. Also engineers have to learn where to find the most up-to-date knowledge regarding the design, which he is making at a certain moment. Because the increased growth in science, rules are outdated faster than in the past; this means that in most cases engineers should not apply the rules they have learned in university. So they should be trained always first to verify if the design method they learned in university is still valid.

Engineers should be trained in a flexible application of the design methodologies they have learned. For design purposes the use of complicated computer software is increasing. The packages are becoming more and more user-friendly, but the insight into the computational process gained is decreasing. This means that the direct link between output and input is less obvious. Engineers have to become more and more aware of the need of checking the output of these programs for inconsistencies (rather than on numerical accuracy). Because of the high quality of presentation methods of modern software, input inconsistencies are often not recognized in time. Engineers have to be trained to identify this problem.[73,74]

Transfer of technology to developing countries can also be a complex matter and needs properly educated engineers. Copying solutions from the Western, industrial countries for application in developing countries is in general not the best solution for solving the problems of developing countries. The main reason for that is that the available resources in the developing world are different from the resources in Western countries. In the industrialized countries there is a strong tendency to solve problems in such a way that the amount of required labor decreases and a capital-intensive solution is sought. The reason for this is the very costly social system and the high standard of living. This causes a big difference between the hourly income and the hourly costs of labor. In developing countries this difference is much less, but on the contrary it is difficult and expensive to import industrial products from elsewhere. Also it is difficult to have sufficient capital available. For those countries, it is more economic to search for solutions, which require hardly any investments, but are relatively labor-intensive. These solutions generally require often more maintenance. However, increased maintenance costs are sometimes very pleasant, provided the initial investment is very low. The cost of the solution can be spread over a longer period without borrowing money for a capital-intensive solution. Maintenance is very important for sustainability of the investment/projects. Maintenance has to be done at the right moment. Also, there has to be a maintaining agency and there has to be a maintenance plan.

In solving these problems, one should always analyze the cause of the problem. Sometimes it is easier to change something in the estuary or river, than to combat the erosion. When it is not possible to take away the cause of the problem, then a number of technical tools are available as discussed in this book. In the design of these methods in most cases a low-investment approach can be followed. Low investment solutions generally require more maintenance than capital-intensive solutions. Therefore the construction has to be designed in such a way that maintenance can be performed easily using local means, that is, with local materials, local people, and with local equipment. These requirements are not very special and one can meet these requirements easily. The main problem is that one has to realize these points during the design phase.

We may conclude from the above that modern engineer must be educated in various technical and nontechnical fields. The task of engineers involved in problems of developing countries should be to adapt (translation) the actual knowledge to appropriate technologies suitable for their problems and their possibilities.

2.8. Conclusions and Recommendations

Problem identification and understanding is very important for a proper choice of solution to water (flood) management and coastal problems. Generally, it may be concluded that there are both physical as well as social aspects to every problem. As a consequence, mere technical solutions often turn out to be mistake. Furthermore, the future use of coast in general should be tailored to fit within the system, whether it is a recreational coast, a wetland, or an ecosystem. For sustainable management, land use planning is required, which does take floodplains and coastal waters into account. Proper quality of environmental boundary conditions defines the quality of design. In many cases, especially in less developed countries, this can be a main obstacle in design of hydraulic and/or coastal structures. Further developments in forecasting and transformation techniques should be stimulated.

There are a large number of hydraulic and coastal structures. For some of them, workable design criteria have been developed in recent years (rubble-mound breakwaters, riprap, block revetments, filter structures, etc.). However, many of these criteria/formulae are still not quite satisfactory, mainly because they lack proper physical background, which makes their extrapolation beyond the present range of experience rather risky. To solve this problem, it will be necessary to continue physical model experiments (on scale and in prototype) to develop, validate, and calibrate new theories. Moreover, there is still a large number of systems with no adequate design techniques, for example, groins, submerged/reef breakwaters, a number of revetment types (gabions, geomattresses), geosystems, open filter, prediction and measures against scouring, etc. However, as opposed to the functional design, the structural design can always be solved by the existing means (design criteria if available or model investigation), assuming availability of funds. Functional design (especially for coastal problems) is one of the most important and most difficult stages in the design process. It defines the effectiveness of the measure (project) in solving a specific problem. Unfortunately, there are still many coastal problems where the present functional design methods are rather doubtful, especially concerning shoreline erosion control measures (i.e., groins, sea walls). Also, adequate measures against lee-side erosion (flanking) deserve more attention.

Alternative (waste) materials and geosynthetics and geosystems constitute potential alternatives for more conventional materials and systems. They deserve to be applied on a larger scale. The geosynthetic durability and the long-term behavior of geosystems belong to the category of overall uncertainties and create a serious obstacle in the wider application of geosynthetics and geosystems and, therefore, are still matters of concern.

The understanding of the coastal responses in respect to the sedimentary coast and its behavior is at least qualitatively available. However, reliable quantification is still lacking, which make functional design of shore erosion control structures very risky. Much mathematical and experimental work is still to be done. Because of scale effects, the experiments will have to be carried out in large facilities or may be verification is even only possible on the basis of prototype observations over a long period. This work is so complicated that international cooperation is almost a prerequisite to achieve success within a reasonable time and cost frame.

Research on hydraulic and coastal structures should benefit from more cooperation among researchers and the associated institutions. Publishing basic information and standardized data would be very useful and helpful in establishing a more general worldwide data bank available, for example, on a website. Systematic (international) monitoring of realized projects (including failure cases) and evaluation of the prototype and laboratory data may provide useful information for verification purposes and further improvement of design methods. It is also the role of the national and international organizations to identify this lack of information and to launch a multiclient studies for extended monitoring and testing programs, to provide users with an independent assessment of the long-term performance of hydraulic and coastal structures, including alternative materials and systems (geosynthetics, geosystems, alternative/waste materials, etc.).

Inventory, evaluation, and dissemination of existing knowledge and future needs, and creating a worldwide accessible data bank are urgent future needs and some actions in this direction should be undertaken by international organizations involved (PIANC, IAHR, UNESCO/WMO, US Coastal Council). It should be recommended to organize periodically (within time span of 5 to 10 years) state-of-the-art reports on various subitems of hydraulic and coastal engineering, which should be prepared by international experts in a given field. It can be organized by creating semipermanent working groups on specific subject and their activities should be paid from a common international fund, which should be established by one of the international organizations. The European projects (MAST, SafeCoast, ComCoast, FLOODsite, etc.) provide good example of international cooperation and dissemination of existing and new knowledge.

Adjustment of the present education system as a part of capacity building for solving future problems should be recognized as one of the new challenges in hydraulic and coastal engineering. It is also important for the proper technology transfer to developing countries and development and maintaining of appropriate technologies for local use. Moreover, solidarity must be found in sharing knowledge, costs, and benefits with less developed countries, which are not able to facilitate the future requirements of integrated coastal management by themselves (on its own) including possible effects of climate change. More attention should be paid to integration of technological innovation with institutional reforms, to rise of awareness to change human behavior, to developing of appropriate technologies that are affordable by poorer countries, to promoting technologies that would fit into small land holdings (local communities), and to capacity building (education), which is needed to continue this process in sustainable way.

Finally, there is a continuous development in the field of hydraulic and coastal engineering, and there is always a certain time gap between new developments (products and design criteria) and publishing them in manuals or professional books. Therefore, it is recommended to follow the professional literature on this subject for updating the present knowledge and/or exchanging new ideas.

References

1. James R. Houston, The coastal structure debate — public & policy aspects, in *Advances in Coastal Structure Design*, eds. R. K. Mohan, O. Magoon and M. Pirrello (ASCE, Reston, WA, 2003).
2. SPM, *Shore Protection Manual*, 4th Edn. (Coastal Engineering Research Centre, US Corps of Engineers, Washington, D.C, 1984).
3. J. Van der Weide, General introduction and hydraulic aspects, in *Short Course on Design of Coastal Structures* (Asian Institute of Technology, Bangkok, 1989).
4. J. P. Van der Meer, Rock slopes and gravel beaches under wave attack, PhD thesis, Delft University of Technology (April 1988) (also available as Delft Hydraulics Communication 396); http://www.vandermeerconsulting.nl/.
5. K. D'Angremond, Coastal Engineering, Achievements and Challenges, *Int. Symp. on Marine Engineering* (Singapore, 2001).
6. RIKZ, Coastal protection Guidelines; a guide to copy with erosion in the broader perspective of Integrated Coastal Zone Management, CZM-C 2001.03, *National Institute for Coastal and Marine Management (RIKZ)* (The Hague, The Netherlands, 2001) (see also http://safecoast. nl/national/index.php?nat=4).
7. CUR, *Beach nourishments and shore parallel structures*, CUR-Report 97–2 (The Netherlands, 1997).
8. N. C. Kraus and O. H. Pilkey (eds.), The effect of seawalls on the beach, *J. Coastal Research* , Special issue No. 4 (1988).
9. K. W. Pilarczyk (ed.), *Coastal Protection* (A. A. Balkema Publisher, Rotterdam, TheNetherlands, 1990) (www.balkema.nl; www.tawinfo.nl).
10. K. W. Pilarczyk and R. Zeidler, *Offshore Breakwaters and Shore Evolution Control* (A. A. Balkema Publisher, Rotterdam, The Netherlands, 1995).
11. H. Hanson, M. M. Thevenot and N. C. Kraus, Numerical simulation of shoreline change with longshore sand waves at groins, *Proc. Coastal Engineering*, Orlando (1996).
12. J. R. C. Hsu, T. Uda and R. Silvester, Shoreline protection methods — Japanese experience, in *Handbook of Coastal Engineering*, ed. J. B. Herbich (McGraw-Hill, New York, 1999).
13. M. Larson, N. C. Kraus and H. Hanson, Analytical solutions of the one-line model of shoreline change near coastal structures, *J. Waterway, Port, Coastal and Ocean Eng.*, ASCE (1997).
14. J. W. Kamphuis, Pushing the Limits of Coastal Engineering, Keynote address and presentation at the Arabian Coasts 2005 Conference, Dubai. Also: Beyond The Limits of Coastal Engineering, International Conference on Coastal Engineering, September 2006, San Diego, CA; http://civil.queensu.ca/people/faculty/kamphuis/publications/.
15. TAW, Fundamentals on water defences (2000). Also: Guide for the Design of Sea and Lake Dikes (1999); Guidelines on Safety Assessment for Water Retaining Structures (1996); Technical Advisory Committee on Water

Defences (TAW), Road and Hydraulic Engineering Division, Delft, the Netherlands; www.tawinfo.nl.

16. R. E. Jorissen and P. J. M. Stallen, *Quantified Societal risk and Policy Making* (Kluwer Academic Publishers, Boston/Dordrecht/London, 1998).

17. RWS, *Design Plan Oosterschelde Storm-Surge Barrier: Overall Design and Design Philosophy*, Ministry of Transport, Public Works and Water Management, Rijkswaterstaat (RWS) (A.A. Balkema, Rotterdam, The Netherlands, 1994).

18. J. K. Vrijling, W. van Hengel and R. J. Houben, Acceptable risk as a basis for design, *J. Rel. Engng System Safety 59* (1998).

19. H. F. Burcharth, State of the art in conceptual design of breakwaters, *Proc. Coastal Structures'99*, Vol. 1, A. A. Balkema, Rotterdam, The Netherlands (1999a).

20. H. F. Burcharth, Reliability Evaluation of a structure at sea, *Proc. Int. Workshop on Wave Barriers in Deepwaters*, Port and Harbor Research Institute, Yokosuka, Japan (1994).

21. H. F. Burcharth, The PIANC safety factor system for breakwaters, *Proc. Coastal Structures'99*, Vol. 2, A. A. Balkema, Rotterdam, The Netherlands (1999b).

22. H. F. Burcharth, Z. Liu and P. Troch, Scaling of core material in rubble mound breakwater model tests, *Proc. COPEDEC V*, Cape Town, South Africa (1999c).

23. H. F. Burcharth, Verification of overall safety factors in deterministic design of model tested breakwaters, *Proc. Advanced Design of Maritime Structures in 21st century*, Port and Harbour Research Institute, Yokosuka, Japan (2001).

24. CUR/TAW, Probabilistic Design of Flood Defences, CUR-Report 141/TAW Guide, (Gouda, The Netherlands, 1990). *Centre for Civil Engineering Research and Codes (CUR), Technical Advisory Committee on Water Defences (TAW)*.

25. PIANC, Analysis of rubble mound breakwaters, Report of Working Group 12, *PIANC, Supplement to Bulletin No. 78/79* (Brussels, Belgium, 1992).

26. P. G. Samuels, *An overview of flood estimation and flood prevention*, Int. Symp. on River Flood Defence (Kassel, Germany, 2000).

27. J. C. Meadowcroft, D. E. Reeve, N. W. H. Allsop, R. P. Diment and J. Cross, Development of new risk assessment procedures for coastal structures, *Proc. Advances in Coastal Structures and Breakwaters*, Thomas Telford, London (1996).

28. M. Koch, Natural hazards and disasters: origins, risks, mitigation and prediction, *Int. Symp. on River Flood Defence*, Kassel, Germany (2000).

29. H. Oumeraci, Conceptual Framework for Risk-based Design of Coastal Flood Defences; a Suggestion, *Proc. Advanced Design of Maritime Structures in 21st century*, Port and Harbour Research Institute, Yokosuka, Japan (2001).

30. R. Jorissen, J. Litjens-van Loon and A. Mendez Lorenzo, Flooding risk in coastal areas: risk, safety levels and probabilistic techniques in five countries

along the North Sea coast, Rijkswaterstaat, Road and Hydraulic Engineering Division, Delft, the Netherlands (2001).

31. K. W. Pilarczyk, Coastal engineering design codes and technology transfer in the Netherlands, *Proc. Coastal Structures'99*, Vol. 2, ed. Inigo J. Losada, A. A. Balkema, Rotterdam, The Netherlands (1999).

32. EAU, *Recommendations of the Committee for Waterfront Structures, Harbours and Waterways*, 8th English Edn. (Ernst & Sohn, Berlin, 1996, 2004).

33. R. E. Fowler and N. W. H. Allsop, Codes, standards and practice for coastal engineering in the UK, *Proc. Coastal Structures'99*, Vol. 2, A. A. Balkema, Rotterdam, The Netherlands (1999).

34. ROM, *Acciones en el Proyecto de Obras Maritimas y Portuarias. Puertos del Estado. Revision for the Technical Committee* (Puertos del Estada, Avda., Madrid, 1999, 2004).

35. OCADI, *The Technical Standards and Commentaries for Port and Harbour Facilities in Japan, The Overseas Coastal Area Development Institute of Japan*, revised English version (Tokyo, 1999/2002).

36. CEM, *Coastal Engineering Manual (CEM)* (US Army Corps of Engineers, Vicksburg, 2006).

37. Rock Manual, *The Rock Manual: The Use of Rock in Hydraulic Engineering*, CIRIA-CUR-CETMEF (2007) enquiries@ciria.org.

38. I. J. Losada (ed.), Coastal Structures'99, *Proc. Int. Conf. Coastal Structures'99, Santander, Spain*, A. A. Balkema, Rotterdam, The Netherlands (2000).

39. CIRIA, Sea Wall; Survey of performance and design practice, Construction Industry Research and Information Association (CIRIA), technical note 125, London (1986).

40. K. W. Pilarczyk (ed.), *Dikes and Revetments* (A. A. Balkema, Rotterdam, The Netherlands, 1998); http://books.google.nl/books?ct=title&q=Coastal +Protection+,+Pilarczyk&lr=&sa=N&start.

41. RWS (Rijkswaterstaat), *A New Coastal Defence Policy for The Netherlands* (Rijkswaterstaat (RWS), Tidal Water Division, The Hague, 1990).

42. MAFF, *Interim Guidance for the Strategic Planning and Appraisal of Flood and Coastal Defence Schemes* (1997). Also: *Flood and Coastal Defence Project Appraisal Guidance; Approaches to Risk* (Ministry of Agriculture, Fisheries and Flood, London, 2000).

43. DELOS, *Environmental Design of Low Crested Coastal Defence Structures*; *D 59 Design Guidelines*, EU 5th Framework Programme 1998–2002, Pitagora Editrice Bologna (2005); www.delos.unibo.it.

44. Y. Goda, *Random Seas and Design of Maritime Structures* (University of Tokyo Press, Tokyo, 1985).

45. J. A. Battjes and J. P. F. M. Janssen, Energy loss and set-up due to breaking of random waves, *Proc. 16th Int. Conf. Coastal Eng., ASCE*, 569–587 (1978).

46. N. Booij, L. H. Holthuijsen and R. C. Ris, The "SWAN" wave model for shallow water, *Proc. 25th Int. Conf. Coastal Engineering*, ASCE, Orlando (1996).

47. M. Borsboom, N. Doorn, J. Groeneweg and M. van Gent, A Boussinesq-type wave model that conserves both mass and momentum, *ASCE, Proc. ICCE*, Sydney (2000). Also: Near-shore wave simulations with a Boussinesq-type model including breaking, *Proc. Coastal Dynamics 2001*, Lund (2001).

48. M. R. A. Van Gent, *Wave Interaction with Permeable Coastal Structures*, PhD thesis, Delft University of Technology, ISBN 90-407-1182-8, Delft University Press (1995). Also: Physical model investigations on coastal structures with shallow foreshores; 2D model test on the Petten Sea-defence, Delft Hydraulics Report H3129-July 1999, Numerical model investigations on coastal structures with shallow foreshores; Validation of numerical models based on physical model tests on the Petten Sea-defence, Delft Hydraulics Report H3351, May 2000, Numerical model simulations of wave propagation and run-up on dikes with shallow foreshores, *Coastal Dynamics'01*. http://www.wldelft.nl/rnd/publ/search.html (insert for author: Gent); http://www.kennisbank-waterbouw.nl/SelfArchiving/CommHydr/CommHydr.htm.

49. Jurjen A. Battjes, Developments in coastal engineering research, *J. Coastal Engineering* **53**, 121–132 (2006).

50. EurOtop Manual, *Wave Overtopping of Sea Defences and Related Structures* (EA Environment Agency, UK, ENW Expertise Netwerk Waterkeren, The Netherlands, KFKI Kuratorium für Forschung im Küsteningenieurwesen, DE, 2007); www.overtopping-manual.com.

51. H. J. Verhagen, G. van Vledder and S. Eslami Arab, A practical method for design of coastal structures in shallow water, *Proc. ICCE'08*, Hamburg, Germany (2008); www.citg.tudelft.nl/live/pagina.jsp?id=71b5a2b8-23b7-474a-9f9b-5c37e622ac33&lang=en.

52. H. F. Burcharth, The lessons from recent breakwater failures — Development in Breakwater Design, *Proc. World Federation of Eng. Organizations*, Technical Congress on Inshore Engineering, Vancouver, Canada (May 1987).

53. H. F. Burcharth, K. d'Angremond, J. W. Van der Meer and Z. Liu, Empirical formulae for breakage of Dolosse and Tetrapods, *Coastal Engineering* **40** (2000).

54. J. P. Ahrens, Large wave tank tests of riprap stability, *C.E.R.C. Technical Memorandum* No. 51 (May 1975).

55. K. W. Pilarczyk, (a) Stability of rock-fill structures, (b) Filters, *in Closure Tidal Basins*, eds. Huis in 't Veld *et al.* (Delft University Press, Delft, 1984); info@library.tudelft.nl; http://repository.tudelft.nl/file/600734/373576.

56. E. Van Hijum and K. W. Pilarczyk, Gravel beaches; equilibrium profile and longshore transport of coarse material under regular and irregular wave attack, Delft Hydraulics Laboratory, Publication No. 274 (July 1982) (also Delft Hydraulics, Publ. no. 293), Delft, The Netherlands.

57. P. Troch, Experimental study an numerical modelling of pore pressure attenuation inside a rubble mound breakwater, *PIANC Bulletin No. 108*, Brussels (2001).
58. ASCE, *Wave Forces on Inclined and Vertical Wall Structures* (Task Committee on Forces on Inclined and Vertical Wall Structures, ASCE, New York, 1995).
59. PHRI, 2001, *Advanced Design of Maritime Structures in 21st Century* (Port and Harbour Research Institute, Yokosuka, Japan, 2001).
60. M. B. De Groot, A. Bezuijen, A. M. Burger and J. L. M. Konter, The interaction between soil, water and bed or slope protection. *Int. Symp. on Modelling Soil-Water-Structure Interactions*, A. A. Balkema, Delft, The Netherlands (1988).
61. CUR, Design manual for pitched slope protection, CUR Report 155, ISBN 90 5410 606 9 (Gouda, the Netherlands, 1995).
62. H. J. Köhler and A. Bezuijen, Permeability Influence of Filter Layers on the Stability of Rip-Rap Revetments under Wave Attack, *Proc. 5th Int. Conf. Geotextiles, Geomembranes and Related Products*, Singapore (1995).
63. M. Klein Breteler, T. Stoutjesdijk and K. Pilarczyk, 1998, Design of alternative revetments, *26th ICCE*, Copenhagen; http://www.wldelft.nl/rnd/publ/search.html (insert author: Breteler).
64. PIANC, Guidelines for the design and construction of flexible revetments incorporating geotextiles for inland waterways, *PIANC, Supplement to Bulletin No. 57*, Brussels, Belgium (1987).
65. PIANC, Guidelines for the design and construction of flexible revetments incorporating geotextiles in marine environment, *PIANC, Supplement to Bulletin No. 78/79*, Brussels, Belgium (1992).
66. G. J. Schiereck, *Introduction to Bed, Bank and Shore Protection* (Delft University Press, Delft, 2001).
67. K. W. Pilarczyk, *Geosynthetics and Geosystems in Hydraulic and Coastal Engineering* (A. A. Balkema, Rotterdam, The Netherlands, 2000).
68. K. W. Pilarczyk, G. J. Laan and H. Den Adel, Application of some waste materials in hydraulic engineering, *2nd European Conf. Environmental Technology*, Amsterdam (1987).
69. G. Mannsbart and B. R. Christopher, Long-term performance of nonwoven geotextile filters in five coastal and bank protection projects, *Geotextiles and Geomembranes*, **15**(4–6) (1997).
70. CEN/CR ISO, Guide to durability of geotextiles and geotextiles related products, European Normalization Committee (CEN), CR ISO 13434, Paris (1998).
71. H. J. Overbeek, J. van der Weide and H. Derks, *Effective Technology Transfer from One Institute to Another* (Delft Hydraulics (for IAHR), 1991).
72. H. J. Verhagen, Education of Coastal Engineers for the 50th ICCE, *Proc. ICCE '96*, Orlando (1996).
73. H. J. Verhagen and W. S. De Vries, Sustainable transfer of coastal engineering knowledge at post gradual level, *Proc. COPEDEC'99*,

Cape Town (1999); www.citg.tudelft.nl/live/pagina.jsp?id=8fefaff8-03a0-442f-9343-b040768ed09d&lang=en.

74. J. W. Kamphuis, Perspective on coastal engineering practice and education, Chapter 41 in *Handbook of Coastal and Ocean Engineering* (World Scientific, Singapore, 1997); http://civil.queensu.ca/people/faculty/kamphuis/publications/.

Chapter 3

Coastal Structures: Action from Waves and Ice

Alf Tørum

Professor Emeritus, Norwegian University of Science and Technology,
Department of Civil and Transport Engineering, Trondheim, Norway
alf.torum@ntnu.no, al-toe@online.no

There are different types of coastal structures such as breakwaters and pile supported structures. Breakwaters are built to protect harbor areas from ocean waves of different kinds, mainly wind-generated waves, but in some areas also against tsunami- or earthquake-generated waves. The design and construction of breakwaters is different in different parts of the world, depending on availability of materials, labor, and material costs and to some extent on tradition and experience. The protection of artificial islands, frequently used by the oil industry, is also designed according to the same principles as breakwaters. Piers and navigation lights are pile-supported structures. In this chapter the emphasis is on wave actions on different types of coastal structures, but ice action on breakwaters is also dealt with. Much attention is paid to the berm breakwater. This is a breakwater developed first in Iceland, but is now used extensively throughout the world, but not much mentioned in the literature. A section on the requirements of stone quality for berm breakwaters and the probability of stone breaking on reshaping berm breakwaters as well as a section on berm breakwater construction is included. Finally a section on the probability of failure of berm breakwaters is also presented.

3.1. Introduction

Until the 1950s, the building of breakwaters was an art based on experience, a trial-and-failure procedure. There was hardly any knowledge of wave climates and on the wave interaction with the breakwater (forces,

run-up etc). But during the Second World War, the science of forecasting
ocean waves was enhanced due to the requirement for forecasting waves for
military landing operation sites, especially in the Pacific Ocean area. The
Norwegian scientist, Professor H.U. Sverdrup, together with the Danish scientist W. Munk, developed a method to forecast waves from a forecasted
wind system, the so-called Sverdrup–Munk method. Although the accuracy
of the method was limited, mainly due to lack of good wave data to calibrate
the method or model, the method was a step forward and was useful for the
war operations. In Norway, the first instrumental wave measurements were
started by the Norwegian Coastal Administration in Bussesundet at Vardø
in 1959, in connection with plans to build a causeway from the mainland to
the Vardø Island, and in 1959 at the fishing harbor Berlevåg, Finmark. The
wave measurements at Berlevåg continued until 1972, by then the longest
continuous wave measurement series in the world.

The first formulae, based on laboratory tests, for evaluating the necessary weight of cover stones on a rubble mound breakwater, came in the
late 1930s.[1] The first laboratory tests were run with regular waves, but
introduction of wave generators for irregular waves in the 1960s made it
possible to carry out tests under more realistic wave conditions. The article
by Carstens et al.[2] was the first international paper to give information of
the stability of rubble mound breakwaters in irregular waves. Gradually the
knowledge of breakwater hydraulics has developed up to the present state
in which the breakwater design is based more on science and less on art.

In Norway, some 600–700 breakwaters have been built, quite many for
fishing harbors, but also for combined fishing and commercial harbors. The
latest large and very exposed breakwater is the one that has been built in
Sirevåg, approximately 50 km south of Stavanger, a very tough and strong
breakwater structure.[3,4] None of the heavy wave-exposed breakwaters in
Norway are subjected to ice loads.

The maximum design-significant wave heights for Norwegian breakwaters are up to approximately 8.0–9.0 m. But throughout the world breakwaters for larger design-significant wave heights have been built. The wave
height record is held by the breakwater at Sines, Portugal, where the design
significant wave height is approximately 11 m and the water depth is 50 m.
A breakwater is under construction at La Coruna, Spain, where the design
significant wave height is approximately 14 m and the water depth is 35 m.
A composite breakwater has been built at Kawaishi, Japan, in maximum
water depth of 65 m (world record?) and for a design-significant wave height
of approximately 8 m.

The state of the art of designing and constructing breakwaters is
reviewed in this chapter, with main emphasis on the design. Most breakwaters are in locations where ice does not influence the design. Therefore

most of the research on breakwater stability has dealt with the stability against waves. However, since there might be an increased need for breakwaters/gravel islands in Arctic areas, a survey on the knowledge on the stability against heavy ice attack on breakwaters has also been included. A chapter on wave induced pore pressures in the bottom soil has been included since this issue has to be taken into account when evaluating the geotechnical stability of a breakwater, especially of breakwaters on weak soils.

For breakwaters on sandy soils, scour protection may be required. The issue of scour and scour protection is not dealt with in this chapter. In spite of much interest on scour so far, more research is still required to understand and predict scour and scour depth at structures. Most of the research on scour at breakwaters and other structures have been through small-scale laboratory tests. In such tests the sand transport is mainly in the bed load mode, while the sand transport under prototype conditions is in the suspension mode. Thus the small scale tests suffer from scale effects.[5] The stability of scour protection layers, or the requirement to stone sizes, has not been systematically investigated, but for site-specific studies the stability of a stone scour protection layer may be investigated.

Although much research have been carried out on the stability of breakwaters and methods have been derived to calculate necessary armor stone weights (rubble mound breakwaters) or wave forces (caisson type breakwaters), it is still recommended to consider site-specific model tests for major breakwaters in heavily exposed areas.

The costs of breakwaters depend very much on local conditions. Cost evaluations and cost considerations have therefore been omitted in this chapter.

The evaluation of the wave and ice climate is not dealt with in this chapter. But it is emphasized that every effort should be made and every available wave and ice data set should be used to evaluate the wave and ice climate to arrive at the design parameters. The uncertainty on the design parameters should be reflected in the design.

Fortunately, the wave heights at a coastal site for breakwaters are smaller than in deep water, due to refraction and especially due to wave breaking.

3.2. Different Types of Breakwaters

Breakwaters may be classified into the following main categories: (1) rubble-mound breakwaters, with subcategories of conventional rubble mound breakwaters and berm breakwaters, (2) Caisson-type breakwaters, with

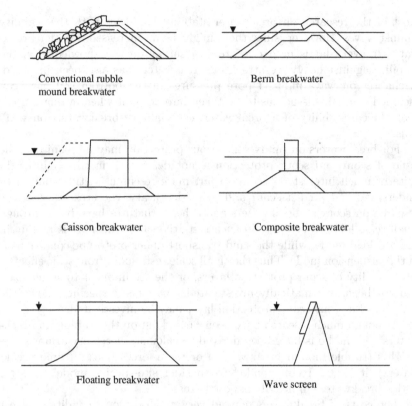

Fig. 3.1.   Main type of breakwaters. Conceptual sketches. Some times a rubble mound, usually from concrete blocks, is placed in front of the caisson breakwater to prevent the caisson from slamming forces from breaking waves.

subcategory composite breakwaters, (3) floating breakwaters, and (4) wave screens (Fig. 3.1).

The floating breakwater and the wave screen are used in wave restricted areas with wave periods less than approximately 3–5 s. They become huge and uneconomical structures if the wave periods become larger, because of their inefficiency of reducing wave heights for longer wave periods. These types of breakwaters are considered not usable in heavy ice-infested areas.

The classical rubble-mound breakwater is covered with stone or concrete armor blocks. Stone armor is used in countries with readily available good quality rock, while concrete armor blocks are used where such stone is not readily available. Several shapes of concrete blocks have been used.

Most frequently, two stone layers are used for the armor. But single-layer breakwaters have been most frequently used in Norway. The reason for this has been the use of simple construction methods. The single-layer breakwater is more vulnerable to damage, although the Norwegian experience is fairly good with respect to maintenance cost, especially in not very exposed areas.

The classical rubble-mound breakwater is the one that is most frequently used. In recent years, the berm breakwater is increasingly used in countries with good rock quality, e.g., Iceland and Norway, but also in many other countries.[6] The advantage of the berm breakwater is that less mass of the individual armor stones is required than for the classical rubble-mound breakwater. The berm breakwater is a "tough" structure compared to the classical rubble mound breakwater, which is a more "brittle" structure.

The caisson-type breakwater is much used in Japan, mainly because of lack of good quality rock.

There are several manuals published on breakwater design, e.g., Refs. 7 and 8. These manuals are detailed, but do not cover the aspects of stability in relation to ice in depth. The berm breakwater is not at all mentioned in Ref. 7. The intention of this chapter is to give a review on breakwater stability, especially the berm breakwater and to compile information on the interaction between breakwaters and ice.

Figures 3.2 and 3.3 show photographs from a rock rubble-mound breakwater with some definitions, in this case the Sirevåg berm breakwater, Norway.

Fig. 3.2.   Sirevåg berm breakwater, Norway.

Fig. 3.3.   Sirevåg berm breakwater. Close up.

3.3. Rubble-Mound Breakwater Hydraulics

3.3.1. *General Discussion on Rubble-Mound Breakwater Stability*

Figure 3.4 shows important failure modes for a rubble-mound breakwater. In this case, the breakwater has a concrete cap or wave screen on top. This is to reduce overtopping and to accommodate traffic on top of the breakwater. In other cases the breakwater is constructed without this concrete wall on top, e.g., the Sirevåg berm breakwater (Fig. 3.3).

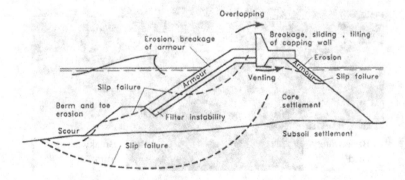

Fig. 3.4.   Important failure modes for conventional rubble mound breakwaters.[9]

The most important failure modes are the failure of the armor stone layer (which is the most important and governs to a large extent the design), including the toe, the failure of the crest structure, and the failure of the rear side due to overtopping of waves.

3.3.2. *Conventional Rubble-Mound Breakwaters with Rock Armor*

The flow of water around the rubble-mound breakwater armor stones during wave run-up and run-down is very complicated. The flow is unsteady and turbulent and there is no information on the detailed flow around the individual stones, which are submerged in a boundary layer. Tørum[10] measured the wave forces on individual armor stones and the water particle velocities adjacent to the "force measuring stone." The results were analyzed in view of the Morison equation.[11] But it is difficult to use the results for the evaluation of the stability of individual armor stones. The formulae for calculating the necessary individual armor stone weight are based on various force and moment equilibrium concepts, leading to formulae with unknown coefficient(s). These coefficient(s) have then to be determined from laboratory wave flume model tests.

3.3.2.1. *Derivation of some breakwater stability formulae*

For better understanding the different formulae for the calculation of the necessary armor stone weight, a brief tutorial review is given on the derivation of a couple of these formulae.

There have been different concepts for the dislocation of the armor stones. Figure 3.5 shows typical armor failure modes.

Figure 3.6 shows schematically the forces that are considered to act on an armor stone. The different concepts on stone dislocation are: (1) rolling, (2) sliding, and (3) lifting, or it could be a combination of all the three.

Iribarren Formula

Iribarren[1] assumed that an armor stone slide along the plane with an angle of inclination, α, when it is removed from its position. It can certainly be argued that in a situation like the one sketched in Fig. 3.6, it is not possible for the stone to slide. It will more likely roll out or be lifted out of its position. However, it may as well be argued that in some time and space there might be a stone that will slide. The concept of Iribarren may then be valid, and for the sake of showing the derivation of a stability formula, with all its shortcomings, we will pursue the Iribarren concept.

A. Tørum

a) *Rocking of unit during up- and down-rush*

b) *Rotation and subsequent down-slope displacement of unit during down-rush*

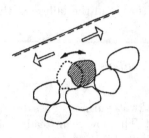
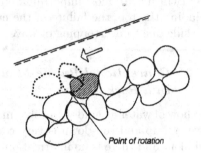

c) *Rotation and subsequent up-slope displacement of unit during up-rush*

d) *Sliding of several armor units (armor layer) during down-rush*

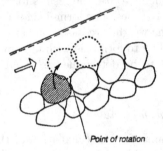
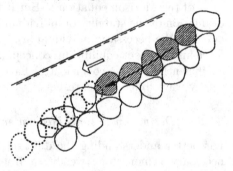

Fig. 3.5.   Typical failure modes of armors.[9]

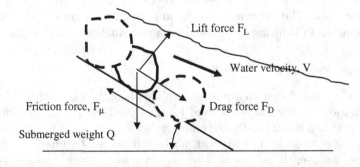

Fig. 3.6.   Conceptual forces acting on a breakwater armor stone.

Sliding will occur when:

$$F_D + Q \sin \alpha \geq F_\mu = \mu(Q \cos \alpha - F_L) \qquad (3.1)$$

where F_D = drag force, F_L = lift force, Q = submerged weight of the stone, μ = friction coefficient and α = inclination angle.

The drag and lift forces are set to:

$$F_D = k_1 \frac{\rho_w}{2} C_D D^2 V^2 \qquad (3.2)$$

$$F_L = k_2 \frac{\rho_w}{2} C_L D^2 V^2 \qquad (3.3)$$

The submerged weight is set to:

$$Q = k_3(\rho_s - \rho_w)gD^3 \qquad (3.4)$$

where k_1, k_2 and k_3 = area and volume coefficients, presently of unknown values, ρ_w = mass density of the water, ρ_s = mass density of the stone, C_D = drag coefficient, C_L = lift coefficient, D = characteristic diameter of the stone, V = water velocity. Note that the inertia force is omitted. This force is generally small for breakwater armor stone.

What should then be considered as the characteristic velocity V? The following reasoning is not precise, but will do for an approximation.

The waves break when the wave height H is roughly equal to the water depth, d. The water particle velocity is then roughly equal to the wave celerity, c, which again is roughly set to $c = V = k_4\sqrt{gd} = k_4\sqrt{gH}$, where k_4 = a velocity coefficient, presently of unknown value. Inserted in Eq. (3.1):

$$k_1 \frac{\rho_w}{2} C_D D^2 k_4^2 gH + k_3(\rho_s - \rho_w)gD^3 \sin \alpha \geq \mu(k_3(\rho_s - \rho_w)gD^3 \cos \alpha$$

$$- k_2 \frac{\rho_w}{2} C_L D^2 k_4^2 gH) \qquad (3.5)$$

or rearranged as:

$$D \geq \frac{\rho_w H \frac{1}{k_3} \left[\frac{1}{2}k_1 C_D k_4^2 + \mu \frac{1}{2} k_2 C_L k_4^2 \right]}{(\mu \cos \alpha - \sin \alpha)(\rho_s - \rho_w)} = \frac{k_5 H}{(\mu \cos \alpha - \sin \alpha)\left(\frac{\rho_s}{\rho_w} - 1 \right)}, \qquad (3.6)$$

where $\frac{1}{k_3}\left[\frac{1}{2}k_1 C_D k_4^2 + \mu \frac{1}{2} k_2 C_L k_4^2 \right] = k_5$

The stone mass is $W = k_3 \rho_s D^3$, which gives

$$W \geq \frac{k_3 k_5^3 \rho_s H^3}{(\mu \cos \alpha - \sin \alpha)^3 \left(\frac{\rho_s}{\rho_w} - 1 \right)^3} = \frac{K \rho_s H^3}{(\mu \cos \alpha - \sin \alpha)^3 \left(\frac{\rho_s}{\rho_w} - 1 \right)^3} \qquad (3.7)$$

All the unknown coefficients are thus lumped together in one unknown coefficient K. This coefficient and the friction coefficient μ have to be determined from model tests.

It is noted that the formula shows that the necessary armor stone weight depends upon the breakwater slope, the mass density of the stone, and the water, the friction angle and the wave height, even to the third power.

This Iribarren formula is a tutorial in respect of some understanding of the mechanism, at least in an approximate way, of the interaction between the waves and the armor stone. But the formula is not much used today. The more frequently used formulae are the Hudson formula[12,13] and the van der Meer formulae.[14,15]

Hudson Formula

Hudson[12,13] did not consider what the armor stone dislocation mechanism was, but assumed that there is force acting on the stone:

$$F_H = k_6 \frac{\rho_w}{2} C_H D^2 k_4^3 g H \tag{3.8}$$

where $C_H =$ force coefficient and $k_6 =$ area coefficient.

According to Hudson, the main stabilization force is the submerged weight of the stone. Hence he set at start of dislocation of the armor stone:

$$k_3 \left(\rho_s - \rho_w \right) g D^3 \leq k_6 \frac{\rho_w}{2} C_H D^2 k_4^2 g H \tag{3.9}$$

or

$$\frac{H}{\left(\frac{\rho_s}{\rho_w} - 1 \right) \left(\frac{W}{\rho_s} \right)^{\frac{1}{3}}} = \frac{k_3^{2/3}}{\frac{1}{2} C_H k_4^2 k_6} = N_s \tag{3.10}$$

N_s is called the stability number. Through a series of model tests, Hudson found that the wave steepness H/L and the depth/wave length relation d/L were not significant parameters in relation to the breakwater stability compared to the breakwater slope (later research shows that this is not always the case). Hudson found that:

$$N_s = (K_D \cot \alpha)^{1/3} \tag{3.11}$$

leading to the Hudson formula:

$$W = \frac{\rho_s H^3}{K_D \left(\frac{\rho_s}{\rho_w} - 1 \right)^3 \cot \alpha} \tag{3.12}$$

The coefficient K_D was obtained from the model tests. Hudson obtained his K_D values, as mentioned, from tests with regular waves, since there were no wave generators available for irregular waves by the time he carried out his tests. Later on recommended values for K_D versus some fractal of H (H_s or $H_{1/10}$) has been given as discussed later.

The stability number N_s may also be expressed as often seen:

$$N_s = \frac{H}{\Delta D_n} \qquad (3.13)$$

where $\Delta = (\rho_s/\rho_w) - 1$, $D_n = (W/\rho_s)^{1/3}$

D_n is thus a nominal diameter equivalent to the edge of a cube with the same mass as the stone, W. In berm breakwater terminology $N_s \equiv H_o$ (see later).

Breakwater Damage

Rubble-mound breakwaters are generally accepted to be slightly damaged by the design waves, i.e., some of the armor stones are allowed to dislocate, but without causing total collapse of the breakwater. There have been several definitions of damage. Burcharth and Hughes[16] list the following:

1. Relative displacement within an area.
   D = number of displaced units/total number of units within reference area.
   Displacement has to be defined, e.g., as position shifted more than a distance D_n, where D_n is stone "diameter," or displacements out of the armor layer. The reference area has to be defined, e.g., as the complete armor area, or as the area between two levels, e.g., SWL $\pm H_s$, where H_s corresponds to a certain damage, or SWL $\pm nD_n$, where $\pm nD_n$ indicates boundaries of armor displacement. SWL = still water level.
2. Number of displaced stones within a strip with width $D_n \cdot N_{od}$ = number of units displaced out of armor layer/(width of tested section/D_n).
3. Relative number of displaced units within total height of armor layer.[15]
   N_{od}/N_a, where N_a is the total number of units within a strip of horizontal width D_n.
   $N_{od}/N_a = D$ if in D the total height of the armor layer is considered, and no sliding $>D_n$ of units parallel to the slope takes place
4. Percent erosion of original volume.[12] D = (average eroded area from profile/area of average profile) $\times$ 100%
5. Relative eroded area,[17] also some times called damage or damage level
   $S = A_c/(D_{n50})^2$, where $D_{n50} = (W_{50}/\rho_s)^{1/3}$ (Fig. 3.7).

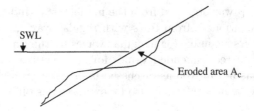

Fig. 3.7.   Eroded area A_c.

In the following some of the most used formulae for calculating the necessary armor stone or concrete unit mass are discussed. These formulae should primarily be used for conceptual designs. Whenever possible, site-specific model tests should be performed, especially for breakwaters subjected to large waves in exposed locations.

3.3.2.2. *Stability of breakwater armor layers*

Most of the general investigations and laboratory tests on the stability of breakwaters have been with the wave direction normal to the longitudinal breakwater axis, Hudson[12,13] and van der Meer.[14,15] Their investigations lead to probably the most known formulae for the calculation of the necessary mass of individual armor units on a straight slope. Recently van Gent *et al.*[18] arrived at a new formula, partly based on reanalysis of data[15] and partly based on new data. Investigations were carried out on the influence on the stability of the wave direction.[19,20]

3.3.2.2.1. Hudson Formula

Hudson used, as mentioned, only regular waves during his model tests because no wave generators for irregular waves were available by the time he carried out the tests. However, there has been tests carried out later to investigate which combination of H and K_D should be used.

Although more sophisticated formulae have been developed (see later), the Hudson formula is still used, at least for conceptual design, because of its simplicity.

Burcharth and Hughes[16] give values of K_D for the Hudson formula, (Eq. (3.12)), as shown in Tables 3.1 and 3.2. The K_D values have been given for a two-layer rock armor, e.g., the armor stones are placed in two layers. Single-layer placements are also used, especially in Norway.[21] But the single-layer rock armor layer is not recommended in areas where ice may be a governing factor for the stability. Single-layer armor is vulnerable to damage by removing only one stone and thereby exposes the filter layer.

Table 3.1. K_D values by SPM[23] H = H_s, for slope angles $1.5 \leq \cot g \, \alpha \leq 3.0$. Head on waves. (Based entirely on regular wave tests.)

Stone shape	Stone placement	Damage[a]			
		5–10%		10–15%	
		Breaking waves[b]	Nonbreaking waves[c]	Nonbreaking waves[c]	Nonbreaking waves[c]
Smooth, rounded	Random	2.1	2.4	3.0	3.6
Rough, angular	Random	3.5	4.0	4.9	6.6
Rough, angular	Special[d]	4.8	5.5		

[a]D is defined according to SPM as follows: The percentage damage is based on the volume of armor units displaced from the breakwater zone of active armor removal for a specific wave height. This zone extends from the middle of the breakwater crest down to the seaward face according to the wave height causing zero damage below still-water level. (See point 4 under the listed damage definitions.)
[b]Breaking waves means depth-limited waves, i.e., wave breaking take place in front of the armor slope. (Critical case for shallow water structures.)
[c]No depth-limited wave breaking take place in front of the armor slope.
[d]Special placement with long axis of stones placed perpendicular to the slope face.
Source: From CEM.[7]

Table 3.2. K_D values by SPM[22]: H = $H_{1/10} \approx 1.27 \cdot H_{1/3}$.

Stone shape	Stone placement	Damage[a]	
		Breaking waves[b]	Nonbreaking waves[c]
Smooth, rounded	Randomly	1.2	2.4
Rough, angular	Randomly	2.0	4.0
Rough, angular	Special	5.8	7.0

[a]D is defined according to SPM as follows: The percentage damage is based on the volume of armor units displaced from the breakwater zone of active armor removal for a specific wave height. This zone extends from the middle of the breakwater crest down to the seaward face according to the wave height causing zero damage below still-water level. (See point 4 under the listed damage definitions.)
[b]Breaking waves means depth-limited waves, i.e., wave breaking take place in front of the armor slope. (Critical case for shallow water structures.)
[c]No depth-limited wave breaking take place in front of the armor slope.
Source: From CEM.[7]

When considering that $H_{1/10} = 1.27 \, H_s$ for Rayleigh distributed wave heights (nondepth-limited waves) it is seen that the recommendations of SPM[22] introduce a considerable safety factor compared to the practice based on Ref. 23.

The coefficient of variation for Eq. (3.11), mainly for $(K_D)^{1/3}$, is estimated to be 0.18 by van der Meer[15] for two-layer rock armor, while Mlakar[24] reported a coefficient of variation for K_D of 0.25 for similar layers.

Jensen *et al.*[25] carried out breakwater stability tests with regular and irregular waves to explore which irregular wave height should be used in the Hudson formula when applying K_D from regular wave tests. Jensen *et al.* found that the $H_{1/20} \approx 1.42\,H_s$ in irregular waves gave the same damage as H in regular waves. They also found that for damage level S = 2 the value of K_D should be set to $K_D = 1.6$.

Mansard *et al.*[26] carried out tests on the stability of two-layer riprap with different gradations. The tests were carried out for slopes 1:1.8 and 1:2.25 and for an Iribarren number $\xi_z \approx 2.5$. The Iribarren number is defined as $\xi_z = \tan\alpha/(2\pi H_s/gT_z^2)^{0.5}$, where T_z is the mean period. For the smallest gradation, comparable with what is found on rubble mound breakwaters, Mansard *et al.* found a $K_D = 1.9$ for damage level S = 2, $K_D = 3$ for S = 4 and $K_D = 4$ for S = 8, when H_s is used in the Hudson formula. Damage levels are further discussed in the Section 3.3.2.2.2 van der Meer formula.

3.3.2.2.2. van der Meer Formulae

Van der Meer[14,15] carried out a comprehensive investigation on the stability of two-layer conventional rubble-mound breakwaters for nonbreaking waves, i.e., the waves do not break in front of the breakwater, but on the breakwater slope. Van der Meer varied the wave parameters as well as the breakwater cross-section construction (Fig. 3.8).

Van der Meer arrived at the following formulae for calculating the necessary armor stone mass:

Plunging waves:

$$\frac{H_s}{\Delta D_{n50}}\sqrt{\xi_z} = 6.2P^{0.18}\left(\frac{S}{\sqrt{N}}\right)^{0.2} \tag{3.14}$$

Surging waves

$$\frac{H_s}{\Delta D_{n50}} = 1.0P^{-0.13}\left(\frac{S}{\sqrt{N}}\right)^{0.2}\xi_z^P\sqrt{\cot\alpha} \tag{3.15}$$

where

$\Delta = ((\rho_s/\rho_w) - 1)$

$\xi_z =$ Iribarren number $= \dfrac{\tan\alpha}{\sqrt{\dfrac{gH_s}{2\pi T_z^2}}}$

$T_z =$ mean zero up-crossing wave period

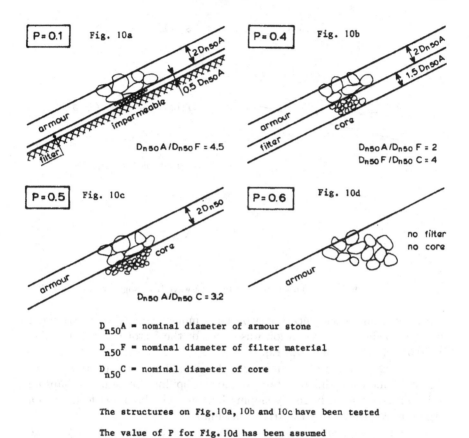

Fig. 3.8. Breakwater cross-sections investigated by van der Meer.[14,15]

S = damage, Fig. 3.7 with definition of damage
N = number of waves
P = "permeability coefficient" (see Fig. 3.8).
α = slope angle.

The "permeability coefficient" P has no physical meaning similar to the permeability coefficient for flow through soil. P was introduced as a notional value to take into account the permeability of the structure.

The transition between plunging and surging conditions is for

$$\xi_{z,\text{trans}} = \left(6.2 P^{0.31} \sqrt{\tan\alpha}\right)^{1/(P+0.5)} \qquad (3.16)$$

Depending on slope angle and permeability the transition lies frequently between $\xi_z = 2.5$ and 4.0, but other values may also be encountered.

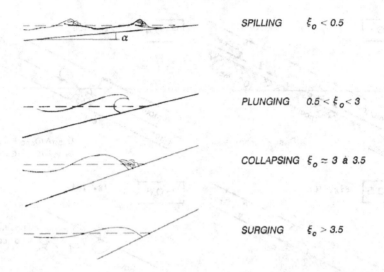

Fig. 3.9.   Three different types of wave breaking.

As a measure of the uncertainty of the formulae, van der Meer estimated that the coefficients of variation was 0.065 for the factor 6.2 in Eq. (3.14) and 0.08 for the factor 1.0 in Eq. (3.15).

Plunging and surging relates to the breaking form of the waves. Figure 3.9 indicates different breaking forms, spilling, plunging, collapsing and surging, on a fairly gently sloping bottom. The breaking form is to a large extent governed by the Iribarren number.

In case of a foreshore where the larger waves break before they arrive at the breakwater, van der Meer proposed the following formulae:

Plunging waves:

$$\frac{H_{2\%}}{\Delta D_{n50}} \cdot \sqrt{\xi_z} = 8.7 P^{0.18} \left(\frac{S}{\sqrt{N}} \right)^{0.2} \tag{3.17}$$

Surging waves:

$$\frac{H_{2\%}}{\Delta D_{n50}} = 1.4 P^{-0.13} \left(\frac{S}{\sqrt{N}} \right)^{0.2} \sqrt{\cot \alpha} \, \xi_z^P \tag{3.18}$$

where $H_{2\%}$ is the wave height exceeded by 2% of the waves in a storm.

It is seen that the formulae of van der Meer are more detailed than the Hudson formula. In addition to the parameters in the Hudson formula, the van der Meer formulae take into account the storm duration (N), the wave period (T_z), the damage level (S), the "permeability" (P) as well as the type of wave breaking. In the van der Meer formulae, the stability number of Hudson, $N_s = H_s/\Delta D_{n50}$, is recognized. Otherwise there are no other

Table 3.3. Lower and upper damage levels.

	Damage level, S	
Slope angle, cot α	Onset of damage	Failure
1.5	2	8
2	2	8
3	2	12
4	3	17
5	3	17

physical meanings attached to the terms in the formulae. The formulae are based on curve fitting to experimental results. That is why there are some "odd" values of coefficients and exponents in the formulae.

Van der Meer[14] stated the damage levels as shown in Table 3.3.

3.3.2.2.3. van Gent et al.'s Formula

Van Gent *et al.*[18] reanalyzed van der Meer's data and added some new experimental data on the stability of the armor layer on conventional rubble-mound breakwaters with stone armor units. Van Gen *et al.* arrived at some minor modifications of van der Meer's formulae. More interesting is that van Gent *et al.* arrived at a new formula. This formula is simpler than Eqs. (3.4), (3.15), (3.17), and (3.18) because of the following reasons:

— There is an influence of the wave period but this influence is considered small compared to the amount of scatter in the data due to other reasons. Therefore the wave period is not used in the formula and there is no separation between "plunging" and "surging" waves.
— There is an influence of the ratio $H_{2\%}/H_s$ but this influence is considered small. Therefore this ratio has been omitted.
— The influence of the permeability of the structure is incorporated in a direct way by using structure parameters, i.e., the diameter of the core material.

The van Gent[18] formula reads:

$$\frac{H_s}{\Delta D_{n50}} = 1.75 \left(\frac{S}{\sqrt{N}} \right)^{0.2} \sqrt{\cot \alpha} \left(1 + \frac{D_{n50,\text{core}}}{D_{n50}} \right) \tag{3.19}$$

3.3.2.2.4. Armor stone mass for rubble mound breakwaters obtained from different formulae

As an exercise, the necessary armor stone mass has been calculated using the Hudson and van der Meer formulae for a breakwater slope of 1:1.5, breakwater "permeability" P = 0.3 (Fig. 3.8), nonbreaking waves (in front of the breakwater), H_s = 7.0 m, N = 3000 waves (approximately 9 hrs sea state duration), damage levels S = 2 and 6 (start of damage and almost failure, Table 3.4). The results are shown in Figs. 3.10 and 3.11. In these figures the necessary stone mass is calculated as a function of the Iribarren number. The slope and significant wave height has.been kept constant (slope 1:1.5 and H_s = 7 m). The Iribarren number then varies only with the wave

Table 3.4.   Table of coefficients to be used in Eq. (3.20)

Armor type	Slope	A	B	C_C	Range of ξ
Stone	1:1.5	0.272	−1.749	4.179	2.1–4.1
Stone	1:2	0.198	−1.234	3.289	1.8–3.4
Dolos	1:1.5	0.406	−2.800	6.881	2.2–4.4
Dolos	1:2	0.840	−4.466	8.244	1.7–3.2

Note: The curves giving the best fit to the data were lowered by two standard deviations to provide a conservative lower envelope to the stability analysis.

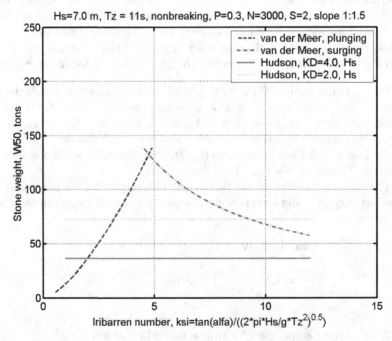

Fig. 3.10.   Necessary amour stone mass versus Iribarren number for slope 1:1.5, breakwater "permeability" P = 0.3, nonbreaking waves, H_s = 7.0 m, N = 3000 waves, damage level S = 2.

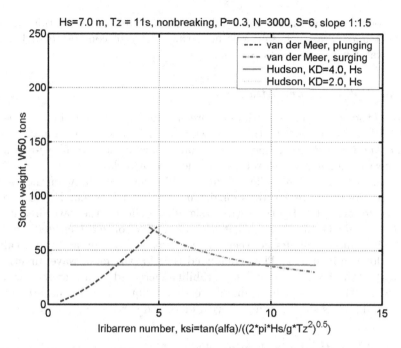

Fig. 3.11. Necessary amour stone mass versus Iribarren number for slope 1:1.5, break-water "permeability" P = 0.3, nonbreaking waves, $H_s = 7.0$ m, N = 3000 waves, damage level S = 6.

period. Since the Hudson formulae have no wave period term, the formulae give results indifferent to the wave period or in this case to the Iribarren number. For a wave period of $T_z = 11$ s, the Iribarren number is $\xi_z = 3.46$. For this Iribarren number the van der Meer formula gives a required armor stone mass $W_{50} = 83$ tons, while the Hudson formula, H_s and $K_D = 2$, gives required stone mass $W_{50} = 72$ tons. However, one has to recognize that there is always some uncertainty on the evaluation of the mean wave period. If the mean wave period is expected to be in the range $T_z = 9-13$ s, the Iribarren number would be in the range $\xi_z = 2.78 - 4.1$ or still in the plunging wave case. In that case the required stone mass is approximately 100 tons for $T_z = 13$ s.

If a higher damage level of S = 6 can be accepted, the van der Meer formulae gives a required armor stone mass of $W_{50} = 42$ tons for the Iribarren number $\xi_z = 4.1$. The Hudson formula does not have information on this high damage level.

Generally speaking, the required armor stone mass for these given conditions are so high that it is very difficult to obtain large enough stones from rock quarries. Under such conditions, other alternative solutions have to be found. Two such solutions are armor units of concrete or a berm breakwater with rock units (see later).

3.3.3. *Oblique Wave Attack*

Most of the general investigations on breakwater stability have been carried out with waves normal to the breakwater axis (2D tests). Galland[20] and Canel and Grauw[19] investigated the stability of conventional rubble-mound breakwaters for normal as well as oblique wave attack.

Galland[20] investigated the stability for different angles between the wave direction and the breakwater's longitudinal axis for irregular long crested waves. He concluded that there was no significant effect of the wave direction.

Canel and Grauw[19] investigated the stability for short crested waves and with different angles between the wave main direction and the breakwater longitudinal axis. They concluded that for the main wave direction normal to the breakwater axis the stability increased with increased wave spreading. For waves with the main direction oblique to the breakwater axis the stability was slightly reduced with increased spreading of the waves.

3.3.4. *Stability of Breakwater Head*

Larger armor units are normally required on the breakwater head. One reason for this is the high cone overflow velocities, sometimes enhanced by wave refraction. Another reason is the reduced support from neighboring units. Figure 3.12 illustrates the critical areas for damage in the round head.

There is much less investigations on the stability of the breakwater head than on the breakwater trunk. Burcharth and Hughes[16] give the following relation for the required unit mass for the head section:

$$\frac{H}{\Delta D_{n50}} = A\xi^2 + B\xi + C_C \tag{3.20}$$

where ξ = Iribarren number, based on the deep water wave length L_o and a characteristic wave height H.

Systematic tests on the stability of breakwater heads have mainly been with regular waves. Burcharth and Hughes[16] give coefficients as shown in Table 3.4 to be used in Eq. (3.20) (Table 3.4).

A limited number of tests in irregular waves produced corresponding results with the peak period T_p equivalent to the monochromatic period and H_{mo} equal to the monochromatic wave height.

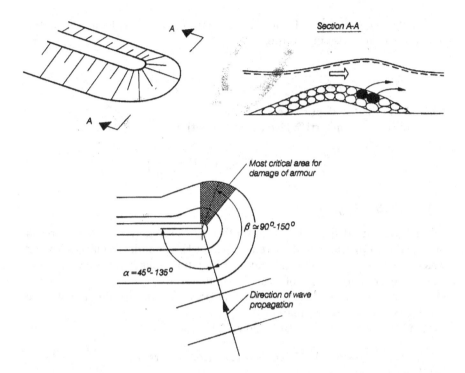

Fig. 3.12.   Illustration of critical areas for damage to armor layers in the round head.[9]

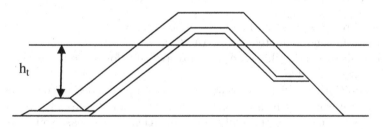

Fig. 3.13.   Cross section of tested breakwaters.[24]

3.3.5. *Stability of the Breakwater Toe*

Sometimes a toe is used as shown in Fig. 3.13 to give support to the armor units and to serve as scour protection.

Meulen *et al.*[27] give the following relation for the calculation of the necessary stone mass in the toe:

$$\frac{H_s}{\Delta D_{n50}} = \left(0.24\frac{h_t}{D_{n50}} + 1.6\right) N_{od}^{0.15} \tag{3.21}$$

where h_t is the toe depth (see Fig. 3.13).

Aminti and Lamberti[28] give another relation:

$$\frac{H_s}{\Delta D_{n50}} = \left(1.1 + 33s_m + 0.15\frac{h_t}{D_{n50}}\right) N_{od}^{0.2} \tag{3.22}$$

3.3.6. *Filter Layers*

Filter layers are defined as layers protecting the underlying base material of soil erosion by waves and currents without the underlayer material being washed out through the armor layer. Filter functions can be achieved by using one or more layers of granulated material such as gravel or small stone of various grain sizes, geotextile fabric, or a combination of geotextile overlaid with granulated material.

A filter layer may have the following functions:

(a) To prevent outwash of finer material from layers beneath
(b) Distribution of weight
(c) Reduction of wave loads on the structures outer stone layer. A granular layer dissipates more energy than a geotextile layer.

Of the three, point (a) is the most important.

The following is quoted from PIANC[6]:

The filter criteria which are generally available for coastal structures are summarized in Table 3.5.[29]

In addition, Rankilor[34] recommended the empirical design curves shown in Fig. 3.14 from which the required grain size ratio D_{15}/d_{15} may be determined as a function of the uniformity coefficient of the base soil, where d is the diameter of the filter (underlayer), while D is the diameter of the overlayer.

Most of the filter criteria presently used in coastal structures is essentially similar to those for steady flow, i.e., they are based on Terzaghi filter rules. However, it has been clearly shown in Ref. 35 that these criteria are rather conservative when applied to steady flow situations. The grain size criterion $D_{15}/d_{15} < 4$ to 5 has been shown to be too safe (with a safety factor of about 2 according to extensive test results). In view of these considerations, the procedure yet adopted in coastal structures appears to be

Table 3.5. Available filter criteria for coastal structures.[29]

No.	Investigators	Filter criteria	Observations
1	Belyashesvskii *et al.*[34]	$D_{60}/D_{10} < 0.2 D_{50}/d_{50}$	for graded rock filters
2	Ahrens[30]	$D_{15}/d_{15} < 4$	for rip-rap revetment underlayers
3	Thompson and Shuttler[31]	$D_{15}/d_{85} < 4$ $D_{50}/d_{50} < 7$ $D_{15}/d_{15} < 7$	for rip-rap revetment underlayers
4	De Grauw *et al.*[29]	$D_{50}/d_{50} < 2$ to 3	for granular filters under strong cyclic (reversing) flow
5	Van Orschot[32]	$D_{50}/d_{50} < 3$ or $W_{50(armor)}^{*)}/W^{*)}50$ (filter) < 25 to 30	for underlayers of rubble mound breakwaters $^{*)}W_{50}$ = Average Stone Weight
		$D_{(armor)}/d_{(under)} < 2.2$	for uniform armor units (breakwaters)
6	SPM[21]	$D_{15}/d_{85} < 5.0$	for graded stone armor, filter blankets/bedding layers
7	Engineer Manual[33]	$D_{15}/d_{85} < 4$ to $5 < D_{15}/d_{15}$	for graded rock filters soil
		$D_{15}/d_{15} < 4$	for underlayer of stone armor

correct. However, the question arises whether the safety factor of 2 is sufficient for the conventional criteria to be adopted in coastal structures. At present, this question is still open, despite the extensive experimental work conducted in the Netherlands[36,37] and in the former USSR[38] on filters subject to cyclic, turbulent flow conditions.

In view of the very complicated flow conditions and transport processes involved in a coastal environmental, rational design rules for filter constructions in coastal structures is only possible if based on the investigation of the actual flow situation and transport mechanisms.

Another problem that has not yet been seriously addressed in coastal structures is the design of the filter thickness. Generally, design criteria for filter thickness are not solely governed by hydrogeotechnical aspects but also by considerations of economy and practicability in the construction process. For instance, a too thin filter layer may be very difficult or impossible to be properly constructed under water. On the other hand, sufficient thickness may reduce risks that would result from possible segregation and/or settlement. Moreover, where suitable filter material is not available, thicker

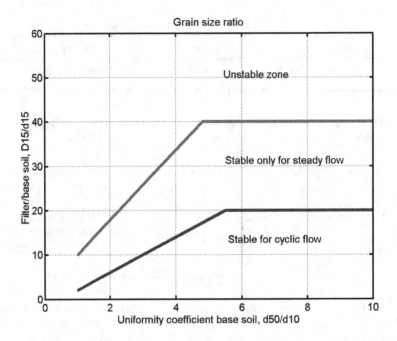

Fig. 3.14.   Filter design curves recommended by Rankilor.[34]

filter layers may be provided to relax grain size criteria. In addition, a
thick filter layer has also a beneficial effect on the stability of the overlying
armor units, as the down rush velocities (which are responsible for most
damages) decrease, i.e., more energy dissipation takes place within the thick
filter layer.

From the hydro-geotechnical view point the filter thickness should fulfill
two criteria, i.e., it should (i) accommodate the time-dependent washing
out of base soil particles into the filter matrix and (ii) provide enough cross-
sectional area to carry out the seeping water without excessive pressure
buildup. Thus, the attempts that have been made yet to investigate the
filter thickness were mainly directed along two distinct paths.

(a) *Seepage analysis*: The main objective of these investigations is to
provide a sufficient cross-sectional area to drain the seeping water
without excessive pressure. Given a filter material with a permeability
k, the required thickness t_f is calculated according to the allowable
discharge q and the allowable hydraulic gradient i ($t_f = q/ki$). This
procedure, however, completely ignores the migration of soil particles
into the filter and its clogging.[39]

(b) *Probability analysis*: The random migration of washed out particles is analyzed and the required filter thickness to arrest washing out of base soil is obtained. Silviers[40] has treated the problem of washing out particles as a stochastic process and proposed an equation to compute the mean distance travelled by a soil particle before clogging, based on the absorbing state of Markov chain. Thanikachalam *et al.*[41] have used queuing theory to describe the random motion of washed out particles and the clogging of filter pores and to formulate a theory for calculating the mean and standard deviation of the washed out particles. The filter thickness is then determined from the computed mean and standard deviation, the number of particles retained per pore opening, and the average grain size of the filter. The results of Thanikachalam *et al.*[41] appear to agree well with those of Silveira.[40] Moreover, they clearly show that increasing the filter thickness beyond a limit does not substantially increase the percentage of trapped particles. However, a much greater thickness is generally required by considering the practical and further aspects mentioned above.

The following concluding remarks are made on filters

- Current filter criteria for steady flow are rather conservative due to the lack of information on the flow and transport mechanisms involved.
- Conventional filter criteria for steady flow cannot readily be applied to cyclic flow conditions.
- Results derived from investigations on uniform filters are normally conservative when applied to broadly graded filters. For instance, a well-graded filter with $C_u = D_{60}/D_{10} = 20$ may catch soil particles of about half the size compared to a uniform filter with the same D_{15}.
- Filter criteria expressed by grain size ratio related to the finer fractions of the filter material like D_{10} and D_{15} and the coarser fractions of the base material (d_{85}) are more reliable.
- To date, broadly graded filter are expected to constitute a more feasible alternative to conventional uniform filters.
- Substantial cost savings may be achieved by accounting for the superimposed loads and for the transport mechanisms beyond the initiation of motion.

However, development of reliable design rules can only be achieved by investigating the (i) mechanics of the initiation of motion and the (ii) transport mechanisms after the initiation of motion.

The latter aspect is of upmost importance with respect to the new era of filters with an allowable material transport rate.

3.3.7. Wave Overtopping

The height of the breakwater crest will to a large extent depend on what wave overtopping can be accepted. If there are frequently used quays just behind the breakwater, not much overtopping should be allowed. For breakwaters that are far from harbor operations, more overtopping may be allowed. The crest height may thus be considered in view of construction costs and operational cost. It is frequently seen that the relative breakwater crest height is approximately $R_c/H_{s,\text{design}} = 1.0 - 1.2$. But this figure varies.

Wave overtopping evaluation is dealt with in more detail Sec. 3.4.

3.3.8. Wave Forces on Wave Crest Screens

Sometimes a wave screen or a crown wall is constructed as a crest structure as indicated in Fig. 3.15.

Pedersen[42] and Burcharth and Hughes[16] have reported on the evaluation of the wave forces, including wave slamming forces, on such screen. Wave-induced forces might cause a crown wall to fail as a monolith by sliding or tilting, or by breaking. With reference to Fig. 3.16, the

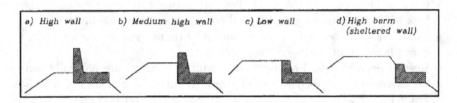

Fig. 3.15.   Typical crown wall configurations.[16]

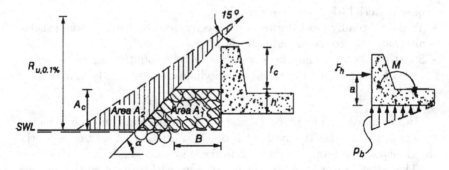

Fig. 3.16.   Notations or calculating wave forces on crown wall.[16]

wave-induced horizontal force and uplift force can be calculated by Pedersen given below.

$$F_{h,0,1\%} = 0,21\sqrt{\frac{L_{om}}{B}}\left(1,6 p_m y_{eff} + A\frac{p_m}{2}h'\right) \tag{3.23}$$

$$M_{0,1\%} = a \times F_{h,0,1\%} = 0.55(h' + y_{eff})F_{h,0,1\%} \tag{3.24}$$

$$p_{b,0.1\%} = 1,00\, A\, p_m \tag{3.25}$$

where

$F_{h,0,1\%}$ = horizontal wave force per running meter of the wall corresponding to 0.1% exceedance probability

$M_{0,1\%}$ = wave generated turning moment per running meter of the wall corresponding to 0.1% exceedance probability

$p_{b,0,1\%}$ = wave uplift pressure corresponding to 0,1% exceedance probability

L_{om} = deepwater wavelength corresponding to mean wave period

B = berm width of armor layer in front of the wall

p_m = $\rho_w g(R_{u,01\%} - A_c)$

$R_{u,0.1\%}$ = wave run-up corresponding to 0,1% exceedance probability

$$R_{u,0,1\%} = \begin{cases} 1,12 H_s \xi_m & \xi_m \leq 1,5 \\ 1,34 H_s \xi_m^{0.55} & \xi_m > 1,5 \end{cases} \tag{3.26}$$

ξ_m = $\tan\alpha / \sqrt{H_s/L_{om}}$

α = slope angle of armor layer

A_c = vertical distance between MWL and the crest of the armor berm

A = $\min\{A_2/A_1, 1\}$, where A_1 and A_2 are areas shown in the Fig. 3.16

y_{eff} = $\min\{y/2, f_c\}$

$$y = \begin{cases} \dfrac{R_{u,0,1\%} - A_c}{\sin\alpha}\dfrac{\sin 15°}{\cos\left(\alpha - 15°\right)}, & y > 0 \\ 0, & y \leq 0 \end{cases} \tag{3.27}$$

h' = height of the wall protected by the armor layer

f_c = height of the wall not protected by the armor layer

The uncertainties of the coefficients in the formulae (3.23)–(3.25) are shown in Table 3.6.

Stability against sliding between the crown wall base and the rubble foundation requires

$$(F_G - F_U)\mu \geq F_H \tag{3.28}$$

Table 3.6. Standard deviation of the coefficients in formulae (3.23)–(3.25).

Coefficients in the formulae	0.21	1.6	0.55	1.00
Standard deviation	0.02	0.10	0.07	0.30

where

μ = friction coefficient for the base plate against the rubble stone
   $(0.5 \leq \mu \leq 0{,}7)$
F_G = buoyancy reduced weight of the structure
F_U = wave-induced uplift force
F_H = wave-induced horizontal force plus force from armor resting
   against the front of the structure

Stability against overturning is maintained if

$$M_{FG} \geq M_{FU} + M_{FH} \qquad (3.29)$$

where

M_{FG} = stabilizing moment of F_G around the heel
M_{FU} = antistabilizing moment of F_U around the heel
M_{FH} = antistabilizing moment of F_H around the heel

A safety factor has to be applied to the right-hand side of Eq. (3.28) and Eq. (3.29) in deterministic design.

Stability against geotechnical slip failures has to be demonstrated. Conventional slip failure calculation methods can be used.

Hydraulic model tests are in general recommended for determination of wave-induced loads on crown walls. From the test series the combinations of simultaneous wave-induced forces that give the minimum crown wall stability must be identified.

3.3.9. *Conventional Rubble-Mound Breakwaters with Concrete Armor Units*

As mentioned in the previous chapter, the required stone mass for a stable rubble-mound breakwater may be so large that it is not possible to obtain such armor stones from quarries. An alternative solution may then be to use concrete armor units. Figure 3.17 shows the most frequently used such units.

The concrete armor units are generally made of un-reinforced concrete, except the multiwhole cubes where fiber reinforcement is sometimes used. The maximum mass of a used concrete unit is 150 tons (parallelepiped)

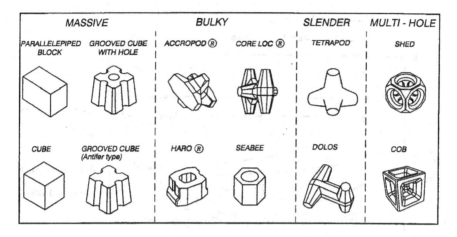

Fig. 3.17. Examples of concrete armor units.[16]

in two layers, slope 1:2, for the Bilbao outer harbor breakwater, Spain. For the new breakwater at La Coruna, Spain, which is under construction, the design wave height is approximately 14 m. The cover layer consists of 150 tons concrete cubes on slope 1:2.

Some of the concrete units are hydraulically very stable (e.g., the Dolos). But this unit has the weakness of easily breaking when it becomes large. This happened during the failure of the Sines breakwater in Portugal in the late 1970s. Several 42 ton dolos units were found broken after the damaging storm waves, which did not exceed the design waves of approximately $H_s = 11$ m. The breakage occurred apparently due to the impact between the units during rocking motions induced by the waves. Since then, more bulky units have been developed, e.g., the Coreloc® and Accropodes®. There are different attitudes toward the different concrete units, based probably on experience and tradition. For example, the Spanish to some extent favor the cube or parallelepiped units, while the Americans favor presently the Coreloc®.

Burcharth and Hughes[16] deal with many details on the concrete armor units, like concrete strength requirement and hydraulic stability. In this chapter only some indicative hydraulic stability coefficients are summarized, mainly from Ref. 16. For more details, Ref. 16 should be consulted, along with Refs. 8 and 43.

For most of the concrete units, the required mass of the units are calculated by formulae different from the Hudson and the van der Meer formulas. But a general term in most of the formulae is the Hudson stability number, $N_s = H_s/(\Delta D_n)$. D_n is the length of a cube having the same mass as the considered concrete armor unit.

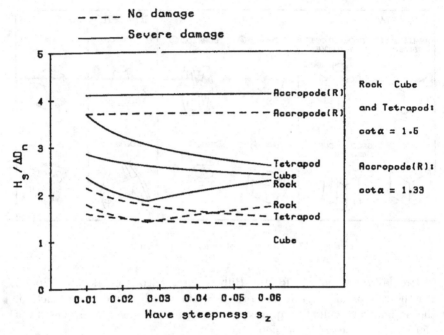

Fig. 3.18. Comparison of stability of different types of breakwater armor blocks. The wave steepness $s_z = s_{om}$

When considering the stability coefficients for concrete units compared to the rock armor stones, it has to be remembered that the mass density of concrete is generally lower than for rock, typically 2200–2400 kg/m^3 for concrete versus 2700 kg/m^3 for rock. This means that a higher K$_D$ value for a concrete unit does not necessarily mean a lower required mass of a concrete armor unit than for a rock armor stone.

Van der Meer[44] indicates stability numbers as a function of the wave steepness for the initiation and failure of rock, cubes, tetrapods, and Accropods[®] as shown in Fig. 3.18. The rock, cubes and tetrapods are placed in two layers, while the Accropods[®] (and Corelocs[®]) are placed in one layer.

3.3.9.1. Coreloc[®], one layer

Burcharth and Hughes[16] recommend for depth-limited plunging to collapsing waves $K_D = 16$ for preliminary design of all trunk sections. This value is considered to be conservative. The Corelock[®] armor unit is thus a very stable armor unit.

Van der Meer[44] discusses the stability of different concrete armor units.

3.3.9.2. *Cubes and tetrapods in two layers*

For cubes in two layers on slope 1:1.5–1:2 van der Meer sets

$$\frac{H_s}{\Delta D_n} = \left[6.7\frac{N_{od}^{0.4}}{N^{0.3}} + 1.0\right] s_{om}^{-0.1} \qquad (3.30)$$

where N_{od} = the actual number of units displayed related to a width (along the longitudinal axis of the structure) of one nominal diameter D_n, N = number of waves, $s_{om} = 2\pi H_s/gT_m^2$, T_m = mean period.

Generally speaking N_{od} is approximately $0.5S$ (S is defined in Sec. 3.3.2.1, Fig. 3.7). N_{od} can easily be compared with the percentage of damage, e.g., the percentage of armor blocks moved out of their initial position. N_{od} gives the actual damage, while percentage damage is always related to the actual structure. The following example may illustrate this.

Assume a breakwater with 15 tons cubes along the slope with $D_n = 1.84$ m and consider a 100 m stretch of the breakwater. Table 3.7 shows the actual number of blocks per 100 m that is being displaced.

If a cross-section slope, one nominal diameter wide, consists of 20 units, $N_{od} = 0.5$ gives $(0.5/20)*100\% = 2.5\%$ damage. If the slope is longer consisting of 40 units, $N_{od} = 0.5$ gives only 1.25% damage.

For tetrapods in two layers, van der Meer sets

$$\frac{H_s}{\Delta D_n} = \left[3.75\left(\frac{N_{od}}{\sqrt{N}}\right)^{0.5} + 0.85\right] s_{om}^{-0.2} \qquad (3.31)$$

For the "No damage" criterion Eqs. (3.30) and (3.31) reduce to:
Cubes, two layers

$$\frac{H_s}{\Delta D_n} = 1.0 s_{om}^{-0.1} \qquad (3.32)$$

Tetrapods, two layers

$$\frac{H_s}{\Delta D_n} = 0.85 s_{om}^{-0.2} \qquad (3.33)$$

Table 3.7.

Damage N_{od}	Number/100 m
0.2	11 units
0.5	27 units
1.0	54 units
2.0	109 units

Whether a breakwater should be designed for no damage or not depends on what type of harbor is behind the breakwater and what the total cost of repair will be, including repair costs and costs incurred by restricted operations in the harbor. Generally speaking the no-damage criterion is a strict criterion, also from an economical point of view. For rock layers, some settlement and small displacement is included in the "start of damage" definition, $S = 2 - 3$. For $N_{od} = 0.5$ a similar situation is found and this may be a more economical situation.

Equations (3.32) and (3.33) give decreasing stability with increasing wave steepness. This is similar to the surging area for rock layers.

3.3.9.3. Accropodes[R], one layer

The Accropodes[R] are normally placed on steep slopes, 1:1.33. From model tests it has been found that

Start of damage, $N_{od} = 0$

$$\frac{H_s}{\Delta D_n} = 3.7 \tag{3.34}$$

and failure, $N_{od} > 0.5$

$$\frac{H_s}{\Delta D_n} = 4.1 \tag{3.35}$$

Equations (3.34) and (3.35) show that start of damage and failure for Accropodes[R] are very close, although at very high $H_s/\Delta D_n$ number. It means that the Accropodes[R] up to a certain wave height level is very stable. But only a small exceedance of this wave height will lead to major damage. The Accropodes[R] and the Corelocs[R] are thus "brittle structures." Van der Meer suggests that the following is used for the design of Accropodes[R]:

$$\frac{H_s}{\Delta D_n} = 2.5 \tag{3.36}$$

There is then a safety factor on the $H_s/\Delta D_n$ value of about 1.5. This is obtained in ideal laboratory testing conditions. But in practice one may perhaps rely on 20–30% safety beyond the given design conditions.

The single layer concrete units require, generally speaking, more precise placements, with the guidance of divers, than the more blocky two layer units. Very experienced contractors for the placement of the one-layer units should be applied.

3.3.10. *Berm Breakwaters*

3.3.10.1. *Introduction to berm breakwaters*

PIANC[6] has issued a report from a working group on State-of-the-Art of Designing and Constructing Berm Breakwaters. The main items for designing a berm breakwater are discussed here.

Berm breakwaters are different from conventional rubble mound breakwaters as indicated in Fig. 3.19.

A conventional rubble mound breakwater is required to be almost static stable for the design wave conditions, while the berm breakwater has traditionally been allowed to reshape into a reshaped static or a reshaped dynamic stable profile as indicated in Fig. 3.19. Nonreshaping static stable berm breakwaters have also lately been considered. Berm breakwaters may thus be divided into three categories:

- Nonreshaped static stable berm breakwater, e.g., only some few stones are allowed to move similar to what is allowed on a conventional rubble mound breakwater.
- Reshaped static stable berm breakwater, e.g., the profile is reshaped into a stable profile where the individual stones are also stable.
- Reshaped dynamic stable berm breakwater, e.g., the profile is reshaped into a stable profile, but the individual stones may move up and down the slope.

The berm breakwater has normally been constructed with a berm that is allowed to reshape. This is because it is presently cheaper to construct the breakwater with an ordinary berm than to construct it according to the final S-shape. The first berm breakwaters designed had a homogenous berm. A more stable design has been developed in Iceland, which is a multilayer berm breakwater, in close cooperation between all partners involved: designers, geologists, supervisors, contractors, and local governments.[45] One reason for this development is the fear that the reshaping process may eventually lead to excessive crushing and abrasion of individual stones as they move on the berm breakwater. However, some of the "old" reshaped berm breakwaters

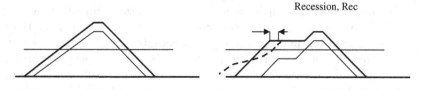

Fig. 3.19.   Conventional and berm rubble mound breakwaters.

have functioned quite well without excessive crushing and/or abrasion of the stones, while others have experienced large reshaping and have been repaired with large blocks on top of the reshaped profile. The question of allowing reshaping or not has obviously to do with the stone quality and the stones ability to withstand impacts crushing and/or abrasion. It is clear though that even a nonreshaping static stable berm breakwater requires armor stones with significantly less mass than cover stones on a conventional rubble mound breakwater.

There are methods available to evaluate the suitability of quarried stones against crushing for reshaped static stable-type berm breakwaters[46,47] (see later). There is less information on how to evaluate the suitability of quarried rock for reshaped dynamic stable berm breakwaters. Hence it is not recommended to design for reshaped dynamic stable conditions.

In many cases, the necessary armor stone mass on conventional rubble mound breakwaters is so large that concrete armor blocks is required if the concept of the conventional rubble mound breakwater is maintained. This is illustrated in Table 3.8, which shows the armor unit mass W_{50} on the Sirevåg berm breakwater (see description of this breakwater later) and the recession as obtained from model tests for the 100-year design wave heights $H_{s,100} = 7.0$ m and for the 10,000-year design wave height $H_{s,10,000} = 9.3$ m. Table 3.8 shows also the necessary armor unit mass for a conventional two-layer rubble mound breakwater with different slopes and with different degrees of damage S. The necessary armor unit mass for the conventional rubble mound breakwater with rock armor has been calculated with the formulae of van der Meer[14] with the assumption of the porosity parameter $P = 0.3$, wave steepness $s_{om} = 0.04$, and the number of waves $N = 2000$. This is a common steepness for high waves outside (Sirevåg).[46]

Table 3.8 shows also the required armor unit mass of cubes and tetrapods for a rubble-mound breakwater with slope 1:1.5. The cube and tetrapod masses have been calculated from the Van der Meer[44] formulae with wave steepness $s_m = 0.04$, slope 1:1.5, damage level $N_{od} = 0.35$, and the number of waves $N = 2000$.

Table 3.8 indicates that it is probably not possible to construct a conventional rubble-mound breakwaters with a reasonable slope from quarried stone for the wave conditions as used in the cases shown. If a conventional rubble-mound structure still is required, concrete armor units will have to be used. However, the mass of these concrete units has to be larger than the mass of the rock units in the berm breakwater. Table 3.8 shows also how "tough" the Sirevåg berm breakwater is, since it can easily withstand the 10,000-year wave event and still be a reshaped statically stable berm breakwater. The probability of failure for the Sirevåg berm breakwater is further dealt with in a later section.

Table 3.8. Comparison of armor unit masses.

(1)	(2) Ho = 2.7 (upper criteria for reshaped static stable) W_{50}, tons	(3) Sirevåg berm breakwater W_{50} = 25 tons Berm width = 19.5 m Recession, m	(4) Conventional two-layer rubble mound breakwater W_{50}, tons	(5) Rubble mound. Concrete cubes two layers W, tons	(6) Rubble mound. Tetrapods two layers W, tons
H_s = 7.0 m	11	4.3	65		
Slope 1:1.5			55	42	30
S = 2			43		
S = 3			35		
Slope 1:2					
S = 2					
S = 3					
H_s=9.3 m	26	8.3		98	69
Slope 1:1.5					
S = 2			156		
S = 3			118		
Slope 1:2					
S = 2			112		
S = 3			78		

Column (1) gives significant wave heights, breakwater slopes, and damage levels.
Column (2) gives the required stone weight for Ho = 2.7 (see later)
Column (3) gives the mean recession for the Sirevåg berm breakwater for H_s = 7.0 m and H_s = 9.3 m.
Columns (4)–(6) give the required armor unit mass for armor layers of rock, concrete cubes and tetrapods, respectively.

It should be added that smaller stones than those used for the Sirevåg berm breakwater may be used (several berm breakwaters have been constructed with relatively smaller stones) and still be considered reshaped statically stable for the 100-year design waves. But the berm breakwater then may become reshaped dynamically stable for the 10,000-year waves. In this case the possibility of stone breakage and abrasion will have to be looked into more carefully.

Tørum et al.[48] showed that the full-scale Sirevåg berm breakwater behaved even better than predicted by the model tests. The main reason for this was thought to be that the stones on the berm were placed randomly in the model, while they were placed orderly in the prototype. However, Tørum et al.[49] showed that there was no significant difference in recession weather the stones were placed orderly or randomly. Tørum et al.[49] concluded that there was a fair agreement between the model tests results and the prototype behavior.

3.3.10.2. Stability and reshaping of berm breakwaters

3.3.10.2.1. General on the stability and reshaping
of the trunk section

The most used parameters in relation to the stability and reshaping of berm breakwaters are the following:

$$Ho \equiv N_s = \frac{H_s}{\Delta D_{n50}}, \text{ stability number,}$$

$$HoTo = \frac{H_s}{\Delta D_{n50}} \sqrt{\frac{g}{D_{n50}}} T_z, \text{ period stability number,}$$

$$To = \sqrt{\frac{g}{D_{n50}}} T_z$$

$$\Delta = \frac{\rho_s}{\rho_w} - 1$$

$$f_g = D_{n85}/D_{n15}, \text{ gradation factor,}$$

where
H_s = significant wave height, W_{50} = median stone mass, i.e., 50% of the stones are larger (or smaller) than W_{50}, T_z = mean wave period, g = acceleration of gravity, ρ_s = density of stone, ρ_w = density of water.

Navid Moghim introduced a modification of the HoTo parameter into a new a parameter.[50,51]

$$Ho\sqrt{To} = \frac{H_s}{\Delta D_{n50}} \sqrt[4]{\frac{g}{D_{n50}}} \sqrt{T_z}$$

Most of the research work on the stability and reshaping of berm breakwaters has been for homogenous berms. But lately some work has also been made on the stability and reshaping of multilayer berm breakwaters. The multilayer berm breakwater allows a better and more economical utilization of the quarry yield than a conventional rubble mound breakwater and it is expected that this will be the future design of berm breakwaters.

Van der Meer[52] developed a computer program BREAKWAT to calculate the reshaped trunk profile of homogenous berm breakwaters for different wave conditions, different breakwater configurations, and different stone sizes under the assumption that the wave direction is normal to the breakwater trunk. The part of BREAKWAT relating to berm breakwaters was based on results obtained by van der Meer[15] for HoTo values above approximately 150, or above what is normally encountered for berm breakwaters. The results of the validation calculations van der Meer,[52] show reasonable agreement with observations.

Van Gent[53] also developed a computer program to calculate the reshaped profile and other items related to berm breakwaters.

Archetti and Lamberti[54] followed a route similar to van der Meer,[15] while Hall and Kao,[55] Tørum,[56] Menze,[57] Tørum *et al.*,[5] have followed a different route, obtaining simple expressions for the recession of the trunk homogenous berm of a berm breakwater. Lamberti and Tomasicchio[58] and Alikhani[59] derived also expressions for the longshore transport of stones of reshaping berm breakwaters. This item is of importance when the wave direction is oblique to the breakwater.

Tørum,[56] Tørum *et al.*,[5,61] and Lykke Andersen[62] followed to some extent the route of Hall and Kao.[55] With reference to Fig. 3.20 the recession, Rec, was analyzed from several model tests at DHI and SINTEF.[56] It was noticed that for a given berm breakwater all the reshaped profiles intersected with the original berm at almost a fixed point A (Fig. 3.20).

There is apparently a difference in reshaping of multilayer berm breakwaters and homogenous berm breakwaters.[63] Hence the stability of two types are treated separately

3.3.10.2.2. Reshaping of multilayer berm breakwaters

As mentioned, berm breakwaters may be designed with a homogenous or multilayer berm breakwater. It has turned out that the multilayer berm breakwater gave less recession than a homogenous berm breakwater, provided the stone dimensions D_{n50} are the same. The possible reasons for this will be discussed later.

Tørum *et al.*[5] and Tørum[64] arrived at the following equation for the mean nondimensional recession for berm breakwaters, with the results from the test on the model of the Sirevåg multilayer berm breakwater as the

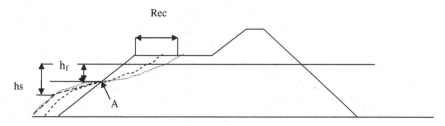

Fig. 3.20. Recession, Rec, of a berm breakwater.

major data basis:

$$\frac{Rec}{D_{n50}} = 0.0000027(HoTo)^3 + 0.000009(HoTo)^2 + 0.11(HoTo)$$

$$- \left(f_{Dn}(f_g) + f_d \left(\frac{d}{D_{n50}} \right) \right) \frac{HoTo}{120} \qquad (3.37)$$

The gradation factor function $f_{Dn}(f_g)$, $f_g = D_{85}/D_{15}$, is given by:

$$f_{Dn}(f_g) = -9.9f_g^2 + 23.9f_g - 10.5 \quad \text{for } 1.3 < f_g < 1.8 \qquad (3.38)$$

and the depth function $f_d(d/D_{n50})$ is given by:

$$f_d(d/D_{n50}) = -0.16 \left(\frac{d}{D_{n50}} \right) + 4.0 \quad \text{for } 12.5 < d/D_{n50} < 25. \qquad (3.39)$$

As an approximation h_f (Fig. 3.20) can be obtained from:

$$\frac{h_f}{D_{n50}} = 0.2 \frac{d}{D_{n50}} + 0.5 \quad \text{for } 12.5 < d/D_{n50} < 25. \qquad (3.40)$$

Figure 3.21 shows the dimensionless recession relation versus HoTo for data from Tørum et al.[49] on tests on the Sirevåg multi layer berm

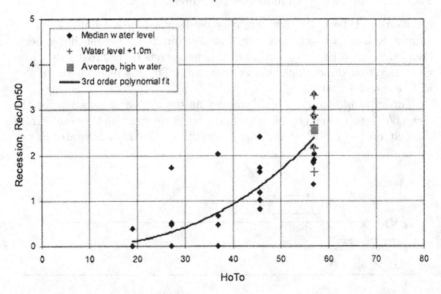

Fig. 3.21. Dimension recession versus HoTo. "Formula by Tørum" is from Ref. 47 and has later been reformulated to cover wider water depth and stone gradation ranges.

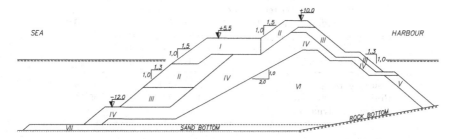

Fig. 3.22. Cross section of the outer end of the Sirevåg berm breakwater. Design waves $H_{s,100} = 7.0$ m, $T_z = 10.6$ s, HoTo $= 48$. Class I stone $W = 20 - 30$ tons, gradation factor $f_g = 1.1$, Class II stones $10 - 20$ tons, Class III stones $4 - 10$ tons and Class IV stones $1 - 4$ tons (Sigurdarson *et al.*[65]).

breakwater (Fig. 3.22). This breakwater was constructed in the time period $1999 - 2001$. The dimensionless recession for multilayer berm breakwaters have been based on D_{n50} for the largest stone class. The depth relation is $d/D_{n50} \approx 8$ for the Sirevåg breakwater.

During a major storm on January 28, 2002 the significant wave height of the waves arriving at the Sirevåg berm breakwater was approximately 8.3 m, measured with a WaveRider buoy 450 m outside the breakwater. There was a refraction effect of the waves from the location of the WaveRider buoy to the breakwater, giving an estimated wave height at the breakwater in the range $H_s = 7.2 - 8.8$ m or approximately the 100-year design wave heights. Tørum *et al.*[48,49] analyzed the wave conditions and the observed behavior of the breakwater during the January 28, 2002 storm and came to the conclusions that there was fair agreement between the model test results and the prototype behavior.

The single stones on the Sirevåg berm breakwater were placed orderly. Tørum *et al.*[41] investigated if there was any effect of orderly placed stones and pell-mell placed stones. They concluded that there was no significant difference between orderly and pell mell placement.

The suggested threshold design criteria for the trunk section, almost head on waves, for different categories of berm breakwaters are shown in Table 3.9.

As mentioned before it is recommended to design for reshaping static stable conditions.

3.3.10.2.3. Reshaping of homogenous berm breakwaters

Lykke Andersen[62] carried out a comprehensive series of 2D tests on a homogenous berm breakwater covering many more variables than ever

Table 3.9. Stability criterion for different categories of berm break-waters for modest angle of attack, $\beta_o = \pm 20°$ (the criterion depends to some extent on stone gradation). Partly based on PIANC.[6]

Category	Ho	HoTo
Non reshaping	<1.75–2.0	<30–55
Reshaping, static stable	1.75–2.7	55–70
Reshaping, dynamic stable	>2.7	>70

before. He arrived at the following equation for the recession:

$$\frac{\text{Rec}}{D_{n50}} = f_{h_b} \cdot \left[\frac{(1 + c_1)\, h - c_1 h_s}{h - h_b} \cdot f_N \cdot f_\beta \cdot f_{HO} \cdot f_{\text{skewness}} \cdot f_{\text{grading}} \right.$$
$$\left. + \frac{\cot(\alpha_d) - 1.05}{2 \cdot D_{n50}} \cdot (h_b - h) \right] \tag{3.41}$$

where

α_d = lower front slope (below berm)
h_b = height of berm. Note that h_b is negative when the berm is above the still water line.
h = water depth
h_s = step height (Fig. 3.20)
$f_{h_b} = 1.18 \cdot \exp\left(-1.64\left(\frac{h_b}{H_s}\right)\right)$ for $h_b/H_s > 0.1$
$f_{h_b} = 1$ for $h_b/H_s < 0.1$.

Note that h_b is negative when the berm is above the still water line and hence f_{hb} is then equal to 1.0.

$c_1 = 1.2$
$f_\beta = \cos(\beta), \beta$ = angle between the wave direction and the breakwater trunk centerline.

$$f_N = (N/3000)^{-0.046 \cdot H_o + 0.3} \quad \text{for Ho} < 5$$

$$f_N = (N/3000)^{0.07} \quad \text{for Ho} > 5$$

N = number of waves in a given storm

$$h_s = 0.65 \cdot H_s \cdot s_{om}^{-0.3} \cdot f_N \cdot f_\beta$$

$$f_{HO} = 19.8 \cdot \exp\left(-\frac{7.08}{H_o}\right) \cdot s_{om}^{-0.5} \quad \text{for To} > T_o^*$$

$$f_{HO} = 0.05 \cdot HoTo + 10.5 \quad \text{for To} < T_o^*$$

$$T_0^* = \frac{19.8 \cdot \exp\left(-\frac{7.08}{H_0}\right) \cdot s_{om}^{-0.5} - 10.5}{0.05 \cdot H_0}$$

$$f_{\text{skewness}} = \exp(1.5 \cdot b_1^2)$$

$$b_1 = 0.54 \cdot Ur^{0.47}$$

$Ur = $ Ursells number $= Ur = \frac{H_s}{2 \cdot h \cdot (k \cdot h)^2} = \frac{H_s \cdot L_p^2}{8 \cdot \pi^2 \cdot h^3}$, where L_p is the wave length corresponding to the peak period.

$$\begin{aligned}
f_{\text{grading}} &= 1 &&\text{for } f_g \leq 1.5 \\
f_{\text{grading}} &= 0.43 f_g + 0.355 &&\text{for } 1.5 < f_g < 2.5 \\
f_{\text{grading}} &= 1.43 &&\text{for } f_g > 2.5
\end{aligned}$$

It was mentioned that there is always inherent scatter in the test results on the stability of rubble mound breakwaters. Figure 3.23 shows the scatter of

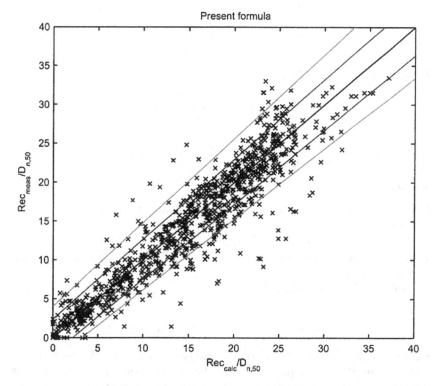

Fig. 3.23. The calculated recession, Rec_{calc}, is calculated according to Eq. (3.5), while Res_{meas} is the measured recession. The outer lines (red) marks the 95% confidence band for all the data he has used, while the inner lines (blue) is the 95% confidence band for his own data.[62]

the test data presented by Lykke Andersen.[62] He has in Fig. 3.23 plotted his own data together with data from Refs. 46, 56, 58 and 66.

Each point in Fig. 3.23 represents the average recession of all the profiles he measured after each test run, while the data points in Fig. 3.21 represents the recession in each of 10 profiles, 7 m apart. Thus Fig. 3.23 represents several hundred test runs.

It should be noted that the width of the berm of the multilayer Sirevåg berm breakwater (Fig. 3.22) and similar berm breakwaters is approximately $10 \cdot D_{n50}$ and that the maximum recession (Fig. 3.21) is approximately $\text{Rec}/D_{n50} \approx 6$ for the 10,000-year waves. This recession is in the lower part of Lykke Andersen[62] results (Fig. 3.23).

Lykke Andersen proposed the following recession taking into account the scatter:

Lykke Andersen, all data:

$$\frac{\text{Rec}_{calc}^{95\%}}{D_{n50}} = 1.080 \cdot \frac{\text{Rec}_{calc}^{50\%}}{D_{nb50}} + 4.00 \tag{3.42}$$

$$\frac{\text{Rec}_{calc}^{5\%}}{D_{n50}} = 0.909 \cdot \frac{\text{Rec}_{calc}^{50\%}}{D_{n50}} - 2.95 \tag{3.43}$$

Lykke Andersen, own data:

$$\frac{\text{Rec}_{calc}^{95\%}}{D_{n50}} = 1.028 \cdot \frac{\text{Rec}_{calc}^{50\%}}{D_{n50}} + 2.38 \tag{3.44}$$

$$\frac{\text{Rec}_{calc}^{5}}{D_{n50}} = 0.961 \cdot \frac{\text{Rec}_{calc}^{50\%}}{D_{n50}} - 2.10 \tag{3.45}$$

3.3.10.2.4. Longshore transport

If the waves approach a berm breakwater at an oblique angle, there is a possibility of transport of stones along the breakwater.

Lamberti and Tomasicchio[50] dealt with long shore transport when the wave attack is oblique to the breakwater axis. Their analysis is very comprehensive and extensive.

Alikhani[51] developed the following simpler expression for the long shore transport:

$$S_N = 0.8 \cdot 10^{-6} \sqrt{\cos \beta_o} (H_o T_{\text{op}} \sqrt{\sin 2\beta_o} - 75)^2 \tag{3.46}$$

where T_{op} signifies that the period parameter is based on the peak period rather than the mean period, β_o is the angle between the mean wave direction and the breakwater axis, S_N is the number of stones transported per wave.

Alikhani[59] recommends the following threshold values for long shore transport:

$$H_oT_{\rm op} = \frac{50}{\sqrt{\sin 2\beta_o}} \quad \text{for the reshaping phase} \qquad (3.47)$$

$$H_oT_{\rm op} = \frac{75}{\sqrt{\sin 2\beta_o}} \quad \text{after the reshaping phase.} \qquad (3.48)$$

where β_o = angle between the mean wave direction and the breakwater axis.

3.3.10.2.5. Stability and reshaping of the berm breakwater head

The head section of a berm breakwater is always of special interest, as it is exposed to 3D flows. Conventional rubble mound breakwater heads are considered less stable than the trunk sections. The main problem caused by deformation at a head section of a berm breakwater is the possible loss of stones by transport of stones away from the profile. Unlike recession at the main trunk section, where reshaping finally will produce an equilibrium profile, the stones from the head may get accreted behind the head and possibly block partly the sailing lane. Once deposited behind the head the stones will not be reactivated by the waves to move back to their initial position. Therefore stone movement at the head section should be limited. During the tests on the Sirevåg berm breakwater,[57] the maximum HoTo-values for the two test set-ups 1 and 2 were 72 and 97, respectively. The reshaping of the Set-up 1 breakwater head was much less than for the Set-up 2 head, although there was no real damage to the breakwater heads for none of the alternative set-ups. The only concern was that more stones were thrown into the area behind the breakwater for Set-up 2 than Set-up 1.

Comparing results of tests by van der Meer and Veldman[66] and Tørum et al.[5] and Tørum[67] it is concluded that if a berm breakwater is designed as a reshaped static stable berm breakwater, HoTo < 70, it seems that by using the same profile of the head as of the trunk the head will be stable, with no excessive movements of the stones into the area behind the breakwater.

3.4. Wave Overtopping

3.4.1. *Wave Overtopping on Conventional Rubble-Mound Breakwaters*

Van der Meer and Jansen[68] arrived at the following relation for the wave overtopping time mean discharge for nonbreaking waves on a rubble-mound slope:

$$\text{Average}: Q_n = \frac{q}{\sqrt{gH_s^3}} = 0.2\exp(-2.6R_n) \qquad (3.49)$$

$$\text{Recommended } Q_n = \frac{q}{\sqrt{gH_s^3}} = 0.2\exp(-2.3R_n) \qquad (3.50)$$

where $q =$ time average overtopping discharge.

$$R_n = \frac{R_c}{H_s}\frac{1}{\gamma_b\gamma_h\gamma_f\gamma_\beta}$$

$\gamma_b =$ reduction factor taking into account a stepped slope, set to 1.0 for a berm breakwater

$\gamma_h =$ depth reduction factor $= 1-0.03(4 - d/H_s)^2$ for $d/H_s < 4$, $d =$ water depth

$$= 1 \text{ for } d/H_s > 4$$

$\gamma_f =$ friction reduction factor

$\gamma_\beta =$ reduction factor for oblique wave attack $= 1-0.0033\beta$, β in degrees

Van der Meer[70] proposed a new formula for the wave overtopping for slopes:

$$Q = \frac{q}{\sqrt{gH_s^3}} = \frac{0.06}{\sqrt{\tan\alpha}}\gamma_b\xi_{op}\exp\left(-4.7\frac{R_c}{H_s}\frac{1}{\xi_{op}\gamma_b\gamma_f\gamma_\beta\gamma_v}\right) \qquad (3.51)$$

but not larger than

$$Q_{\max} = \frac{q_{\max}}{\sqrt{gH_s^3}} = 0.2\exp\left(-2.3\frac{R_c}{H_s}\frac{1}{\gamma_f\gamma_\beta}\right) \qquad (3.52)$$

where $\xi_{op} = \tan\alpha/(2\pi H_s/(gT_p^2))$, Iribarren number.

$\gamma_v =$ a reduction factor due to a vertical wall (if any) on the slope

It should be mentioned that there generally is a tremendous spread in experimental data on wave overtopping.[68,72]

The results of de Rouck et al.[69,71] indicate that the runup in scale models is smaller than in full scale, especially for low waves. However, the reason for this apparent discrepancy is not fully resolved yet.

3.4.2. Wave Overtopping on Berm Breakwaters

The breakwater crest should be high enough to prevent excessive overtopping. The acceptable rate of overtopping depends on the design of the

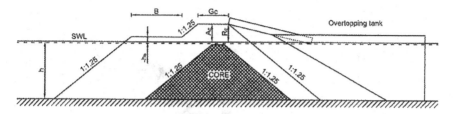

Fig. 3.24.   Test set-up.[62]

rear side and location of harbor facilities in relation to the location of the breakwater(s).

The most extensive investigation of wave overtopping on berm breakwaters has been carried out by Lykke Andersen.[62] Figure 3.24 shows his test set-up.

Lykke Andersen[62] proposed the following equation to calculate the mean overtopping discharge (mean of several waves):

$$Q_* = 1.79 \cdot 10^{-5} \cdot (f_{HO}^{1.34} + 9.22) \cdot s_{\text{op}}^{-2.52} \cdot \exp(-5.63 \cdot R_*^{0.92}$$
$$- 0.61 \cdot G_*^{1.39} - 0.55 \cdot h_{b*}^{1.48} \cdot B_*^{1.39}) \tag{3.53}$$

where

$$Q_* = \frac{q}{\sqrt{g \cdot H_{mo}^3}}$$

where q is the average overtopping discharge per meter structure length (width) [m^3/s/m]. H_{mo} is the significant wave height at the toe of the structure (frequency domain parameter), and g is the acceleration due to gravity.

$$R_* = \frac{R_c}{H_{mo}}, \quad R_c = \text{breakwater crest height.}$$

$$G_* = \frac{G_c}{H_{mo}}, \quad G_c = \text{breakwater crest width}$$

$$B_* = \frac{B}{H_{mo}}, \quad B = \text{berm width}$$

$$h_{b*} = \frac{3 \cdot H_{mo} - h_b}{3 \cdot H_{mo} + R_c} \quad \text{for } h_b < 3 \cdot H_{mo}$$

$$h_{b*} = 0 \quad \text{for } h_b \geq 3 \cdot H_{mo}$$

Figure 3.25 shows the calculated overtopping according to Eq. (3.53) versus the measured overtopping. Lykke Andersen considers that the

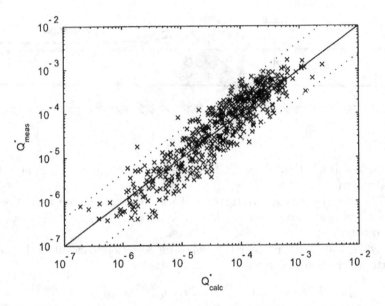

Fig. 3.25. Evaluation of the overtopping formula (Eq. (3.53)) against data related to reshaped profiles. Dashed lines show the 95% confidence band, i.e., only 5% of the data are outside this band. Note that the axes are on log scale.[62]

performance of the formula seems good considering the amount of scatter normally related to wave overtopping tests.

The overtopping formula give only overtopping discharges for "solid water" and spray is not included because spray does not reproduce well in a laboratory scale test.

A group of European researchers have issued a report on wave overtopping based on available overtopping data.[72] This report is probably the most comprehensive report available on wave overtopping.

3.4.3. *Rear Side Stability of Berm Breakwaters*

The rear side stability is to a large extent depending on the overtopping discharge. Lykke Andersen[62] gives the following formulae for the damage on the rear side with a slope of 1:1.25, assuming minor reshaping of the berm:

$$\text{No damage: } \frac{q}{\sqrt{g \cdot \Delta \cdot D_{n50}^3}} < 4 \cdot 10^{-5} \qquad (3.54)$$

$$\text{Start of damage: } 4 \cdot 10^{-5} \le \frac{q}{\sqrt{g \cdot \Delta \cdot D_{n50}^3}} < 3 \cdot 10^{-4} \qquad (3.55)$$

$$\text{Moderate damage: } 3 \cdot 10^{-4} \leq \frac{q}{\sqrt{g \cdot \Delta \cdot D_{n50}^3}} < 7 \cdot 10^{-4} \qquad (3.56)$$

$$\text{Severe damage: } 7 \cdot 10^{-4} \leq \frac{q}{\sqrt{g \cdot \Delta \cdot D_{n50}^3}} \qquad (3.57)$$

Lykke Andersen has in a personal communication with Alf Tørum emphasized that his criteria for rear side damage are probably conservative.

3.5. Caisson-Type Breakwaters

3.5.1. *General on Caisson-Type Breakwaters*

Caisson-type breakwaters are normally constructed as reinforced concrete caissons, built in a dry dock and towed to its position and lowered to a prepared stone/gravel bed. Japan[74] is probably the country where most caisson-type breakwaters have been built, but there has been research on the wave forces and the structural and foundation response in several other countries, although such breakwaters have not been built there (e.g., Germany).

The wave forces on the vertical wall breakwater are integrated wave pressures from a standing like wave and impulsive pressures from breaking waves. There is frequently in Japan placed concrete armor units in front of the caisson to mitigate impulsive pressures from breaking waves.

Most caisson-type breakwaters are built with straight walls, extending from the lower edge to the top of the breakwater. But caisson with inclined walls on the upper part and with perforated/slotted wall has also been built to reduce wave forces and to reduce reflections (perforated/slotted walls).

Figure 3.26 shows different "forms" of wave induced pressures versus time on a vertical wall.

3.5.2. *Extended Goda Formula[73] for Wave Actions on Main Body of a Caisson Breakwater with Vertical Walls*

Wave pressure exerted upon a front wall of the vertical or composite breakwater is assumed to have a linear distribution as shown in Fig. 3.27.

The elevation to which the wave pressure is exerted, denoted with η^* is given by

$$\eta^* = 0.75 \, (1 + \cos \beta) \lambda_1 H_D \qquad (3.58)$$

in which β denotes the angle between the direction of wave approach and a line normal to the upright wall, and H_D is the wave height to be used in

a) Non-breaking wave

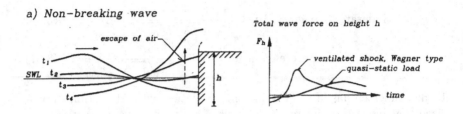

b) Breaking (plunging) wave, almost vertical front

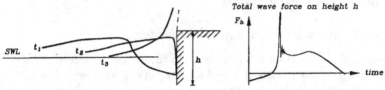

c) Breaking (plunging) wave, large air pocket

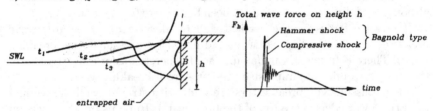

Fig. 3.26.   Different types of wave induced pressures on a vertical wall breakwater.[16]

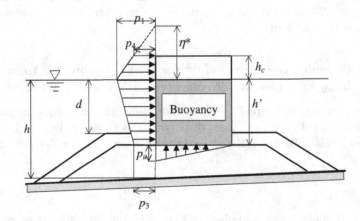

Fig. 3.27.   Distribution of wave pressure and uplift exerted on the main body of breakwater.

calculation as specified later. The wave direction should be rotated by up to 15° toward the line normal to the upright wall from the principal wave direction in consideration of inaccuracy in defining the wave direction.

The pressures are given by

$$p_1 = 0.5 \left(1 + \cos\beta\right)(\alpha_1\lambda_1 + \alpha_2\lambda_2 \cos^2\beta)\rho_w g H_D \tag{3.59}$$

$$p_3 = \alpha_3 p_1 \tag{3.60}$$

$$p_4 = \begin{cases} p_1\left(1 - \dfrac{h_c}{\eta^*}\right) & : \eta^* > h_c \\ 0 & : \eta^* \le h_c \end{cases} \tag{3.61}$$

in which α_1, α_2, and α_3 are the coefficients given below, λ_1 and λ_2 are the pressure modification factors, ρ_w is the density of seawater, g is the acceleration of gravity, and h_c is the crest height of front wall above the still water level.

$$\alpha_1 = 0.6 + \frac{1}{2}\left[\frac{4\pi h/L}{\sinh(4\pi h/L)}\right]^2 \tag{3.62}$$

$$\alpha_2 = \min\left\{\left(\frac{h_b - d}{3h_b}\right)\left(\frac{H_D}{d}\right)^2, \ \frac{2d}{H_D}\right\} \tag{3.63}$$

$$\alpha_3 = 1 - \frac{h'}{h}\left[1 - \frac{1}{\cosh(2\pi h/L)}\right] \tag{3.64}$$

where $\min\{a, b\}$ denotes the smaller one of a or b and

d = water depth on top of the armor units (or foot protection blocks)
h = water depth at the location of the front wall
h' = water depth at the toe of the front wall
h_b = water depth at an offshore distance of 5 times the significant wave height from the front wall
L = wavelength at the water depth h

The uplift exerted on the bottom of the main body is assumed to have a linear distribution with the maximum intensity of the following:

$$p_u = 0.5(1 + \cos\beta)\alpha_1\alpha_3\lambda_3\rho_w g H_D \tag{3.65}$$

The buoyancy is applied to the immersed part of the main body regardless of wave overtopping.

The pressure modification factors λ_1, λ_2, and λ_3 are given the values of 1.0 for standard vertical or composite breakwaters, but they are assigned smaller values for composite breakwaters covered with wave-dissipating concrete blocks (see Ref. 73, p. 109 and Tables VI-5-58 and VI-5-59 of Burcharth and Hughes[16]).

The wave height H_D is the height of the highest wave, which is taken as 1.8 times the design significant wave height $H_{1/3}$ when the breakwater is located outside the surf zone. Inside the surf zone, H_D should be evaluated by taking the random wave breaking process into consideration. The wave period for evaluation of the wavelength L is the significant wave period of the design wave, which is approximately equal to $0.9T_p$ and $1.2T_m$ for wind waves.

For the cases in which considerations are needed for the exertion of impulsive breaking wave pressures, the coefficient α_2 is to be replaced by $\alpha^* = \max\{\alpha_2; \alpha_I\}$, where α_I is the coefficient for impulsive breaking wave pressure and defined below.

$$\alpha_I = \alpha_{IH}\alpha_{IB} \tag{3.66}$$

where

$$\alpha_{IH} = \min\{H/d; 2, 0\} \tag{3.67}$$

$$\alpha_{IB} = \begin{cases} \cos\delta_2/\cos\delta_1 & : \delta_2 \leq 0 \\ 1/(\cosh\delta_1 \ \cosh^{1/2}\delta_2) & : \delta_2 > 0 \end{cases} \tag{3.68}$$

$$\delta_1 = \begin{cases} 20\delta_{11} & : \delta_{11} \leq 0 \\ 15\delta_{11} & : \delta_{11} > 0 \end{cases} \tag{3.69}$$

$$\delta_{11} = 0.93\left(\frac{B_M}{L} - 0.12\right) + 0.36\left(0, 4 - \frac{d}{h}\right) \tag{3.70}$$

$$\delta_2 = \begin{cases} 4.9\delta_{22} & : \delta_{22} \leq 0 \\ 3.0\delta_{22} & : \delta_{22} > 0 \end{cases} \tag{3.71}$$

$$\delta_{22} = -0.36\left(\frac{B_M}{L} - 0.12\right) + 0.93\left(0.4 - \frac{d}{h}\right) \tag{3.72}$$

where B_M is the width of the berm of the rubble mound in front of the main body. The above formulas are due to Takahashi *et al.*[75]

Figure 3.28 shows failure modes for caisson breakwaters. Goda[73] gives the following formulae for the total horizontal force and the overturning moment this force gives around the rear lower corner (the heel):

$$P = \frac{1}{2}(p_1 + p_3)h' + \frac{1}{2}(p_1 + p_4)h_c^* \tag{3.73}$$

$$M_P = \frac{1}{6}(2p_1 + p_3)h'^2 + \frac{1}{2}(p_1 + p_4)h'h_c^* + \frac{1}{6}(p_1 + 2p_4)h_c^{*2} \tag{3.74}$$

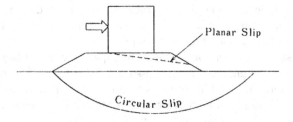

(A) Sliding

(B) Overturning

(C) Failure of Foundation

Planar Slip

Circular Slip

Fig. 3.28. Failure modes of caisson breakwaters.[75]

in which

$$p_4 = p_1(1 - h_c/\eta^*) \quad \text{for } \eta* > h_c$$

$$p_4 = 0 \quad \text{for } \eta* \le h_c$$

$$h_c^* = \min\{\eta^*; h_c\}$$

The total uplift pressure and its moment around the heel of the upright section (Fig. 3.27) are calculated with:

$$U = \frac{1}{2}p_u B \tag{3.75}$$

$$M_u = \frac{2}{3}UB \tag{3.76}$$

Goda[73] give formulas for safety factors for sliding and overturning, which has been modified to:

$$\text{Against sliding: } SF_{\text{sliding}} = \frac{\mu(M \cdot g - U - Bouy)}{P} \qquad (3.77)$$

$$SF_{\text{overturning}} = \frac{M \cdot g \cdot t - Buoy \cdot t - M_u}{M_p} \qquad (3.78)$$

where

M = mass of the caisson in air
μ = coefficient of friction
t = the horizontal distance between the center of gravity and heel of the upright section. For a square formed caisson t is the same for M and for buoyancy.
Buoy = buoyancy of the caisson
B = width of the caisson

The bearing capacity of the foundation is to be analyzed by means of the methodology of foundation engineering for eccentric inclined loads. Such analysis has not been shown in this chapter.

In the design of vertical breakwaters in Japan, the safety factors against sliding and overturning must not be less than 1.2. The coefficient of friction between concrete and rubble stones is usually taken as $\mu = 0.6$.[73]

The Goda formula tends to overestimate the total wave loading on composite breakwaters by about 10% with the coefficient of variation of 0.1 or so (see Takayama and Ikeda[76] and Table VI-5-55 of Ref. 16). The bias and uncertainty of the Goda formula should be taken into consideration in probabilistic design of composite breakwaters.[77]

3.5.3. Wave Forces on Caissons with Inclined Walls and with Circular Cylinders

3.5.3.1. Wave forces on caissons with inclined walls

Takahashi et al.[78] carried out a series of wave force tests on caisson breakwaters with cross-sections as shown in Fig. 3.29. They developed a calculation method for the wave forces on a sloped caisson. The method is a modification of the Goda formula (Fig. 3.30).

The base pressure diagram and the front pressure diagram are the diagram obtained by the Goda method (Fig. 3.27), assuming a vertical wall. The forces F_p (and the forces F_{SH} and F_{SV}) and F_v are then modified. Figure 3.30 shows how the modifications were done at an early stage.

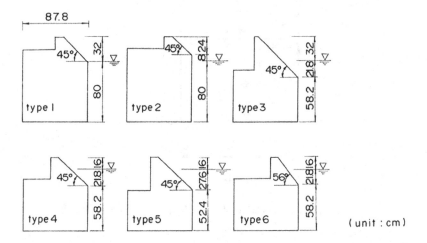

Fig. 3.29.   Caisson cross sections tested by Takahashi *et al.*[78]

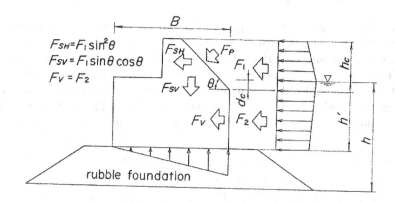

Fig. 3.30.   Sketch of the wave forces on inclined caisson.[78]

However, Takahashi *et al.*[78] introduced, based on laboratory experiment, modifications to these forces.

$$F_{SH} = \lambda'_{SL'} F_1 \sin^2 \theta \tag{3.79}$$

where

$$\lambda'_{SL} = \min\{\max\{1.0, -23(H/L)\tan^{-2}\theta + \sin^{-2}\theta\}, \sin^{-2}\theta\}$$

and

$$F_{\mathrm{v}} = \lambda_{\mathrm{v}} F_2$$

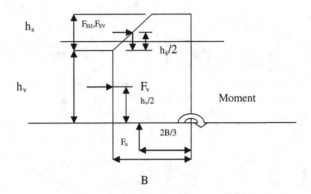

Fig. 3.31. Location of forces F_{SH}, F_{SV} and F_V.

where

$$\lambda_v = \min\{1.0, \max\{1.1, 1.1 + 11d_c/L\} - 5.0\,H/L\}$$

Takahashi *et al.*[78] give no information on moments and uplift wave pressures. For the uplift pressures, the same pressures as for the vertical wall may be assumed in the calculations. Due to lack of knowledge about the exact wave pressure distributions on the slope and on the vertical lower part, the horizontal and vertical forces F_{SH} and F_{SV} and the horizontal force F_v are assumed to be located as shown in Fig. 3.31. The inclined caisson gives some reduction in wave forces compared with the vertical wall caisson.

3.5.3.2. *Wave forces on caisson with circular cylinders*

Caissons with circular cylinder forms have been used at Hanstholm, Denmark and Brighton, UK. The circular cylinder form may be an advantage from the structural point of view. However, there are few investigations on the wave forces on caissons with circular cylinders.

Figure 3.32 shows the pressure distribution around cylinder walls as obtained by Khaskhachich and Vanchagov.[79] This indicates a horizontal wave force reduction of 10–15% compared to a vertical wall.

Inspired by the design of the large concrete platform in the North Sea/Norwegian Sea in the 1970s and 1980s (CONDEEP etc.) a caisson breakwater with circular cylinders has been proposed, as indicated in Fig. 3.33. There are two rows of cylinders and the diameter of each cylinder is approximately 20 m. Jahr[81] carried out laboratory model tests on caissons as indicated in Fig. 3.33 for water depth 50 m and caisson depth 25 m. Her results support that the force reduction was approximately 10–15% of the wave forces on a vertical wall caisson.

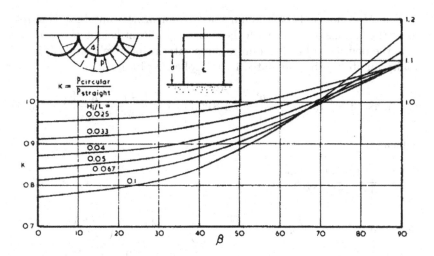

Fig. 3.32.   Distribution of pressure around cylindrical wall.[79,80]

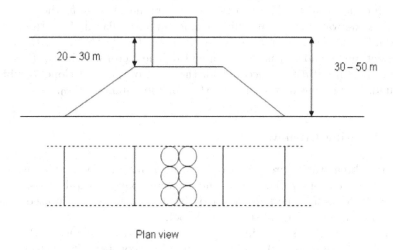

Plan view

Fig. 3.33.   Proposed composite caisson breakwater with circular cylindrical caissons.

3.5.4.   *Formulae for Minimum Mass of Armor Units of Rubble Foundation for a Caisson Breakwater*

One of the available formulas for designing armor units of the rubble foundation of a composite breakwater is due to Tanimoto *et al.*[82] and expressed as below. For other formulas, see Tables VI-5-45 to 48 of Ref. 16.

$$M = \frac{\rho_s}{N_s^3 (\rho_s/\rho_w - 1)^3} H_{1/3}^3 \qquad (3.80)$$

in which M denotes the minimum mass of an armor unit that is stable against the actions of waves with the design significant height $H_{1/3}$, ρ_s and ρ_w are the densities of armor unit and seawater respectively, and N_s is the stability number to be calculated by the following formula:

$$N_s = \max\left\{1.8; \left(1.3\frac{1-\kappa}{\kappa^{1/3}}\frac{h'}{H_{1/3}} + 1.8\exp\left[-1.5\frac{(1-\kappa)^2}{\kappa^{1/3}}\frac{h'}{H_{1/3}}\right]\right)\right\} \quad (3.81)$$

in which

$$\kappa = \frac{4\pi h'/L'}{\sinh(4\pi h'/L')}\kappa_2 \quad\quad (3.82)$$

$$\kappa_2 = \max\left\{0.45\sin^2\beta\cos^2\left(\frac{2\pi x}{L'}\cos\beta\right); \; \cos^2\beta\sin^2\left(\frac{2\pi x}{L'}\right)\right\}:$$
$$0 \le x \le B_M \quad (3.83)$$

The function $\max\{a; b\}$ denotes the larger one of a or b, the term h' denotes the water depth at which armor units are placed, L' is the wavelength for the design significant wave period at the depth h', β is the wave incident angle, and B_M is the berm width. The factor 0.45 in Eq. (3.83) is due to Kimura et al.[83] to account for the effect of the front slope of rubble mound. It is not clear if M means $M_{\min}$ or the minimum W_{50}.

3.6. Floating Breakwaters

Floating breakwaters are primarily used in relatively sheltered areas, e.g., small boat harbors in fjord areas and sheltered bays. Several cross-section shapes have been suggested and partly been used. A very often used cross-section is shown in Fig. 3.34 (a box shape).

The breakwater shown in Fig. 3.34 was tested in the Ocean Basin at Marintek.[84] Figure 3.32 shows a two-element breakwater, with each element being 22 m long. But it is referred to test results from a test layout with 6 elements, each element being 22 m long. The tests were carried out with model scale 1:10. In addition to measurements of waves in front and behind the breakwater, mooring forces were also measured in several of the mooring lines. The tests were carried out with long-crested and short-crested waves. Tests were carried out with two configurations of the floating breakwater lay-out: (1) All the six elements were rigidly connected as a long beam (Stiff model). (2) Each element was loosely connected to the neighbor element with chains and fenders (Fendered model).

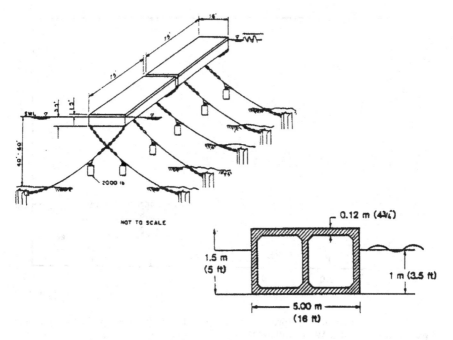

Fig. 3.34.   Floating breakwater.[74]

Figures 3.35 and 3.36 show wave measurement results. $H_{mo,i}$ is the significant wave height of the incoming waves (H_{s}.in) H_{mo} is the significant wave height behind the breakwater, in this case at wave staff 18 located approximately 10 m (full scale) behind and in the middle of the breakwater.

Figures 3.35 and 3.36 shows that there are no significant differences between long- and short-crested waves (directional waves) and there are no significant differences between "Stiff" and "Fendered" model. It is also seen that the wave height reduction factor is sensitive to the peak wave period.

As a general conclusion from these tests and similar other tests on the efficiency of the floating breakwater is that in order to give adequate wave damping they have to be relatively wide in relation to the wave length. Thus they will be uneconomical to use on open coasts for waves with periods 10–15 s. Only in restricted waters with large depths they may become economical.

Mooring forces were also measured. The mooring forces are governed by the slow drift oscillations and not by the motion of the floating breakwater for each individual wave. At the moment there is no good method to calculate the slow drift oscillations of floating breakwaters.[85,86] The findings from the experiments of Stansberg *et al.*[84] are (1) the slow drift oscillations

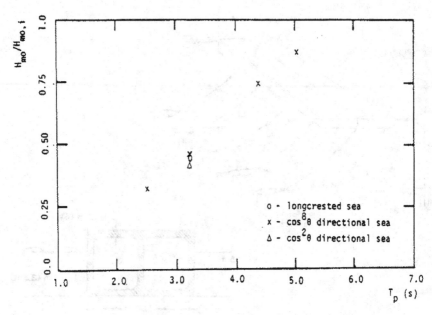

Fig. 3.35.   Wave reduction factor, $H_{mo}/H_{mo,I}$ ($H_s/H_{s,i}$) versus peak period T_p of input wave. Fendered model. Wave staff 18.[84]

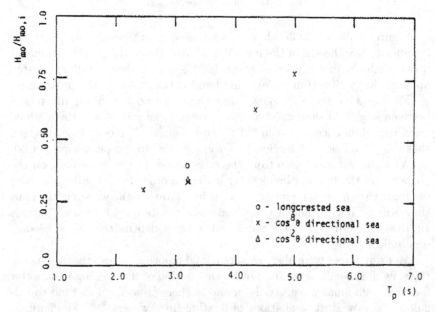

Fig. 3.36.   Wave reduction factor, $H_{mo}/H_{mo,I}$ ($H_s/H_{s,i}$) versus peak period T_p of input wave. Stiff model. Wave staff 18.[84]

are dominant with respect to mooring forces; (2) the slow drift motion is larger on a long floating breakwater than on a short one, and (3) the slow drift sway motion gets significantly reduced in multidirectional irregular waves compared to those in unidirectional irregular waves.

3.7. Wave Screens

3.7.1. *Wave Damping Effects of Wave Screens*

Similar to floating breakwaters, wave screens have primarily been used in relatively sheltered areas (e.g., small boat harbors in fjord areas and sheltered bays). Figure 3.37 shows a definition sketch of a wave screen.

$Himo$ = significant height of incoming wave, $Htmo$ = significant height of transmitted wave, $Hrmo$ = significant height of reflected wave, Fmo = significant wave force on the screen.

The transmission coefficient $KTmo$ is defined as:

$$KTmo = \frac{Htmo}{Himo}$$

Figure 3.38 shows the transmission coefficient as obtained by Kriebel[87] for different ratios of submergence w and wave length L_p (corresponding to the wave length for the peak period) and for different w/d ratios, $0.4 < w/d < 0.7$.

3.7.2. *Wave Forces on Wave Screens*

For an impermeable wave screen with a small gap at the bottom, an approximation to the wave pressure exerted on the front surface can be made by applying the Goda formula for vertical caissons, excluding the pressures

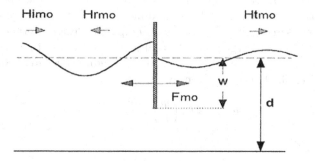

Fig. 3.37.   Definition sketch for a generic wave screen or wave barrier.[87]

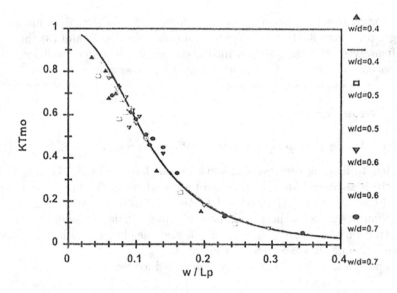

Fig. 3.38.   Measured and predicted zero-moment wave transmission coefficients.[87]

below the submergence w, Fig. 3.37. It provides an upper limit to the actual wave pressure, because it neglects a reduction of pressure caused by partial wave reflection and by the bottom gap. The maximum horizontal force is then obtained from the vertical integration of the wave-induced pressures along the height of the wall. The maximum overturning moment about the seafloor by wave actions is similarly obtained. Some design diagrams are available for the maximum forces and moments for wave screens. Kriebel[88] and Bergmann and Oumeraci[89] provide information on wave loads for full-depth permeable (slotted) wave screens. References 87, 90–94 provide information on wave loading on partial-depth impermeable wave screens. For wave screens of other configurations, hydraulic model tests are required for evaluation of wave pressures, forces and moments. For the case of an impermeable wave screen CEM[7] presents an empirical design method for computing wave forces. Large-scale physical model test data of Kriebel et al.[90] were used to develop an expression for the significant force. Results are shown in Fig. 3.39 and are given by

$$F_{MO} = F_0 \left(\frac{w}{h}\right)^{0,386(h/L_p)^{-0,7}} \tag{3.84}$$

where

$$F_O = \rho g H_{MO} \frac{\sinh(k_p h)}{k_p \cosh(k_p h)} \tag{3.85}$$

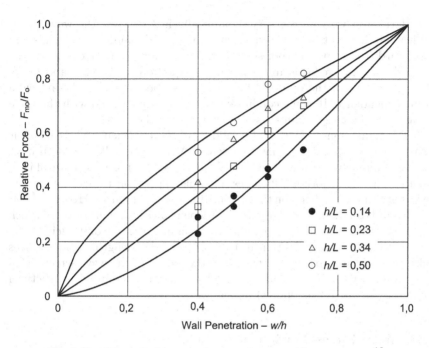

Fig. 3.39. Dimensionless wave forces on impermeable wave screen.[16]

and where

F_{MO} = significant force per unit width on partial-depth impermeable wave screen

F_O = significant force per unit width on full-depth wall

H_{MO} = significant wave height

h = water depth

w = wave screen draft or penetration

L_p = wavelength associated with peak spectral wave period, T_p

k_p = wave number associated with peak spectral wave period, $k_p = 2\pi/L_p$

ρ = water mass density

g = acceleration of gravity

The CEM[16] method should be used within the limits of data for which it was developed: $0,4 < w/h < 0,7$ and $0,14 < h/L_p < 0,5$.

The design force per unit width on the wave screen should be the load corresponding to the design wave height of $1.8\,H_{MO}$ recommended by Goda.[73] The appropriate design force is then given by

$$F_{\text{Design}} = 1,8F_{MO} \qquad (3.86)$$

Horizontal wave forces vary dynamically in time over the wave cycle. When an incident wave crest is at the wall, dynamic wave loads are landward (toward the harbor side). When an incident wave trough is at the wall, dynamic wave loads are seaward. The equations and figures above give the magnitude of the dynamic force. This can be coupled with sinusoidal time dependence. In some conditions, wave loads associated with the wave trough can be larger than those associated with the crest.

The expressions above are also valid for 0° wave incidence where incident wave crests are parallel to the wave screen. Waves generally approach wave screens obliquely and wave loads are then reduced for nonzero angles of incidence. The Goda formula includes a simple expression to account for the effect of oblique incidence on the pressure intensity. Gilman and Kriebel[91] and Grune and Kohlhause[92] suggest a cosine dependence in oblique waves in which wave loads at 90° incidence are reduced to one-half of those for 0° incidence.

In addition to a reduction in the force per unit length, oblique waves also cause a variation in loadings along the length of the wave screen. This longitudinal variation should be considered when designing the structural support (pier or pile supports) for the wave screen.

3.8. Wave Forces on Circular Cylinders

3.8.1. *Wave Forces on Slender Circular Cylinders*

Wave forces from nonbreaking waves on a slender vertical pile are commonly calculated according to the Morison equation[11] (Fig. 3.40):

$$dF = dF_D + dF_M = 0.5\rho_w C_D D u|u|dz + \rho_w C_M \frac{\pi D^2}{4}\frac{\partial u}{\partial t}dz \qquad (3.87)$$

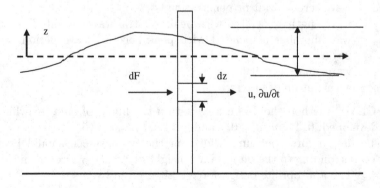

Fig. 3.40. Wave forces on a slender vertical pile.

where ρ_w is the mass density of water, C_D is the drag coefficient, C_M is the inertia coefficient, D is the pile diameter, u is the water particle velocity, and t is time.

By integrating Eq. (3.87) from the bottom to the surface elevation, the total wave force is obtained. The Morison equation is to a large extent used in the oil industry and much research has been carried out on the appropriate values of the drag coefficient, C_D, and the inertia coefficient, C_M.

Eddies can be shed from a cylinder alternatively on either side, thus inducing a transversal force normal to the direction of the particle velocity direction. The eddies are shed at twice the frequency of the wave. The lift force thus induced is expressed as follows:

$$dF_L = 0.5\rho_w C_L u_{\max}^2 \sin 2\omega t \cdot dz \qquad (3.88)$$

where $u_{\max}$ is the maximum water particle velocity in a wave cycle, ω is the radian frequency of the wave $= 2\pi/T$, and $C_L =$ lift coefficient.

ISO 21650[95] give guidance on the choice of C_D, C_M, and C_L coefficients and the following is quoted from ISO 21650,[95] partly based on other manuals etc.:

"The drag, inertia and lift coefficients have been obtained from analysis of results from experimental work. The values of the drag and inertia coefficients depend on the Keulegan–Carpenter (KC) number, the Reynolds number and the cylinder surface roughness."

The Keulegan–Carpenter number (KC) and the Reynolds number (Re) are defined as:

$$KC = \frac{u_{\max}T}{D}, \quad \mathrm{Re} = \frac{u_{\max}D}{\nu}$$

where $\nu =$ kinematic viscosity. One of the most extensive investigations on drag, inertia, and lift coefficients was carried out by Sarpkaya.[96,97] The experiments were performed in an oscillating U-tube water tunnel for a range of Reynolds numbers up to 700,000 and Keulegan–Carpenter numbers up to 150. Relative roughness (k/D) for the cylinders varied between 0.002 and 0.02. The roughness k is the average height of the roughness elements glued to the cylinders.

Marine growth is considered to contribute mostly to the roughness of a structural element in the ocean.

The cross-sectional dimensions of structural elements shall be increased to account for marine growth thickness. Marine growth will depend on the location. The NORSOK standard[98] gives recommendation for marine growth roughness as shown in Table 3.10. The marine growth roughness may be different in other locations.

Table 3.10. Thickness of marine growth. The water depth
refers to mean water level.[88]

Water depth	Altitude 56° − 59°	Altitude 59° − 72°
Above +2 m	0	0
+2 m −40 m	100 mm	60 mm
Under −40 m	50 mm	30 mm

ISO CD 19902[100] cautions that it takes very little roughness to make
the rough value of C_D (see later) realistic. Site-specific data shall be used
to reliably establish the extent of the hydrodynamically rough zones. Otherwise the structural elements shall be considered rough down to the sea
floor.

Sarpkaya carried out his tests under idealized conditions with rectilinear
oscillating flow. For "real" waves, there is still some controversy on the
choice of relevant C_D and C_M values. Different standards, codes, and guidelines for the offshore oil industry give different values (see Refs. 98–101).
Moe and Gudmestad[103] and Gudmestad and Moe[102] discuss the coefficients
and recommend the American Petroleum Institute's recipe on selection of
coefficient values.[101] The following is quoted from[100]:

"For typical design situations, global hydrodynamic loading on a structure
can be calculated using Morison's equation with the following values of the
hydrodynamic coefficients for unshielded circular cylinders:

smooth $C_D = 0.65$ $C_M = 1.6$

rough $C_D = 1.05$ $C_M = 1.2$"

NORSOK[98] elaborate somewhat more on the issue of selecting C_D and C_M
values and the following is quoted from Ref. 98, which again adhere mainly
to API[101] for slender tubular structural elements.

For structures with small motions, the wave actions can be calculated as
follows:

a) If the Keulegan–Carpenter number (KC) is less than 2 for a structural
   element, the actions may be found by potential theory:

aa) If the ratio between the wave length L and the tubular diameter D is
    greater than 5, the inertia term in the Morison formula can be used
    with $C_M = 2.0$.

ab) If the ratio between L and D is smaller than 5, the diffraction theory
    should be used.

b) If KC is greater than 2, the wave action can be calculated by means
   of the Morison formula, with C_D and C_M given as functions of the
   Reynolds number, Re, and the Keulegan–Carpenter number KC and

relative roughness. It should be noted that Morison formula ignores lift forces, slamming forces and axial Froude–Krylov forces.

c) For surface piercing framed structures consisting of tubular slender members (e.g., conventional jackets) extreme hydrodynamic actions on unshielded circular cylinders are calculated by Morison formulae on the basis of:

— Stokes 5th order or stream function wave kinematics and a kinematics spreading factor on the wave water particle velocity, which is 0.95 for North Sea conditions. This kinematics spreading factor is introduced in the regular wave approach to account for wave spreading and irregularity in real sea states.
— Drag and inertia coefficients equal to

— $C_D = 0.65$ and $C_M = 1.6$ for smooth members
— $C_D = 1.05$ and $C_M = 1.2$ for rough members.

These values are applicable for $u_{max} T_i/D > 30$, where u_{max} is the maximum horizontal particle velocity at storm mean water level under the wave crest, T_i is the intrinsic wave period and D is the leg diameter at the storm mean level.

d) For (dynamic) spectral or time domain analysis of surface piercing framed structures in random Gaussian waves and use of modified Airy (Wheeler) kinematics with no account of kinematics factor, the hydrodynamic coefficients should in absence of more detailed documentation be taken to be:

CD = 1.0 and CM = 2.0

These values apply both in stochastic analysis of extreme and fatigue action effects. If time domain analysis is carried out with non-symmetry of wave surface elevation properly accounted for, the hydrodynamic coefficients in item c) could be applied.

Taken into account that the codes for offshore oil industry structures are calibrated, which is not the case for structures in shallow water and in the coastal zone, it is recommended to apply the Stokes 5th order wave theory, or some other high order theory, together with $C_D = 1.4$ (see Ref. 105) and $C_M = 2.0$ for structures in these areas. For square formed piles, C_D values, which are approximately 30% higher than for the circular pile, are recommended. Note that possible wave slamming actions should be considered (see later).

All wave theories predict that the maximum water particle velocities are maximum at the crest height. The Morison equation thus also predicts that the maximum wave force also occurs at the crest height. Dean et al.[106] and Tørum[107] found a slight modification to this due to surface effects (Fig. 3.41).

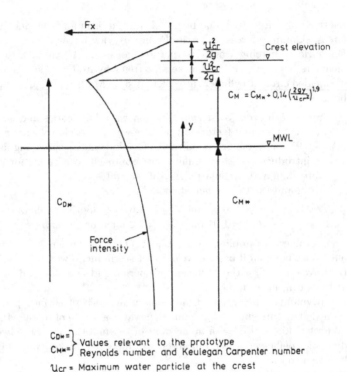

Fig. 3.41. Recommended design time invariant C_D and C_M values in the surface zone area.[107]

C_{D*} and $C_{M'} = $ values relevant to the prototype Reynolds and Keulegan–Carpenter number. $U_{cr} = $ maximum water particle velocity at the crest.

3.8.2. *Wave Forces on Large Diameter Cylinders*

Wave action on large volume bodies should be calculated on the basis of wave diffraction theory. The incoming waves are reflected or scattered and the wave potential is expressed as a sum of the potential of the incoming wave and the potential of the scattered wave.

For simple structural shapes like a vertical circular cylinder resting on the sea bottom, analytical solution proposed by MacCamy and Fuchs[104] can be applied. This solution shows that there will be a phase angle between the maximum force and the wave as the wave passes the cylinder. This phase angle depends on the D/L ratio. Comparing the MacCamy and Fuchs analytical solution to the general formulae for the inertia force, dF_h on a

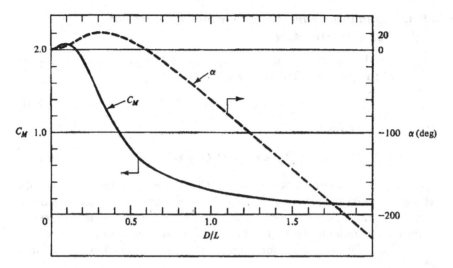

Fig. 3.42. Variation of inertia coefficient C_M and phase angle α of maximum force with parameter D/L.[108]

unit length of the cylinder:

$$dF_h(z) = \rho_w C_M \frac{\pi D^2}{4} \frac{\partial u(z)}{\partial t} \qquad (3.89)$$

where $\partial u/\partial t$ is calculated at the centre of the cylinder, C_M and phase angles as shown in Fig. 3.42 are found.

For general structures consisting of several large volume components, boundary element or finite elements should be used.[95]

MacCamy and Fuchs solution and some computer programs based on boundary elements methods are based on linear wave theory. However, in the vicinity of large bodies, the free surface elevations may be increased due to motions, diffraction, radiation, wave/current interaction effects, and other nonlinear wave effects (e.g., shallow water effects). These should be accounted for in the wave action calculation and used to estimate deck clearance and freeboard.

Ground *et al.*[110] carried out tests on wave forces for a 100 m diameter circular cylinder in 25 m water depth and for waves with periods in the range 7.7–27.3 s and wave heights in the range 4.8–17 m. They also considered the set-down effect due to the influence of radiation stress. The waves with longer periods and greater heights were highly nonlinear with regard to water surface elevations. The comparison with the MacCamy and Fuch's theoretical results showed a fair agreement for the horizontal force, especially when the set-down effect was included. The overturning moment was significantly underestimated by the MacCamy and Fuch's theory.

3.8.3. Slamming Wave Forces on Slender Circular Cylinder Structures

If the waves break against the pile (Fig. 3.43), a slamming force may occur on part of the pile, $\lambda\eta_b$. The total force is then

$$F = F_D + F_M + F_s \tag{3.90}$$

The slamming force is written as:

$$F_s = 0.5\rho_w C_s D C_b^2 \lambda\eta_b \tag{3.91}$$

where C_s is a slamming force factor, C_b is the breaking wave celerity (the water particle velocity is set equal to the wave celerity at breaking), λ is the curling factor, which indicates how much of the wave crest is active in the slamming force. The nature of the slamming force is indicated in Fig. 3.44. The slamming force may have a short duration τ_p but high intensity.

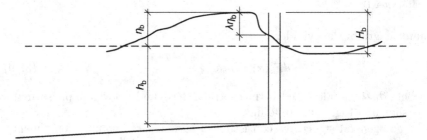

Fig. 3.43.   Breaking wave leading to possible slamming force.

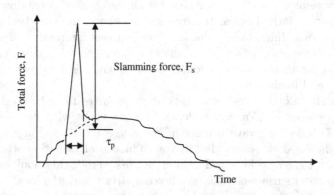

Fig. 3.44.   The nature of the slamming force.

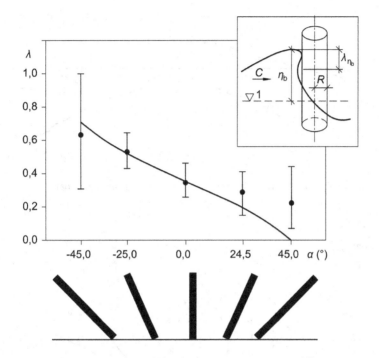

Fig. 3.45.   Curling factor versus pile inclination.[112]

Different values of C_s have been obtained by different researchers. One of the latest investigations on wave slamming forces on cylinders in a large model scale set-up have been carried out by Wienke[111] and Wienke and Oumeraci.[112] They set $C_s = 2\pi$ and obtained values of λ as shown in Fig. 3.45 for different inclination of the pile.

When a wave breaks over a three-dimensional shoal the conditions may be as indicated in Fig. 3.46. The shoal may go "dry" for part of the time when the wave passes over.[113] The flow conditions are then complicated and no wave theory has been established to calculate the waver particle velocities under such conditions. Grønsund Hanssen[114] and Grønsund Hanssen and Tørum[115] measured the water particle velocities as a wave broke over a specific shoal. Figure 3.47 shows the measured peak velocities for different wave heights.

Goda,[116] Hovden,[117] Hovden and Tørum,[118] and Kyte and Tørum[119] investigated the wave forces on a single vertical cylinder erected on a shoal. Later Grønsund Hanssen and Tørum[115] investigated the wave forces on a tripod (Fig. 3.48) located at the shoal Knubbehausen off the coast at Kragerø, Norway. The water depth on the shoal is approximately 7.5 m

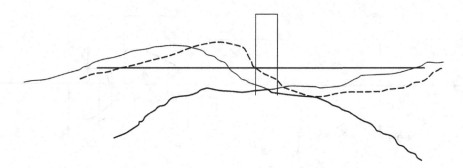

Fig. 3.46.   Wave breaking over a three-dimensional shoal.

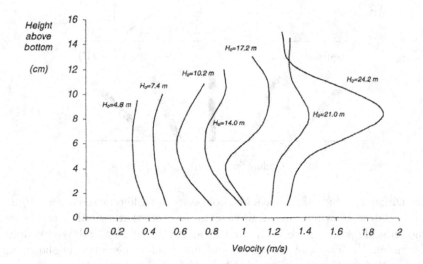

Fig. 3.47.   Measured maximal water particle velocities over a specific shoal.[115]

and the depth increases abruptly to more than 40 m around the shoal. The shoal is thus like a peaky underwater mountain. The height of the waves approaching the shoal may be more than 20 m. They will thus break heavily on the shoal, giving rise to slamming forces on the structure.

Grønsund Hanssen and Tørum[115] formulated a curling factor λ as follows:

$$\lambda = \frac{F_{\text{measured}}}{0.5\rho_w C_{ss} U^2 \sum_i D_i l_i} \qquad (3.92)$$

where $U =$ mean water particle velocity above the shoal (Fig. 3.47), $C_{ss} = \pi$ (note that this is different from $C_s = 2\pi$ as used by Wienke and

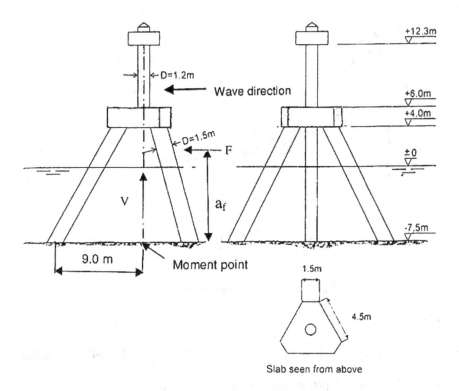

Fig. 3.48.   Tripod (light tower) on the shoal Knubbehausen.

Oumeraci).[112] D_i and l_i are the diameter and the length of the elements that will be subjected to wave forces.

It was assumed that for the tripod on Knubbehausen the two front legs, the horizontal plate and the single vertical cylinder carrying the light were subjected to slamming forces simultaneously. The triangular horizontal plate were made equivalent to a cylinder with the same area as the triangle, $D = 6$ m.

Grønsund Hanssen and Tørum[115] arrived at the following expression for the mean water particle velocity over the shoal:

$$U = \sqrt{gd\left(1 + \frac{\eta_{\text{maks}}}{d}\right)}$$

$$\cdot \left(8.717\left[\frac{H_0}{gT^2}\right]^{\frac{gT^2}{2000d}} + 0.470\sqrt{\frac{gT^2}{d}} - 11.300\right) \qquad (3.93)$$

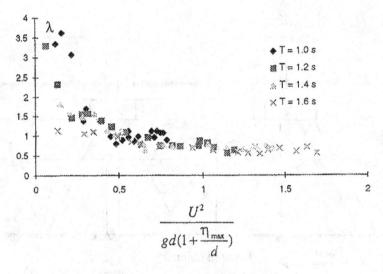

Fig. 3.49. λ versus velocity parameter.[115]

where

$$\frac{\eta_{maks}}{d} = 8.283 \left[\frac{H_o}{gT^2}\right]^{\frac{gT^2}{2000d}} + 0.430\sqrt{\frac{gT^2}{d}} - 10.467 \qquad (3.94)$$

where d is water depth on the shoal, T is wave period, H is wave height, η_{max} is maximum crest elevation and g is acceleration of gravity.

Figure 3.49 shows the value of λ versus velocity parameter.

Grønsund Hanssen and Tørum[115] also obtained also overturning moments for the tripod. The moments were much influenced by the vertical forces exerted on the horizontal slab.

When a wave crest arrives at a pile there will be a run-up on the pile, Op (Fig. 3.50) considerably above η_{max}. This run-up may have to be considered, especially if the will be a navigation light on top of the structure. Tørum[107] estimated the runup to be:

$$Op = \frac{u^2}{2g} \qquad (3.95)$$

where u is the water particle velocity at the wave crest. In the case of a vertical cylinder on a shoal, u is taken as the average velocity U (Eq. (3.93)).

The duration of a slamming force, τ_p (Fig. 3.44) is in the range

$$\tau_p = 0.25\frac{D}{u} - 0.5\frac{D}{u} \qquad (3.96)$$

If a natural frequency of oscillation of the structure is in this range or close to it, there might be an amplification of the response of the structure.

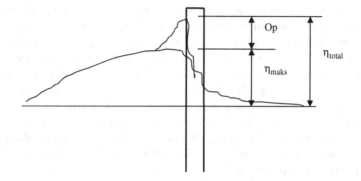

Fig. 3.50.   Run-up on a pile.

3.9. Wave-Induced Pore Pressures in the Sea Bed and Geotechnical Stability

3.9.1. *General on Geotechnical Stability Analysis*

The overall slope stability of any breakwater has to be checked. The bearing capacity of the rubble mound and the seabed foundation can be analyzed with slip surface calculations (Fig. 3.51). There are different methods and approaches to this (e.g., the simplified Bishop method,[74] for circular slip surfaces and Janbu[120] for noncircular slip surfaces). Such an analysis has to include the wave-induced pore pressures in the breakwater trunk and in the bottom soil. De Rouck and Troch[121] refer to sensitivity analysis showing the paramount importance of the pore pressure along a potential slip surface; they are as important as the shear resistance characteristics.

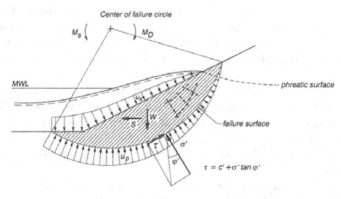

Fig. 3.51.   Simplified cross section of a rubble mound breakwater with a possible slip circle.[61]

The analysis of the pore pressures in the breakwater trunk may be carried out with some numerical program.[121] Similarly the program may also be used for the sandy bottom. The breakwater trunk material is generally so coarse that there will most probably not be any influence on the pore pressure from air/gas in the water. The sea bed may consist of fine sand and clay. Only a small content of air/gas in the soil has a significant influence on the wave induced pore pressures in a fine soil.

But there has been some debate if there might be some air/gas in the pores of the bottom soil of sand, silt, or clay. An overview and discussion on the issue of air/gas content in the soil and its effect on wave-induced pore pressures is therefore included.

3.9.2. *Wave-Induced Pore Pressures*

Wave-induced pore pressures in a seabed of sand or clay has been an item for extensive research. One of the major difficulties in analyzing and interpreting both laboratory and field data has been related to the question of air/gas content in the pores of fine soil. There is presently no commercial instrument available to readily measure the small quantities (some few percent) of air/gas that may be present, although some development of such instruments has been in progress.[123]

3.9.2.1. *De Rouck's[109] investigations on wave-induced pore pressures*

Some of the most advanced wave induced pore pressure measurements in the field have been carried out by Julien de Rouck,[125] de Rouck and van Damme,[125] and de Rouck and Troch.[121] These measurements were carried out in connection with the planning of the extension of the Zeebrugge harbor, Belgium, and the results were used to evaluate the geotechnical stability of the new breakwaters. De Rouck also developed an analytical model, based to some extent on two-dimensional consolidation theory. We will compare the results of de Rouck's measurement with a more advanced theory developed by Mei and Foda.[126]

Mei and Foda developed their two-dimensional model based on the mechanics of continuous porous media. The comparison with the de Rouck's measurements show that the agreement between the Mei and Foda model and the field measurements is poor when full saturation is assumed. A saturation degree of 97% (3% air/gas) gives a better agreement with the measurements.

Reference is made to de Rouck,[124] de Rouck and van Damme,[125] and de Rouck and Trock[121] regarding the details of the measurements.

Wave-induced pore pressures were measured for varying water depths (approximately 5–10 m) and to depths of approximately 18 m below the sea bottom.

3.9.2.2. *Mei and Foda's theory*

Mei and Foda derived a theory, following to some extent Prevost,[127] on wave-induced pressures in a sea bed of soils. Mei and Foda's theory is also described in Mei.[113]

Mei and Foda arrived at the following expression for the wave-induced pore pressure ratio from waves travelling over a horizontal bottom with an infinite thick soil layer:

$$\frac{u(z)}{u(0)} = \left[\frac{1}{1+m} e^{-\frac{2\pi z}{L}} + \frac{m}{1+m} e^{\frac{(i-1)z}{\delta\sqrt{2}}} \right] e^{i\left(\frac{2\pi x}{L} - \frac{2\pi t}{T}\right)} \tag{3.97}$$

where

$u(z)$ = pore pressure at depth z below the sea bed

$u(0)$ = wave induced pressure at the sea bed

(note that in soil mechanics u is used for pressure, while p is used for pressure in hydrodynamics)

$$m = \frac{n}{1-2\nu} \frac{G}{K_w} \tag{3.98}$$

where n = porosity of the soil, G = shear modulus of the soil, ν = Poisson's ratio, K_w = compression (bulk) modulus of water/gas mixture,

$$\delta = \left(\frac{k}{\rho_w g} \frac{TG}{2\pi} \right)^{1/2} \left[\frac{nG}{K_w} + \frac{1-2\nu}{2(1-\nu)} \right]^{-1/2} \tag{3.99}$$

where k = permeability coefficient of the soil and T = wave period.

To be able to compare with de Rouck's results, we have to apply a value of G that corresponds to the value of oedometer Young's modulus, E_{oed}, obtained by de Rouck from piezo-cone measurements. The following relations between G, E (Youngs modulus) and E_{oed} apply:

$$G = \frac{E}{2(1+\nu)} \tag{3.100}$$

$$E = E_{oed} \frac{(1-2\nu)(1+\nu)}{1-\nu} \tag{3.101}$$

$$G = E_{oed} \frac{(1-2\nu)(1+\nu)}{2(1-\nu^2)} \tag{3.102}$$

3.9.2.3. *Variation of bulk modulus with air/gas content*

The bulk modulus K_w varies with the fraction of air/gas in the soil water. The relation between the saturation degree and the bulk modulus is set equal to:

$$\frac{1}{K_w} = \frac{S}{K_{ow}} + \frac{1-S}{p_a} \tag{3.103}$$

where

K_{ow} = bulk modulus of fully saturated water
S = degree of saturation, $S = 1.0$ is fully saturated
p_a = absolute water pressure

Figure 3.52 shows how the bulk modulus varies with the fraction of air for 13 m water depth plus 5 m depth into the soil (approximately the average depth for the location of the measurements of de Rouck) making the absolute water pressure $p_a = 0.28 \cdot 10^6 \, \text{N/m}^2$. The bulk modulus of fully saturated water ($S = 1.0$, fraction of air/gas = 0) is set to $K_{ow} = 2.3 \cdot 10^9 \, \text{N/m}^2 = 2300 \, \text{MPa}$.

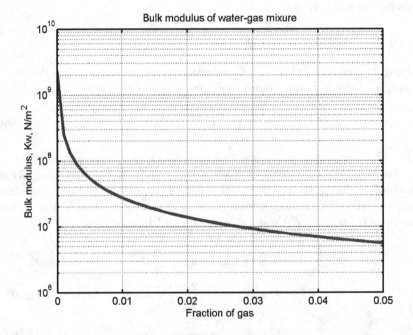

Fig. 3.52. Bulk modulus of water versus gas content. $p_a = 0.28 \, \text{MPa}$.

Table 3.11. Soil characteristics and wave periods, Campaign 4.[121,124]

	k m/s	E_{oed} MPa	Wave period, T, s
Tides			
Stiff Bartonian clay	$5.5 \cdot 10^{-8}$	78.4	44700
Sand	$2.5 \cdot 10^{-4}$	61.1	44700
Waves			
0–10 m depth	$2.5 \cdot 10^{-4}$	61	6.25
0–4 m depth	$2.5 \cdot 10^{-4}$	45	6.25

3.9.2.4. *Determination of E_{oed} and k*

De Rouck[109] determined values of E_{oed} indirectly from field piezo-cone measurements. The permeability coefficient k was determined from pump tests in the field. For sand and clay de Rouck arrived at the values shown in Table 3.11.[106]

The profile from which these data were obtained was not homogenous and the assumption of a homogenous indefinitely thick soil layer, as assumed in the theoretical analysis, may be questioned. However, for the present calculation it has been assumed that the soil is homogenous and isotropic.

If we assume Poisson ratio $\nu = 0.3$ we obtain $G = 0.28E_{oed}$ or $G = 16.8 \times 10^6$ Pa when using $E_{oed} = 60 \times 10^6$ Pa as found by de Rouck.

3.9.3. *Comparison between Theoretical and Pore Pressure Measurement Results*

Several comparisons were made of the pressure amplitudes at different locations below the bottom with Mei and Foda's model and with the de Rouck's measurements, measurement campaign 4 and to some extent from measurement campaigns 2 and 3.

Figure 3.53 shows comparison with fully saturation, $S = 1.0$, while Fig. 3.54 shows comparison with $S = 0.97$, both for sand and storm waves, $T_z = 6.25$ s. The calculations with the Mei and Foda model has been carried out with $G = 23 \times 10^6$ Pa, slightly larger than evaluated from de Rouck's measurements, $G = 16.8 \times 10^6$, if we apply the Poisson's ratio $\nu = 0.3$ for the E_{oed} found by de Rouck. We see that the comparison between measurements and theory indicate that there is some 3% air/gas in the soil pores.

De Rouck measured also pressures in the soil induced by tides (Figs. 3.55–3.58). The tide is not significantly attenuated in sand, while the tide may be significantly attenuated in clay. Again we see that the comparison between the measurements and the theory indicates, at least for Bartonien clay case, that there is some 3% air/gas in the pores of the soil.

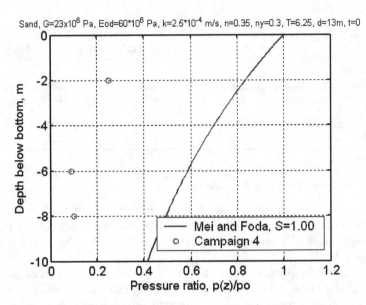

Fig. 3.53. Wave induced pore pressure amplitude ratios in sand. $G = 23\,\text{MPa}$, $E_{\text{oed}} = 60\,\text{MPa}$, $k = 2.5 \times 10^{-4}\,\text{m/s}$, $n = 0.35$, $\nu = 0.3$, $T = 6.25\,\text{s}$, $S = 1.0$ (fully saturated).

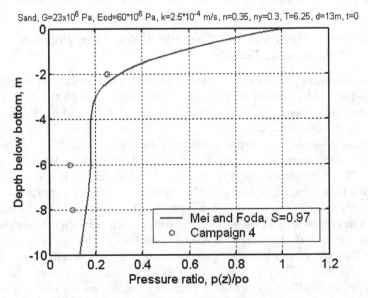

Fig. 3.54. Wave induced pore pressure amplitude ratios in sand. $G = 23\,\text{MPa}$, $E_{\text{oed}} = 60\,\text{MPa}$, $k = 2.5 \cdot 10^{-4}\,\text{m/s}$, $n = 0.35$, $\nu = 0.3$, $T = 6.25\,\text{s}$, $S = 0.97$.

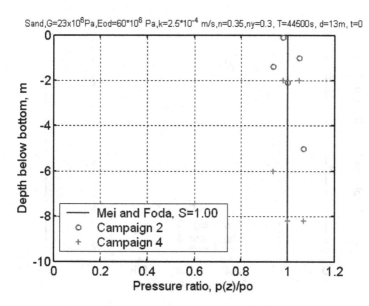

Fig. 3.55.   Tide-induced pore pressure amplitude ratios in sand. $G = 23\,\mathrm{MPa}$, $E_{\mathrm{oed}} = 60\,\mathrm{MPa}$, $k = 2.5 \cdot 10^{-4}\,\mathrm{m/s}$, $n = 0.35$, $\nu = 0.3$, $T = 44700\,\mathrm{s}$, $S = 1.0$ (fully saturated).

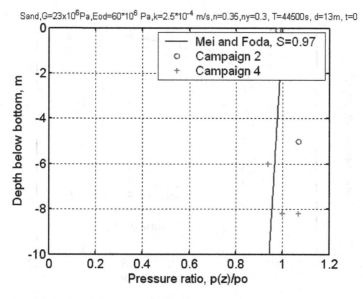

Fig. 3.56.   Tide induced pore pressure amplitude ratios in sand. $G = 23\,\mathrm{MPa}$, $E_{\mathrm{oed}} = 60\,\mathrm{MPa}$, $k = 2.5 \cdot 10^{-4}\,\mathrm{m/s}$, $n = 0.35$, $\nu = 0.3$, $T = 44700\,\mathrm{s}$, $S = 0.97$.

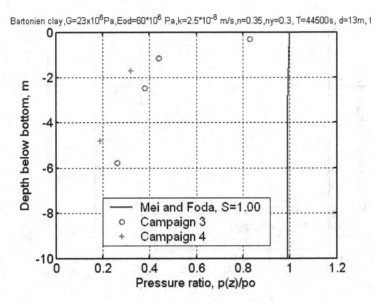

Fig. 3.57.   Tide-induced pore pressure amplitude ratios in Bartonian clay. $G = 23\,\text{MPa}$, $E_{\text{oed}} = 60\,\text{MPa}$, $k = 2.5 \times 10^{-8}\,\text{m/s}$, $n = 0.35, \nu = 0.3, T = 44700\,\text{s}$, $S = 1.0$ (fully saturated).

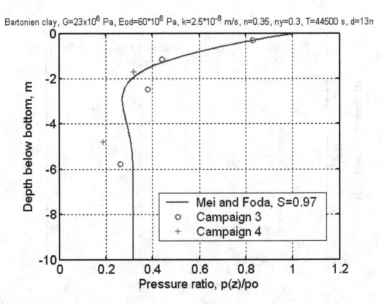

Fig. 3.58.   Tide induced pore pressure amplitude ratios in Bartonian clay. $G = 23\,\text{MPa}$, $E_{\text{oed}} = 60\,\text{MPa}$, $k = 2.5 \cdot 10^{-8}\,\text{m/s}$, $n = 0.35, \nu = 0.3, T = 44500\,\text{s}$, $S = 0.97$.

3.9.4. *Discussions on Wave-Induced Pore Pressures*

From the calculations, it is obvious that some air/gas has to be introduced into the Mei and Foda model in order to arrive at some agreement with the measurements. $S \approx 0.97$ seems to give a reasonable fit (Figs. 3.54 (sand) and 3.58 (day)).

Air/gas content in the soil pores has been an issue of debate. No instrument was available to measure small air content in the soil pores yet, but within the EU research project LIMAS, the Department of Civil and Transport Engineering, NTNU, Trondheim, such an instrument has been developed and has successfully been used in the tidal zone at a beach in France.[123] But it remains to have an instrument to measure air content in the soil in deeper waters.

Bonjean *et al.*[129] report on air content in sand in the depth range 0–0.5 m in a tidal range area. They found by geoendoscopic tests an air content of up to 5% during the LIMAS project.

It seems that wind waves and tidal waves may be significantly attenuated with depth in the bottom soil if there is gas/air in the soil pores. This leads to increased effective stresses in the soil. De Rouck and Troch[121] give an interesting account of this. The intake of water to the LNG terminal in Zeebrugge consists of a reservoir (30×30 m) with reinforced concrete walls of 1.00 m thickness. Vertical movements of the reservoir during spring tide have been measured. Unexpectedly the reservoir settled at high tide and emerged at low tide, although the buoyancy force should have "lifted" the reservoir at high tide. The intake is founded at -7.00 m. Soil layers below -7.0 m are as follows:

-7 to -25 m: dense sand
-25 to -37 m: Bartonian clay
below -37 m: very dense tertiary sand.

From Figs. 3.55 and 3.56 it is seen that the attenuation of the tide in sand is very limited. Moreover, the compressibility of sand is very low, so vertical movements due to compression or decompression of the sand is negligible.

On the other hand, Fig. 3.57 shows a clear attenuation of the tide in the Bartonian clay layer. Due to this attenuation, the variation of the effective stress at a certain level is given by de Rouck and Trock[121]:

$$\Delta\sigma' = \rho_w g H_{\text{tide}}(1 - |e^{-Az} \cos Az|) \tag{3.104}$$

where H_{tide} is the tidal variation and A is a coefficient obtained by de Rouck and Trock,[121] $A \approx 0.34 \, \text{m}^{-1}$ for tides and Bartonian clay, $A \approx 0.008 \, \text{m}^{-1}$ for tides and sand, $A \approx 0.60 \, \text{m}^{-1}$ for waves and sand.

De Rouck and Troch carried out settlement analysis with the following results when moving from low tide to high tide:

Settlement due to increased effective stresses in the clay due to tide attenuation: $s_1 = 6.42$ mm
Rising due to increased buoyancy: $s_2 = 0.46$ mm
Resulting settlement: $s = s_1 - s_2 = 6.42 - 0.42 = 5.96$ mm.

The measured resulting movements were 5.7 mm (bolt 41) and 5.4 mm (bolt 42). There is thus a fair agreement between the measured and calculated settlements.

Hovland *et al.*[130] discuss effects on slope stability of gas hydrates and seeps and remark that many of the known historical large-scale underwater slides have occurred in areas with gas hydrates. Could the tidal variations or some other large-scale sea surface variations (storm surge, tsunami) through the mechanism discussed above be a contributing mechanism to slope failure?

3.10. Ice Interaction with Breakwaters

3.10.1. *General on Ice Interaction with Breakwaters*

Most of the built breakwaters are in non-Arctic areas. There has thus been little research on the interaction between ice- and rubble-mound breakwaters and on the requirement to the mass of the armor stones to sustain the ice attack compared to the research on the wave attack. Ice forces on different other type of structures in arctic areas, especially for the oil industry structures (caisson, towers etc.) have been investigated extensively and the knowledge on ice forces on these types of structures may to some extent be used for the caisson-type breakwaters, although different ice loading codes give different ice loads.

There has been some more research on pile-up and ride-up of ice on slopes, beaches etc. These items are important with respect to the crest height of breakwaters. But this research is not reviewed in this chapter.

Most of the literature gives some general and qualitative information and viewpoints on the design of structures in arctic areas and the required armor stone mass for structures like gravel islands etc.[131–133]

There has been some investigation on the interaction between ice and breakwaters/riprap with respect to requirement to the armor stone size. These studies have been carried out as laboratory or model tests, mainly in ice tanks. Such tests may be subjected to scale effects, which are discussed, for example, by Refs. 134, 135 and 144.

We will in the following discussion review some of the work that has been done on ice action on breakwaters with respect to (a) requirement to the mass of the stones and (b) requirement to the breakwater crest height in order to prevent ice overriding the breakwater. But first some brief about the property of ice.

3.10.2. *Ice Properties in Arctic Areas*

The property of sea ice varies. The ice thickness has been a significant parameter in the few studies on the armor unit stability against ice attack on rubble-mound breakwaters. As background information the following is quoted from Løset et al.[136] on level (first year) ice for different areas:

> Level ice thickness increases during the winter season and reaches its maximum in spring or beginning of summer. Monthly level ice thickness for several seas is compared in Table 3.12.
>
> The extreme level ice thicknesses of these areas are: 1.8–2.1 m in the Beaufort Sea,[137] 2 m in the Kara Sea[142] and 1.8 m in the Barents Sea.[138] In the spring, when the ice reaches the maximum thickness, the ice cover is generally not homogeneous. In the Kara Sea for instance, it contains equal parts of autumn ice (1.2–1.8 m), winter ice (0.8–1.4 m), late winter ice (0.4–0.8 m).[139] Table 3.12 shows that the first year level ice thickness is approximately the same in the Beaufort and the Kara Seas. This is mainly due to an equal temperature regime. However, other features differ significantly.

3.10.3. *Ice Interaction with Rubble-Mound Breakwaters*

As mentioned, few studies have been conducted on the interaction between rubble-mound breakwaters and ice. The following discussion reviews some of these studies.

Table 3.12. Level ice thickness (m).

Reference	Sea	Month							
		11	12	1	2	3	4	5	6
Sanderson[140]	Beaufort	0.48	0.80	1.10	1.34	1.50	1.65	1.74	1.70
Croasdale[141]	Beaufort	0.30	0.60	0.90	1.20	1.40	1.60	1.70	1.75
Mironov et al.[142]	West Pechora	0.40	0.60	0.80	1.00	1.00	1.10	1.10	1.00
average	East Pechora	0.40	0.70	1.00	1.20	1.30	1.40	1.45	1.30
Riska[143]	Pechora	0.30	0.50	0.70	0.90	1.10	1.20	1.20	—
	Kara	0.60	0.90	1.20	1.40	1.60	1.70	1.80	—
Vinje[138]	North Barents[a]	0.40	0.45	0.55	0.70	0.80	1.00	1.05	1.10

[a]Latitude: about 77°N (Hopen).

3.10.3.1. *Port at Nome, Alaska*

Ettema *et al.*[144] and Perdichini *et al.*[145] report on a study of ice attack on
the rubble-mound breakwater for the port development at Nome, Alaska.
Figure 3.59 shows the planned layout of the port while Fig. 3.60 shows
a typical cross-section through the causeway. The design significant wave
height for the causeway was $H_s = 5.1$ m with a peak period of $T_p = 8.9$ m.[127]

The waves governed the size of the armor rock, 20 tons with a mass
density of 2650 kg/m^3, on the causeway (it is not stated whether the given
20 tons is median, maximum or what, but it is believed that it is the

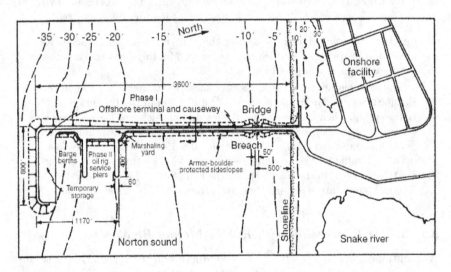

Fig. 3.59.   Layout of the causeway at Nome, Alaska.[144]

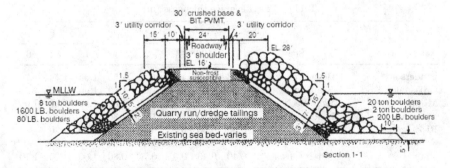

Fig. 3.60.   Typical cross-section of the causeway at Nome.[144]

mean mass of the armor units and that the variation of the mass of the individual armor units is within 20 ± 3 tons or $D_{n50} = 1.96$ m). The principal objectives of Ettema *et al.*'s study was to identify, evaluate, and provide guidance for alleviation of deleterious interaction of sea-ice with the causeway structure. Ettema *et al.* carried out physical model tests on 1:20 and 1:30 scale, using urea ice in the model. This is a model ice that has been found suitable for model testing.

The sea ice in the Norton Sound, where Nome is located, undergoes large-scale movements from November until late May. Two design ice sheets were chosen for the evaluation of ice loads: 3-foot (0.90 m) thick mobile, cold ice and 4.5-foot (1.35 m) thick ice undergoing ductile crushing during late winter, or, equivalently, the same thickness of warm mobile ice at spring break-out. Crushing and flexural strength up to 435 psi (≈ 3.0 MPa) and 102 psi (0.7 MPa) respectively were adopted. Mobile ice sheets were concluded to move at velocities ranging from a mean value of 0.70 fps (0.21 m/s) to a maximum of 2.5 fps (0.75 m/s).

Figure 3.61 shows schematically ice over-ride of an armor boulder protected side slope while Fig. 3.62 shows model ice-sheet impact with the armor-boulder protected side slope.

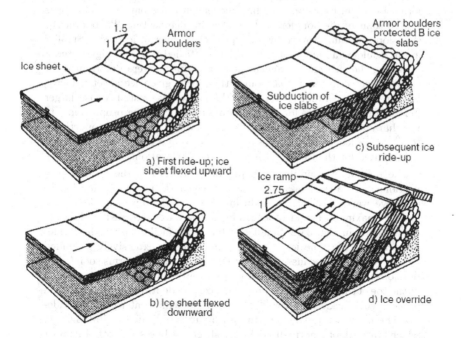

Fig. 3.61. Schematic of ice over-ride of an armor boulder protected side slope.[144]

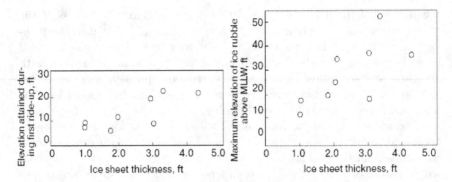

Fig. 3.62. Max. height of ice above water level during first ride-up event and max. height of ice above water level during an entire test. After Ettema *et al.*[144]

Ettema *et al.* discuss side slopes as follows:

Ice impact on the side slopes.

Model tests showed that impacting ice sheets with thickness in excess of 2 ft (0.60 m) can drive ice rubble over major portions of the port. An important characteristic of ice-sheet impact with a structure is the formation of an ice ramp. The first movement of ice onto the side slopes caused some of the boulders to be rocked in their seating. Thereafter the side slopes became covered with a protective layer of ice slabs. Smaller armor boulders about 4 tons would have satisfied the ice design criteria. However, the size and arrangement of the rip-rap were determined for wave impact criteria. It was found that for a given ice sheet velocity, thicker ice more rapidly over-rides the causeway crest, because larger blocks of ice rubble are formed (the fracture width being proportional to the characteristic length l, of the ice sheet, which in turn is proportional to $h^{0.75}$, h is ice-sheet thickness (see Fig. 3.63) and fewer flexural failures are required for the ice to reach a given crest elevation.

The maximum, equilibrium depth of ice rubble on the crest is proportional to the thickness of the parent ice sheet. The maximum height of above water level attained during the first ride-up of an ice sheet, and the maximum height of ice rubble above water recorded in the tests are depicted in Fig. 3.62. Smaller water depth facilitates ice sheet ride-up, provided that the ice sheet does not become extensively grounded before fracturing, because a submerged toe of the ramp forms sooner and develops more rapidly. In deep water, ice rubble will not become grounded at the toe of the side slope and beyond a certain depth, which is a function of ice sheet thickness, water depth plays no significant role of ice ride-up. The influence of water depth on ride-up is indicated in Fig. 3.63, in which water depth normalized by ice-sheet thickness is plotted against the time normalized by a characteristic time T_l, required for ice rubble

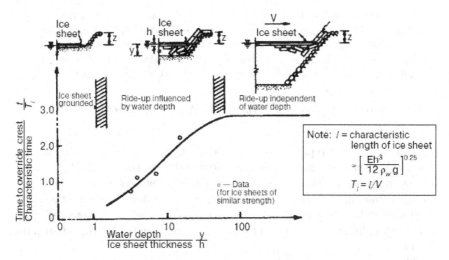

Fig. 3.63. Influence of water depth and ice-thickness on the time for ice rubble to surmount a side slope.[144]

to overtop the causeway; T_l is the characteristic length of an ice sheet, l, divided by its velocity, v, and is a measure of the period of ice fracturing.

The maximum horizontal forces, per unit width of ice sheet, recorded during the first ride-up event and maximum horizontal forces per unit width of ice sheet during an entire test are plotted against ice-sheet thickness in Fig. 3.64. Also plotted in Fig. 3.64 (to the right) are the forces

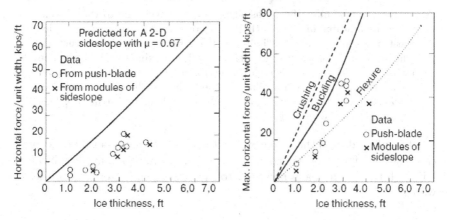

Fig. 3.64. Maximum horizontal force per unit width of ice sheet during first ride-up event (to the left) and max. horizontal force per unit width of ice sheet during an entire test.[144]

estimated for an ice sheet to buckle and crush against a wide structure. It is evident from Fig. 3.64 that the forces exerted on side slopes by ice sheets are smaller than those predicted for buckling or crushing failure of ice. This is due to the nonsimultaneous and irregular nature of ice-sheet failure that occurs once ice rubble has accumulated along a side slope.

3.10.3.2. *Breakwaters at North Bay, Lake Nipissing, Ontario, Canada*

MacIntosh *et al.*[146] report on the Canadian experience with ice and armor stones. Of particular interest is their description of the North Bay breakwaters on Lake Nipissing. The studies of the breakwaters at North Bay had the advantage in that there was a conventional rubble mound and a berm breakwater close to each other. MacIntosh *et al.* does not, however, give any detailed information on the breakwaters (e.g., water depth, armor unit size etc.).

The following is quoted from MacIntosh *et al.*:

> MacIntosh[147] believes that the flatter slopes at the waterline of a berm breakwater reduce the likelihood of rock movement by ice. In addition, the berm itself can ground ice rubble for protection during the winter. He noted that during break-up, the ice melted away from the rocks before it moved off, so there was no potential for rafting or plucking. He believes that the presence of ice in the interstices between the rocks may act as cement to help resist external ice forces. It is nevertheless worth noting that Lake Nipissing is nontidal, so that these breakwaters were subjected to moving ice only during freezing and breakup.

> During one winter there was significant damage done to the conventional breakwater in the North Bay region. In this case, a moving ice sheet physically moved, or "bull-dozed" a large section of the structure. The armor rock on the breakwater weighed approximately 1 tonne. The annual ice thickness in this region is approximately 0.7 m. Information on the speed of the ice is not available. (These photos) show that bulldozing can occur and cause extensive damage to a rubble mound structure. This type of failure, however, is quite rear.

It is inferred from the paper by MacIntosh *et al.* that a berm breakwater can resist ice forces better than a conventional rubble-mound breakwater, although there is no basis presently to say how much better a berm breakwater is.

3.10.3.3. *Ice action on Riprap: CRREL report 96-12*

Sodhi *et al.*[148] investigated by small-scale tests ice action on riprap. The objective of the study was (1) to conduct small-scale tests of ice shoving

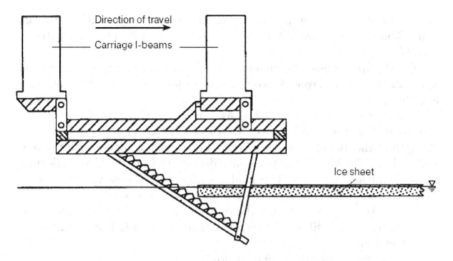

Fig. 3.65.   Test set-up.[148]

against a sloped surface protected with a model riprap, (2) to measure the forces generated during the ice-riprap interaction during the tests, and (3) to determine the conditions when ice showing damages a riprap cover layer.

Figure 3.65 shows the test set-up. The riprap was placed on a sloping board, which again was attached to a carriage spanning the refrigerated basin at CRREL's Ice Engineering Facility, and pushed against model ice sheets grown in the basin. The slope of the board varied as 1:3, 1:2 and 1:1.5 (steepest). There was instrumentation to measure velocity and horizontal and vertical forces between the riprap and the model ice. The linear scale of the tests was 1:8, based on typical ice flexural strength.

The model riprap was scaled down stones in approximate gradation as suggested by the US Army Corps of Engineers. The stone shape should be blocky, not elongated, to enable the stones to "nest" together more effectively to increase their resistance to movement. Stones within a certain range of shape factor, which was defined roughly as the ratio of the longest dimension to the shortest dimension of the rock, were specified to limit the number of rocks that were too elongated to be effective as riprap. A ratio of 2.5 should not be exceed by more than 30%, while a ratio of 3.0 should not be exceed by more than 15%.

Three different riprap sizes and gradations were used during the tests. The gradation, D_{85}/D_{15}, of the riprap was 2.7, 2.0, and 2.0 respectively. D_{100} (e.g., the largest stone size) was used as the characteristic stone size. Note that in this study D was defined as $D = (6W/(\pi \rho_s))^{1/3}$, or the equivalent diameter of a sphere. D_{100} was 127, 76, and 38 mm, respectively, for the three tested ripraps, corresponding to 1.00, 0.60, and 0.30 m full scale. The mass density of the stones was 2600 kg/m^3. The rock porosity was 0.3.

The riprap was placed with a layer thickness of approximately 2D$_{100}$ on a 38 mm thick sand layer. A geotextile was placed between the sand layer and the riprap.

The model ice was established as a urea ice (e.g., a solution of 1% of urea in the water and an ice growth at $-10°C$). The desired flexural strength of the model ice sheet was 87.5 kPa, which compare with the full-scale value of 700 kPa. There were some deviations from this desired flexural strength during the tests. The model ice sheet thickness varied between 24 and 65 mm, corresponding to 0.192 and 0.51 m full scale.

The carriage speed varied in the range 0.20–0.80 m/s, but was for most of the 35 test runs 0.40 m/s, corresponding to 1.13 m/s full scale, except for three test runs.

The failure levels were defined as follows:

A failure level designated by 0 was assigned to tests when no significant displacement of rocks was observed.

A failure level designated by 1 was assigned to slight damage during tests in which small areas, on the order of 20 cm^2, of geotextile filter fabrics was observed. A failure level 2 was assigned to damages were moderate areas, 20–65 cm^2, of the geotextile filter fabrics was observed. Lastly, a failure level of 3 was assigned to damages in those tests in which large areas riprap gouged from the bank, resulting in exposed bank areas of 65 cm^2, corresponding to 5.2 m^2 full scale, or more.

The data as given in Ref. 148 are plotted in Fig. 3.66, showing the damage level versus ratio of stone diameter and ice thickness, h/D$_{100}$. As expected, there is scatter in the data due to the randomness of the ice hitting the riprap slope and the randomness of the size and placement of the riprap stones.

The following is quoted from Ref. 148 on the observation of ride-up and pileup (*references to photos omitted*):

> We observed ice ride-up only during two tests (D$_{100}$ = 38 mm and h = 40 mm, D$_{100}$ = 127 mm and h = 54 mm). The slope angle of the bank in both these tests was 18.4° (1V:3H). During an ice ride-up event, we observed that the ice sheet was being pushed up the riprap slope and continually slid up and over the top of the model and back into the tank (*no information can be found in Sodhi et al. on the height of the top of the model*). During the ice ride-up events, the ice sheet dislodged some rocks upon initial impact, but generally did not displace enough rocks to expose the bank. The action of the ice sheet sliding on top of the riprap shaved the underside of the ice sheet and filled in the interstitial spaces

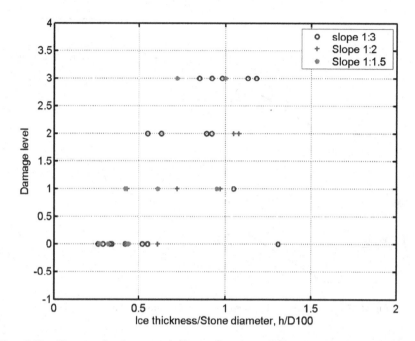

Fig. 3.66. Damage level versus h/D_{100}. Based on Ref. 148. The small damage for $h/D_{100} = 1.3$ for slope 1:3 is for a high ice velocity, carriage speed 0.80 m/s, corresponding to 2.25 m/s full scale.

between the rocks, causing the riprap surface to become smooth. There was very little displacement of the riprap.

During the initial stages of a test with a pileup, an ice sheet would contact the riprap, break into pieces, and ride on top of the riprap. Although the ride continued for a long time in a few tests, the ice would usually pile up above the water line and offer resistance to the incoming ice pieces. As the pile grew larger, there were instances when we observed an ice sheet pushing under the pile, gouging the model riprap, and bringing up small stones.

We observed the rocks being transported from the bed to the surface of the ice pile during most of the tests in which the ice piled up on the bank. At times, the ice sheet would buckle one or more times during a test. As discussed later, the forces associated with buckling represent the maximum force that a model ice sheet could impact on the model riprap. Buckling events represents large scale failure of the model ice sheet, and this did not cause large scale failure of riprap in our tests. However, the model ice sheet caused small scale damage by removing individual stones one at a time.

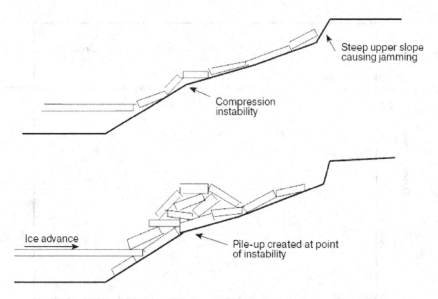

Fig. 3.67. Possible beach design to resist ice ride-up. After Crosadale.[149]

3.10.3.4. *Ice action on revetments/breakwaters*

Croasdale *et al.*[149] proposed a beach design as shown in Fig. 3.67 to resist ice ride-up. The idea is that the bump in the slope will cause high bending stresses in the ice, which will eventually break and start to pile up at this bump. Tørum[150] suggested that a static stable berm breakwater, with such a bump built into it, would be suitable in arctic areas where the structure is subjected to ice attack. This concept was used to develop the shoulder ice barrier (SIB),[151,157,158,162] which will be discussed later.

Recently Daly *et al.*[152] investigated through model tests, scale 1:20, the ice impact on armor stone revetments at Barrow, Alaska. The revetment is planned to protect the city of Barrow from ice shoves, which previously have reached up to approximately 75 m inland. Figure 3.68 shows an ice shove in January 2006. Figure 3.69 shows the cross-section of the tested revetment model. The water depth is 10 m. The berm width is 5.50 m and the height and width of the crest are 10 m and 5.5 m, respectively. The armor thickness is 5.5 m.

Sodhi *et al.*[148] observed that damage to riprap appears to take place during pile-up events when the incoming ice sheet is forced to go between the riprap and the pile-up ice bringing with it rocks from the riprap layer to the surface of an ice pile.

Daly *et al.* performed four test series at scale 1:20. The ice in all tests was 1.5 m (prototype) thick with a flexural strength of 600 kPa. Figure 3.70

Fig. 3.68.　Ice rubble pile formed in January 2006 at Barrow, Alaska, as a result of ice shove. (www.gi.alaska.edu/snowice/sea-lake ice/images/ice_events/html)

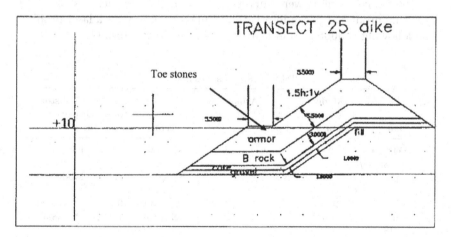

Fig. 3.69.　Cross section of the tested revetment at Barrow, Alaska. Measures in mm.[152]

shows the test set-up in the test basin. Two sections were tested at the same time, designated Model North and Model South. Each section covered a length along the barrier corresponding to approximately 61 m. The size of the revetment stones varied between W = 1.7 tons and W = 7.2 tons. It is

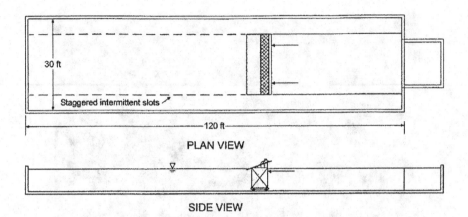

Fig. 3.70. Layout of the ice test basin.[152]

not clearly stated by Daly *et al.* if this corresponds to W_{50} or not, but it is believed that the given weight is W_{50}. Some tests were carried out with the stones placed randomly, while in other tests the stones were placed orderly or as designated by Daly *et al.* as selectively. Table 3.13 gives an overview of the tests. Figures 3.71 and 3.72 show photographs from the tests.

The tested sections were surveyed before and after each test 3 and 4 by taking profiles at 30 cm intervals along the sections with a laser profiler. The following is quoted from Daly *et al.* on the Test 4 results:

Table 3.13. Test results.[152]

Test	Date	Model speed (cm/s)	Test section	Armor layer stones W_{50} (?) tons	Stone placement	Toe stone tons	Total stone displacement (m³) prototype
1	February 28, 2007	20		1.5	Random	—	Extensive
2	April 19, 2008	10	N	2.7–3.6	Selective	7.2	Moderate
			S	3.6	Random	10	Extensive
3	July 18, 2008	10	N	3.6	Selective	7.2	−258.6
			S	3.6	Selective	11.8	−104.3
4	August 29, 2008	10	N	3.6	Selective	11.8	54.4
						18.1	25.1
			S	7.2	Selective	11.8	132
						7.2	−117.1

Fig. 3.71. Photograph of the test set-up of the South model prior to Test 2.

Fig. 3.72. Test 2. The difference between the North section (to the right) and the South section is displayed. The removed material on the randomly placed stones on the South section can be seen.

In this test, both the North and South models used selective placement of carefully graded armor stone: 3600 kg armor stone on the North model, and 7200 kg armor stone in the South model. In addition, the toe stone was varied for each model, resulting in four separate tests listed in Table 3.13. It is interesting that 3600 kg armor stone resulted in less total displacement than the 7200 kg armor stone. Visual inspection of the revetments after the test suggests that the large majority of 7200 kg were not lost from the revetment but a few were moved around and several of the large 11,800 kg toe stones were rolled up the revetment on the south side. The large volume of the individual may have contributed to these results.

3.10.3.5. Comparison between the results of Sodhi et al.,[148] Ettema et al.,[144] and Daly et al.

Figure 3.66 shows that there is no failure when h/D_{100} is smaller than approximately 0.5, while there is almost always failure if h/D_{100} is larger than 1.0.

Ettema et al.,[144] who tested the causeway at Nome, Alaska, with rock size W_{100} of approximately 22 tons or $D_{100} \approx 2.00$ m, stated that the first movement of ice onto the side slope caused some of the stones to be rocked into their seating. Thereafter the side slope became covered with a protective layer of ice slabs. The Nome causeway was tested for ice thicknesses up to 1.35 m, in which case h/D_{100} was up to 0.7. The results of Ettema et al. is thus in agreement with the results of Sodhi et al. However, Ettema et al. states that "smaller armor boulders about 4 tons ($D_{100} = 1.43$ m, $h/D_{100} = 0.93$), would have satisfied the ice design criteria; however, the size and arrangement of the riprap was determined for wave impact." But Ettema et al. do not mention the source for the stated ice design criteria. Sodhi et al.'s results indicate that riprap with 4 tons rock would have failed for the ice attack of 1.35 m thick ice.

Daly et al. state that the required stone size from the wave point of view is the size used during Test 1 or 1.5 tons. This stone size is too small to resist the ice forces from 1.5 m thick ice. For 1.5 m thick ice, the tests of Sodhi et al. indicate that a stone size of $D_{100} \approx 3$ h, where h = ice thickness, or $D_{100} = 4.5$ m for ice thickness of 1.5 m., corresponding to $W_{100} = 246$ tons. To quote from Daly et al.: "Stone of this size would be very large, very expensive, and costly to transport and install. The laboratory tests described here suggests that 3500 kg (4 tons) armor stone, which would have a large dimension of roughly 2 m, could resist ice shoves very well if selectively placed and due care was taken in the design and placement of toe stones. Selective placement is a more expensive method of construction than random placement, but the benefits are all smaller stone size requirement and greater resistance to ice shoves."

3.10.4. *Interaction between Breakwaters and Drifting Ice Set in Motions by Waves*

There have been some few studies on the interaction between breakwaters and drifting ice set in motions by waves. Kitamura *et al.*[153] reports on such a study of the interaction between drifting ice and doubly peaked submerged breakwaters (Fig. 3.73).

Kitamura *et al.* carried out laboratory scale tests in a wave flume on a scale of 1:30. The front slope of the submerged breakwaters was 1:3 and the rear slope was 1:1.5. The mass of the armor stones corresponded to 3,800 kg. Kitamura *et al.* used polypropylene with mass density of 900 kg/m^3 to model the ice sheets. The modulus of elasticity of polypropylene is greater than that of sea water ice. Therefore the experimental results are considered by Kitamura *et al.* to be conservative.

The dimensions of the model ice floes used in the experiments were based on the results of surveys carried out by Kunimatsu.[154] The side length of the rectangular ice floes were in the range corresponding to 0.90–6.5 m and the thickness of the ice floes was in the range corresponding to 0.50–1.45 m. The wave periods used in the tests corresponded to 6 and 10 s, while the

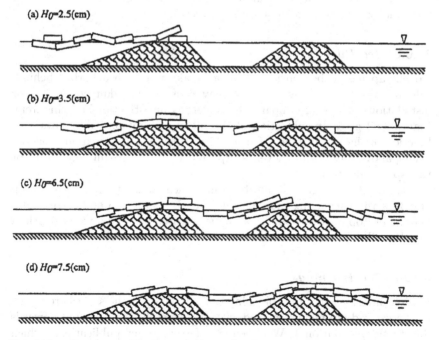

Fig. 3.73. Conceptual illustrations of ice floes overtopping a doubly submerged breakwater.[153]

wave heights corresponded to the range 0.60–2.70 m. The water depths corresponded to the range 3.9–5.1 m and the submergence of the breakwater crest corresponded to the range 0–1.2 m.

Generally speaking there was not much damage to the armor stone layer when W_i/W_s was less than 5. W_i is the mass of the ice floe and W_s is the mass of the armor stone.

3.10.5. *Design of Ice Protection Rock Rubble-Mound Barriers in the Caspian Sea*

Lengkeek *et al.*[155] dealt with ice attack on ice protection barriers in the Caspian Sea. Their primary concern was the geotechnical stability of the rubble barrier. They state however that the cover layer should be at least one time the design ice thickness. They also state that the rock diameter should be about half the ice thickness, which means that the cover layer should consist of two layers or more. This is somewhat contradictory to the results of Sodhi *et al.* (Fig. 3.66). The results of Sodhi *et al.* indicate that the rock size diameter should be 1–2 times the ice thickness. This is, may be, an indication of the poor knowledge that exists on the requirement to the rock size on rock slopes subjected to ice loads.

3.10.6. *Ice Pile-Up*

Ice pile-up has to be considered for a breakwater with a harbor behind. Pile-up over the breakwater crest may cause severe damage to harbor installations. There are apparently numerous investigations on this item. Figure 3.62 gives an indication of the pile-up height on a breakwater as function of the ice thickness. For a 1.35 m thick ice, the maximum pile-up height above mean water level is expected to be approximately 15 m for the Nome causeway.

Barker *et al.*[156] give results from numerical simulations on ice pile-up of 0.30 m thick ice along shorelines. For a shore slope 1:3 and a horizontal plateau on top of the slope 2 m above still water line and a water depth of 6 m, the pile-up was estimated to approximately 7 m above sea level.

3.10.7. *Ice Plucking*

Limited information on ice plucking is available (e.g., the stones are frozen to the ice and are being removed with the ice if the ice is shifted upwards due to tidal variations). Wuebben[158] refers to other publications, which indicate that to avoid plucking, the stone size D_{n50} should at least be equal to the ice thickness. It requires a certain ice sheet size to bear this stone. We consider a stone diameter $D_{n50} = 1.5\,\text{m}$ with a mass density of $2700\,\text{kg/m}^3$, or $W_{50} = 9112\,\text{kg}$, and a $1.5\,\text{m}$ thick ice with mass density $950\,\text{kg/m}^3$ the required ice sheet area to carry this stone is approximately $186\,\text{m}^2$. Such an ice flow would necessarily be in contact with several stones and it is questionable if plucking may occur for such stones.

3.10.8. *Ice and Wave Forces on a Specially Designed Caisson*

3.10.8.1. *General*

Discoveries of hydrocarbons in shallow ice-infested waters (e.g., the Northern Caspian Sea) raise the need for intensive drilling activities during the ice season. Experience with mobile drilling units in the seasonally ice-infested waters in the Northern Caspian Sea solely originates from the current Sunkar drilling barge at Kashagan and Kalamkas. However, with increased drilling activities upcoming, innovative drilling concepts are desirable due to the objective of maintaining drilling operations during the ice period with conventional non-ice-resistance drilling platforms. Inspired by the suggestion of Croasdale *et al.* (Fig. 3.67 by the experience with the berm breakwaters to apparently withstand ice forces,[133,150] an external removable stand alone gravity based caisson structure, the shoulder ice barrier (SIB) was suggested.[158,159] Figure 3.74 shows the concept of the SIB for shallow water drilling, while Fig. 3.75 shows an illustration of an assumed maximum rubble-pile at a SIB for shallow water. It should be

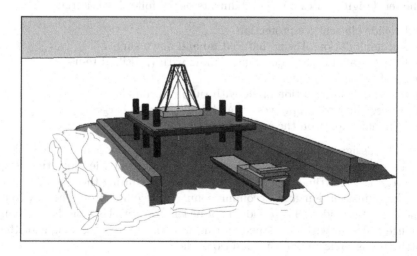

Fig. 3.74. The concept of the SIB as an ice barrier.

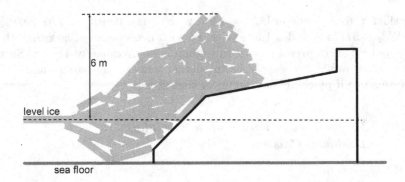

Fig. 3.75.   Concept of rubble pile up.

noted that the tests of Daly *et al.* to some extent confirm that a bermed shape is advantageous for ice barriers.

Some results are presented in the following from ice model testing in the large ice tank of at the Hamburgische Schiffbau–Versuchsanstalt (HSVA) for a SIB in 6 m water depth, Gürtner *et al.*[159,163]

3.10.8.2. *Design basis*

As a first design case, it has been suggested that the ice protection shelter is designed for a potential use at the area of the offshore situated hydrocarbon field Kashagan in the north-eastern Caspian Sea at a depth of 4–6 m. The shelter design focuses on the fulfilments of the following criteria:

- Provide efficient ice protection
- Accommodation of wind induced up and down surges
- Easy installation and decommissioning with (possible) re-use
- Self-floating during transport
- Successive stabilization inline with rubble accumulation
- Accessibility by supply vessels and liquid storage barges
- Minimal impact on the environment

The design conditions of offshore structures in the Northern Caspian Sea are primarily driven by extreme shallow water, water level fluctuations, ice regime wintertime, and hydrodynamic regime during the fall season.

Easy installation and decommissioning is required from the environmental point of view. There should not be left any gravel bed on the bottom. Hence a design with skirts penetrating into the soil by the weight of the structure is preferred, if soil conditions allow.

3.10.8.3. *Ice conditions in the Caspian sea*

According to Lengkeek *et al.*,[155] ice occurs in the North Caspian Sea for about 3–4 months each winter. Average thermal ice growth is usually less than 0.5 m, but in severe winters, a sheet ice thickness up to about 0.9 m can occur. Rafted ice is common and creates extreme thickness for the design of wide structures at about 1.4 m (although locally thicker rafting has been measured). Ice salinity is low (1–2 ppt). Ice is wind driven at speeds, which can be over 1 knot. For a review on the ice conditions encountered in the Northern Caspian Sea, see Gürtner *et al.*[159] and the references therein.

3.10.8.4. *Ice force tests*

The ice force tests were carried out in the large ice tank of the Hamburgische Schiffbau–Versuchsanstalt (HSVA), as mentioned. The results of the tests are fully reported by Gürtner.[160] Some results are given here.

Model testing was carried out according to Froude's scaling of 1:20. The water depth was set to 6 m and the velocity of ice to the SIB was set to 0.5 m/s. All model tests were conducted in natural grown columnar grained level ice. The method for preparation of the ice was according to description of Evers and Jochmann.[160] The ice thickness varied during different test runs, and ranged between 0.24 and 0.96 m. The target flexural strength was 750 MPa at an ice temperature above the freezing point. For details on model scaling in ice the reader is referred to Schwarz.[141] The SIB models were fabricated from plywood. The surfaces were coated with industrial varnish to provide an authentic surface structure for the experiments. Dynamic friction measurements of ice against the SIB surface gave friction coefficients for dry and wet friction of 0.14 and 0.12, respectively.

The experimental tank at HSVA is 78 m long, 10 m wide, and 2.5 m deep. For the purpose of simulating shallow water conditions, a retrievable frame was installed on which the SIB, in relation to the tank width, was centrally mounted, Fig. 3.76. The length of the SIB corresponds to 30 m and the width corresponded to 22 m. In two of the test runs, the initial model base structure of 30 m length was equipped with dummy side panels to investigate the influence of three dimensional effects of ice bypassing.

3.10.8.5. *Ice force test results*

Figure 3.77 shows a summary plot of the global loads per meter structural width during test run #3100 when the base structure was equipped with side panels (Fig. 3.76b).

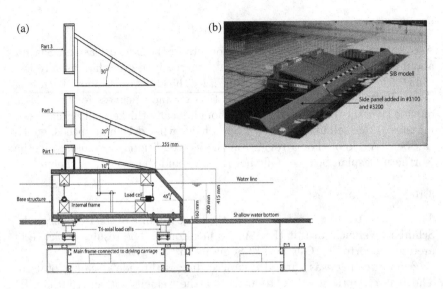

Fig. 3.76.   Section view of model SIB and the model mounted in the ice basin.

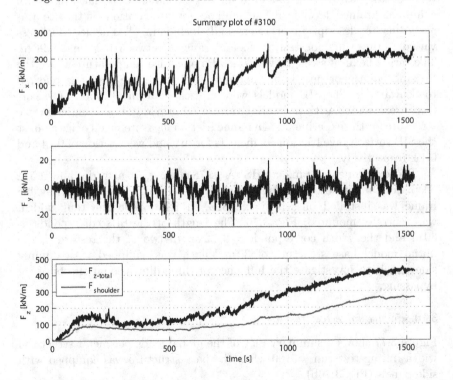

Fig. 3.77.   Summary plot of loads to the SIB with ice thickness of $h_i = 0.64$ m.

It can be seen that the horizontal force F_x steadily builds up until 200 s (full scale) into the run, where after the force trace shifts character to a series with fluctuating load peaks of short duration. A fast buildup of the force level precedes the force peaks, whereas, when the ice fails, the horizontal force typically releases abruptly. About 1000 s into the test run the horizontal force become stable. The vertical force F_z is steadily building up toward a constant maximum force of about 415 kN/m at the very end of the run. The vertical shoulder force $F_{shoulder}$ follows the same trend as F_z. The vertical force is in the same order of magnitude halfway into the run, whereas in the second half of the run the global vertical force is about twice the horizontal load.

After each test run where the condition of steady-state ice accumulation was reached, the ice rubble profiles were measured *in-situ*. Plots comparing the two-dimensional side view of the rubble profiles are shown in Fig. 3.78.

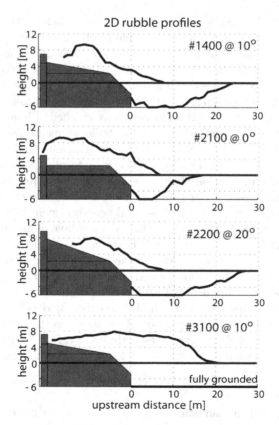

Fig. 3.78. Summary plots of ice rubble profiles where a steady state accumulation process was reached. Note that the set-up #3100 included the dummy side panels.

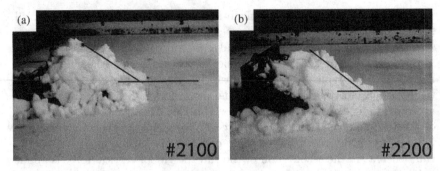

Fig. 3.79.   Comparison photographs showing the conditions where the rubble inclination of approximately 30° is reached to fail rubble towards the upstream side in run (a) #2100 and (b) #2200.

It is noted that the pile-up of ice in general follows the anticipated pile-up (Fig. 3.75). Concerning the rubble mechanism observed during the model tests, it turned out to be a process in which the structural dependence, in terms of the shoulder inclination, vanishes after exceeding rubble pile steepness in excess of 30°. That is, the rubble mechanism started developing toward the upstream side, rather than contributing toward increasing the rubble height when this steepness was reached. Figure 3.79 shows these observed conditions on the rubble building mechanism. Note that the structural shape of the SIB becomes insignificant after this condition is reached. Nevertheless, this condition is naturally reached faster for steeper shoulder inclination than for shallower ones. After this condition was reached, no leeward over-riding of ice could be observed.

The event resulting in the highest load peak was identified to be caused by ice pushing through the unconsolidated ice rubble at the structure and directly failing by crushing on the SIB surface. Due to the constraint of gravity and buoyancy on either side of the ice sheet, forces are higher than observed in typical out-of-plane breaking of the ice sheet.

No full-scale measurements of ice forces on offshore structures of either sloped or vertical type with a length of 30 m exist. Bjerkås[162] presented the following relation to define upper bound pressures on the basis of full-scale measurements for offshore structures within $5\,\text{m} < D < 165\,\text{m}$, where D is the diameter or length of the structure, for a quasi static loading:

$$p_G = 2.05 \cdot D^{-0.06} \tag{3.105}$$

where $p_G =$ global pressure in MPa, and D is the structure diameter or length in meters.

An estimate of the total global ice force is given by, Gürtner[151]:

$$F = p_G h_i D \qquad (3.106)$$

where h_i is the ice thickness.

The analyzed results indicates that Eqs. (3.105) and (3.106) give reasonable relation also for the SIB, when D is set to 30 m and for the special condition of ice being pushed through unconsolidated ice rubble as discussed by Gürtner and Gudmestad.[163] Thus, in case of a narrow structure, the referred special condition may be relevant for the design. With increasing structural dimensions, however, this condition is not considered being able to proceed over the complete contact area and thereby reducing the expected forces on sloping structures compared to estimates based on Eqs. (3.105) and (3.106).

3.10.8.6. *Wave force tests on SIB*

As mentioned, wave loads are not crucial in the Northern Caspian Sea for the SIB. However, in deeper water, 15–20 m (e.g., in the shallower parts of the Beaufort Sea, the Barents Sea and the Pechora Sea), if the SIB is utilized as a breakwater at ice-infested harbour locations, the wave forces may govern the design loads for the stability of the SIB. Hence the wave forces were investigated on a "deeper" SIB model structure in a wave flume[164,165] (Fig. 3.80). The model scale was 1:100, the width of the SIB model corresponded to 27 m, the length 20 m, and the total height to 34 m, and the water depth corresponded to 16 m. The model was designed with no wave-induced water pressure on the underside, simulating the conditions if the SIB, with skirts, is placed directly on a sandy sea bed with no prepared gravel bed, similar to the CONDEEP concrete platforms in the North Sea.

Figures 3.81 and 3.82 show the measured horizontal wave overturning moments and wave horizontal forces. The measured horizontal wave forces were compared with the Goda formula for vertical wall caissons and for

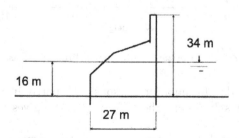

Fig. 3.80.   The wave force tested SIB as breakwater and shoulder ice barrier.

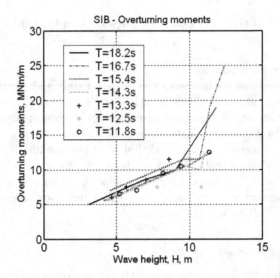

Fig. 3.81.   Overturning moment due to horizontal wave forces. Prototype values.

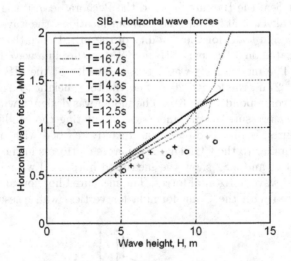

Fig. 3.82.   Horizontal wave forces. Protype values.

the Takahashi et al.[78] formula for caissons with partly inclined front wall
(Fig. 3.83). The result is that the SIB of Fig. 3.80 gives lowest value for
the forces.

It can be concluded that the wave forces and wave overturning moments
on the SIB are significantly reduced compared to a vertical wall. The

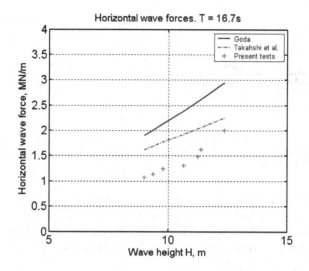

Fig. 3.83. Horizontal wave forces per meter length compared with calculation models. Prototype values.

horizontal ice forces on a SIB are normally much smaller than on a vertical wall, but may under certain conditions locally be of the same magnitude as on a vertical wall. The ice will provide a stabilizing vertical force on the SIB, which will not be present on a vertical wall structure. Hence it is concluded that the SIB seems to be an attractive concept for the Northern Caspian Sea as well as for other Arctic areas.

3.11. Feasibility Study for a Breakwater/Artificial Island Under Arctic Conditions

3.11.1. *General*

Timco *et al.*[133] refer to a proposed artificial island for the Pechora Sea in approximately 20 m water depth (Fig. 3.84).

The proposed rock protection layer has a slope of 1:6. The proposed mass of the armor stone units is 10 tons. The small slope is apparently based on the experience from the gravel islands in the Beaufort Sea. The experience from the Beaufort Sea area has, according to Timco *et al.*, clearly shown that the development of a grounded rubble field around an artificial island effectively protects it throughout the winter months.

Timco *et al.* discuss the action of moving ice cover on the proposed island concept with the main emphasis of the analysis focused on the ability of

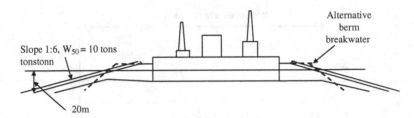

Fig. 3.84. Side-view of a proposed island for the Pechora Sea. After Timco *et al.*[133]
An alternative solution with a berm breakwater concept as wave and ice protection is
indicated by the author.

the moving ice to dislodge the armor stones protecting the island. However,
there is little quantitative information available on this issue.

Malyutin *et al.*[166] discuss various concepts for platforms that have been
reviewed for the "Prirazlomnoye" field in the Pechora Sea:

- Truss platform piled to the soil
- Reclaimed "artificial" island
- Gravity base platforms with reinforced concrete and steel substructure

Malyutin *et al.* states the following on the artificial island:

Construction of an artificial island would require dredging of a huge volume
of weak soil (to a depth of 15–20 m below the seabed) and delivery of
refill materials for the island from remote borrows areas, including those in
Norway. It is not feasible to complete the island within one summer season
and the earlier placed fill would be scoured during the off season time and
the fill material would mix with the weak soil. As a result the facility would
be unserviceable.

The proposed artificial island Timco *et al.* discuss is apparently also
located on weak soil (clay) and it may not be cost effective. But from an
exercise point of view with respect to wave and ice conditions and forces on
an artificial island, we stick to the artificial island as shown in Fig. 3.84. The
discussions are valid for rubble-mound breakwaters as well as for artificial
islands and may thus be relevant for an artificial island on less weak soil
conditions.

3.11.2. *Wave and Ice Conditions at the Proposed Island for the Pechora Sea*

Lavrenov *et al.*[167] discusses the extreme wave heights in the Pechora Sea.
The wave heights given in Table 3.14 are inferred from this paper.

Table 3.14. Wave parameters at 20 m water depth in the southeast Pechora Sea as inferred from Ref. 167.

Return period (years)	Significant wave height H_s(m)	Maximum wave height $H_{max,m}$	Mean wave period T_z(s)	Period of maximum wave, $T_{H_{max}}$,(s)
20	4.7	8.7	8.5	10.1
50	5.1	9.1	9.1	10.8
100	5.4	9.5	9.5	11.3

The given mean period gives a wave steepness, $s_{mo} = 2\pi H_s/(gT_z^2)$, of approximately 0.04. Since T_z may vary, we have also made calculations of the required cover stone size for $s_{mo} = 0.06$ or for $T_z = 7.6$ s.

It has to be realized that there are uncertainties on these values. May be the "true" values of the wave heights are within ±15–20%.

Timco et al.[133] give the following account on the ice conditions in the Pechora Sea:

This area is normally ice-covered from approximately mid-November to late June. As such the island must be designed to withstand the effects of the ice environment. In November, the ice is thin and only partially covers the sea. In December and January, the ice cover continues to grow in terms of thickness and extent, with ice concentrations typically ranging from eight-tens to ten-tens (m). The ice cover is most severe in March and April. An extreme ice thickness is 1.2 m, but usually the maximum offshore level ice thickness is in the range of 0.6 to 0.9 m. The Pechora Sea is open to the west and northwest, and most pressure events occur when ice moves from west to east, compressing the ice cover against coastal constraints. The ice is very dynamic, and approximately 50% of the ice cover is rough and hummocked by mid-winter. In this region, the ice floes are normally hundreds of meters in size, although ice floes a few kilometers in extent may sometimes interact with the island. Although landfast ice forms in the shallow waters, a moving pack ice cover is characteristic at the island location.

This information compares well with the information in Ref. 136.

It is also emphasized that for important structures with severe exposure to waves and ice, model tests should be carried out.

3.11.3. *Stability of Armor Layer of the Proposed Artificial Island*

The required cover stone size to protect the slope of the artificial island (Fig. 3.76) has been calculated with the Hudson formula (Eq. (3.12)) and

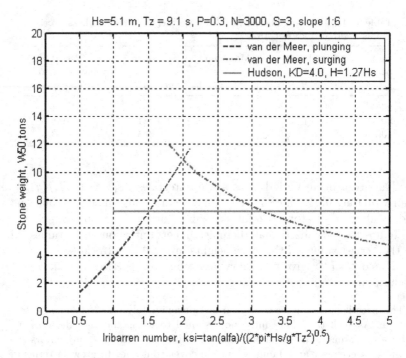

Fig. 3.85. Required mass of the armor stones on the 1:6 slope of the gravel island (Fig. 3.84).

the van der Meer formulae (Eqs. (3.14) and (3.15)), as shown in Fig. 3.85. Two layers of armor rock have been assumed. The mass density of the armor stones has been assumed to be 2700 kg/m^3.

The Iribarren number is in the case of wave steepness $s_{mo} = 0.04, \xi = \tan \alpha/(2\pi H_s/(gT_z^2))^{1/2} = 0.166/(2\pi \cdot 5.1/(9.81 \cdot 9.1^2))^{1/2} = 0.83$ and in case $s_{mo} = 0.06, \xi = 0.70$. It is seen from Fig. 3.85 that the required stone mass W_{50} is approximately 4 tons according to van deer Meer, and approximately 7 tons according to Hudson. So it seems that the proposed stone size of 10 tons should be adequate to protect the slope against waves.

With respect to stability against ice, Timco et al.[133] discusses basically various mechanisms for the movements of the armor rocks: plucking, sliding and bulldozing. The ice forces on the island and the amount of ice ride-up and pile-up was dealt with other places.

Timco et al. conclude from theoretical considerations that the armor protection was adequate to resist the ice forces. However, Timco et al.[133] did not have the results of Sodhi et al.[148] Sodhi et al.[148] did tests on ice attack on riprap on slopes 1:3, 1:2 and 1:1.5. According to Fig. 3.66 the damage level was not very sensitive to the slope angle. If we assume that the results presented in Fig. 3.66 is also representative for slopes 1:6, the ratio ice thickness/stone diameter, h/D_{100}, should be in the range of 0.5–1.0. If we set the ice thickness $h = 1.2$ m, D_{100} should be in the range of 1.2–2.4 m or the mass of the armor stones in the range $W_{100} = 2.5$–19.5 tons. This implies that the median mass should be in the approximate range $W_{50} = 2$–16 tons. We believe that not much damage should be done to the cover layer by the ice, because the layer should be able to withstand wave attacks in subsequent summer/fall seasons. Hence the proposed cover layer with stone size $W = 10$ tons may not be adequate to resist ice attack. But more research is needed to arrive at proper design criteria on ice attack on the cover layers such as the ones proposed on this artificial island.

3.11.4. *Stability of a Berm Breakwater Concept Alternative*

As indicated in Fig. 3.84 one may think of a berm breakwater concept as an alternative to the uniform slope. The advantage of the berm concept is that it requires less material than the uniform slope.

The requirement to the stone size is shown in Table 3.15.

Figure 3.86 shows the recession for a berm versus significant wave height for $W_{50} = 4.3$ tons, Eq. (3.6), assuming the wave steepness $s_{om} = 2\pi H_s/gT_z^2 = 0.04$ and $s_{om} = 0.06$, while Fig. 3.87 shows the same for $W_{50} = 10$ tons.

The proposed cross section for a static stable berm breakwater against wave attack is shown is shown in Fig. 3.88. Note that the berm breakwater is considered to be a tough structure and can easily be built to survive larger waves than the design waves.

There is virtually no quantitative information on the requirement of the stone size for ice attack on a berm breakwater, except the work of Daly

Table 3.15. Requirement to stone the maximum size in a berm breakwater concept for the gravel island shown in Fig. 3.84. $H_s = 5.1$ m, $T_z = 7.3$–9.1 (wave steepness $s_{om} = 0.06$–0.04).

Category	Ho	HoTo	W_{50}, tons	Recession. Rec, m
No-reshaping	2.0	45	10	≈4
Reshaping static stable	2.7	71	5	≈5

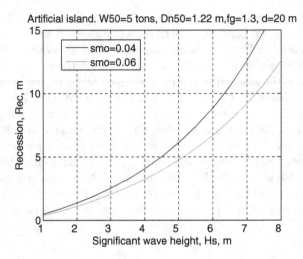

Fig. 3.86. Mean recession for a proposed berm breakwater concept for the artificial island. $W_{50} = 5.0$ tons.

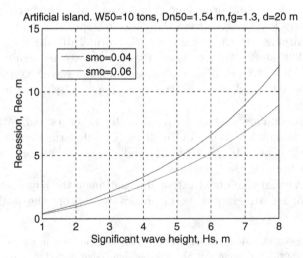

Fig. 3.87. Mean recession for a proposed berm breakwater concept for the artificial island. $W_{50} = 10$ tons.

et al. who carried out tests on a rubble mound cross section resembling a berm breakwater. The results of Daly et al.[152] as described previously indicate also that a berm structure is beneficial from the ice resistance point of view.

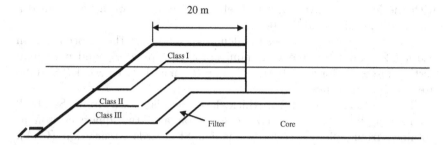

Fig. 3.88. Conceptual design of the cross section of the berm for the artificial island (Fig. 3.84). 20 m water depth. $H_s = 5.1$ m, $T_z = 7.6$–9.1 s. Nonreshaping. Class I stones: 8–12 tons, Class II stones: 4–8 tons, Class III stones: 1.5–4 tons.

3.11.5. *Wave and Ice Forces on a Vertical Wall Caisson-Type Breakwater*

We have made a brief estimate of the wave and ice loads on a caisson-type breakwater as shown in Fig. 3.89. The same design wave heights as for the calculations related to the artificial island have been used (Table 3.14).

The height of this caisson is taken higher than normally used for caisson-type breakwaters. But it is believed that a fairly large height is required for protection of ice pile-up. The height is somewhat arbitrarily chosen such that there will be no significant wave overtopping on the caisson for the 50-year design wave height, $H_{\max} = 9.1$ m, $T_{\max H} = 10.8$ s. To compare

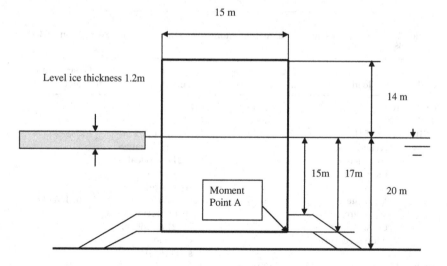

Fig. 3.89. Conceptual design of a Caisson-type breakwater.

with the ice forces given by Sandwell Engineering[168] we have assumed a length of the caisson to be 100 m.

The wave force has been calculated following the procedures in Sec. 3.5.2. The calculated maximum horizontal force, F_h, and maximum vertical force, F_v, for wave height $H_{\max} = 9.1$ m and wave period $T = 10.8$ s are as shown in Table 3.16.

The ice loads for 1.2 m thick level ice is taken from Sandwell Engineering,[168] who compared ice loads on a 100 m square caisson as obtained from different codes: Canadian Standards Association (CSA), American Petroleum Institute standards (API), and the Russian SNiP and VSN codes. Sandwell Engineering[168] calculated also the ice forces for a first year ridge of total thickness 10 m interacting with the same structure. A keel-to-sail ratio of 4.4, a consolidated layer thickness of 1.5 m, and a width of 23 m was assumed. The ridge was assumed to be embedded in an ice sheet with the same characteristics as for the level ice of 1.2 m thickness.

Table 3.17 shows the calculated ice forces on a 100 m square caisson. The overturning moments relates to Point A (Fig. 3.89), and has been obtained by multiplying the ice forces with the depth to point A (17 m).

Table 3.16. Wave forces and overturning moment, Point A, on caisson type breakwater of 100 m length (Fig. 3.89) [Wave height $H_{\max} = 9.1$ m, wave period $T = 10.8$ s].

Horizontal force, MN	Vertical force, MN	Overturning moment, MNm
163	36	2574

Table 3.17. Ice forces and overturning moment on a 100 m square caisson, "deep" water.[168]

Code	1.2 m level ice load (MN)	Overturning moment (MNm)	10 m ridge ice load (MN)	Overturning moment (MNm)
CSA	100	1700	127	2159
API	180	3060	239	4063
SNiP	597.6 if the structure is considered as a monopod	10159	896.4 to 1195.2 (using formulae (121) applicable to monopod structure)	15238–20318
	159.4 if the structure is considered as a wide structure	2710	239.1 to 318.8 (using formulae (122) applicable to a wide structure)	4064–5420
VSN	171.6	2917.2	257.4	4375.8

Table 3.17 shows that the different ice load codes give significantly different load. The given ice loads tends to be generally larger than the wave loads, so a much wider and heavier caisson is required for the ice loads. But it seems feasible to design a caisson breakwater such that it can withstand the ice forces for ice conditions as used for the calculations, depending on the code to be used. Thicker ice will necessarily give higher ice forces and this requires then an even wider and heavier caisson. A caisson with a sloping front (Fig. 3.80), may be an attractive solution.

3.12. Rock Properties Requirements for Berm Breakwater Designs

3.12.1. *General on Rock Properties for Berm Breakwater Construction*

PIANC[6] gives an account of rock type, rock quality and quarry yield for berm breakwaters. The following is quoted from PIANC[6]:

> Smarason *et al.*[169] gives an account on how rock quality is considered in Icelandic berm breakwater design. Their account may also be useful as guide for others for their quarry investigations. The following is an excerpt of the paper by Smarason *et al.*[169]
>
> Commercial rock quarries are relatively few in Iceland. New quarries are therefore often needed for new harbor projects, although existing one can sometimes be used. Selection of suitable quarries begins with inspection of geological maps and aerial photographs of the area adjacent to the planned breakwaters and the area is widened until successful. The search for suitable armor stone quarries is initially directed at any prominent thick lava flows, which may be accessible on low ground or in accessible benches or hillocks. Promising sites are visited and inspected visually for geological features such as rock type, weathering forms, pores, pore fillings (amygdales), alteration and joint density. Further investigation is performed through pneumatic drilling and core sampling at promising sites before bidding. Quarries may be located right by the breakwater structure or up to 40 km away. However the transport distance is commonly 5–15 km. Core material is sometimes produced more economically in poorer quarries closer to the structure or dredged from sediments on the sea floor. Possible quarry yield and rock quality is weighted against transportation distance in each case, to optimize cost effectiveness and strength and durability of the structure. Environmental impact assessment is carried out for quarries exceeding 150,000 m^3 in total volume in situ rock, or 200,000 m^3 of blasted rock.
>
> Scan lines at horizontal and vertical rock exposure are used to measure the fracture density of the rock. Two scan lines, at right angles to each other, are measured with a tape measure on the upper surface and also in vertical

sections if possible. Pneumatic drilling is carried out to give an idea of the thickness of the rock, possible size of quarry and an idea of the soundness and alteration grade of the rock. Core drilling is usually carried out to give further information regarding spacing of discontinuities (joint and fractures), rock quality, specific gravity, absorption, point load index, freezing/thaw resistance and optical inspection in thin sections. Measurements of discontinuity spacing in scan-lines and cores are used to establish an idea of the possible size distribution of blasted material from the rock. It is assumed, in the interpretation that the shape of the stones is on average cubic.

The conventional rubble mound breakwaters are usually built as rubble mound structures consisting of an inner core of fine material of quarry run, covered with an armor layer of stones. The armor layer of a conventional rubble mound structure usually consists of two layers of armor stones, or concrete armor units if armor stones are not readily available, and a filter or an underlayer to prevent the finer material from being washed out. The armor layer extends about 1.0 to 1.5 times the design significant wave height above the design water level and to the same distance below the lowest water level. The size of the armor stone needed to resist the wave energy is proportional to the wave height in third power. This means that for conventional rubble mound breakwaters very large stones (Table 3.8) are often needed in large quantities. This design method can be characterized as a "demand-based design".[170]

The design philosophy of berm breakwaters aims at optimizing the structure not only with respect to wave load but also possibly yield from an armor stone quarry, which can be characterized as "supply-based-design." The initial idea of berm breakwaters was that they should be wide voluminous structures, built of two stone classes with a wide size gradation, allowing considerable reduction of armor stone size.

An "Icelandic type" of berm breakwaters has been developed, where the structure is less voluminous and more stable than the original berm breakwater concept anticipated. An emphasis is put on maximizing the outcome of the armor stone quarry and utilizing this to the benefit of the design. The goal of the Icelandic type berm breakwater is that it shall be "non-reshaped statically stable."

It is advisable to design a berm breakwater as a "non-reshaping static stable" berm breakwater. But since it is a minor probability of crushing of good quality rock by the impact when the stones are moving through reshaping (see Sec. 3.12.2), it is possible to design for "reshaping, static stable" conditions.

Reverting to Smarason *et al.*,[169] "Close collaboration between the designer and geologist in the preparation of berm breakwater projects has proven very effective over the years. This has resulted in better designs and better use of the quarried material. This collaboration gives the designer the chance to fully utilize material from the quarry down to the smallest possible stone size and have often resulted in 100% utilization of the quarries.

Close cooperation between the geologist and the project supervisor with the contractor is often necessary to achieve maximum results in the quarry. Blasting and sorting of armor stone are by no means an easy task and slight alteration of spacing and tilt of drill holes may at times help to improve blasting results. It has to be realized that the contractor and the buyer should work as a team aiming for the same goal. Experienced contractors rely on the predicted yield curves in their bidding."

Figure 3.90 shows as an example the quarry yield prediction mass distribution curve for the quarries at Sirevåg, Norway, for building the Sirevåg berm breakwater (Fig. 3.2). From this prediction curve the stone classes and quarry yields were determined as shown in Table 3.18 and Fig. 3.22.

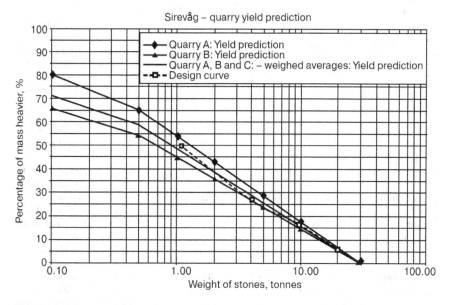

Fig. 3.90. Quarry yield prediction and design curve for the Sirevåg berm breakwater. After Sigurdarson *et al.*[4]

Table 3.18. Stone classes and quarry yield for the Sirevåg berm breakwater.

Class	$W_{min} - W_{max}$ (tons)	W_{mean} (tons)	W_{max}/W_{min}	Expected quarry yield (%)
I	20–30	23.3	1.5	5.6
II	10–20	13.3	2.0	9.9
III	4–10	6.0	2.5	13.7
VI	1–4	2.0	4.0	19.3

Table 3.19. Guidelines for quality control of armor stone of igneous rocks (After Smarason et al.[169] Modified by Smarason[173]).

Test	Excellent (A)	Good (B)	Marginal (C)	Poor (D)	Comments
Rock type	Gabbro Granite Dolerite Anot- thosite Porh. basalt	Ol.-tholeiite Alkali basalt Quartzite Gneiss	Tholeiite basalt Andsite	Rhyolite Dacite Hyalo- clatite	Guidelines for rock types without correlation to rock density
Specific gravity (SSD) Tons/m^3	>2.9	2.65–2.9	2.5–2.65	<2.5	Density of rock is a good indicator of hydraulic stability in a breakwater
Water absorption (%)	<0.5	0.5–1.0	1.0–2.0	>2.0	Important indicator of alteration and resistance to degra- dation, especially in cold climate
Freeze/thaw tests Flaking in kg/m^2	<0.05	0.06–0.10	0.11–0.20	0.21–0.50	Swedish standard SS 137244 in a 3% NaCl solution for concrete
Point Load Index $I_{s(50)}$ (MPa)	>8.0	5.0–8.0	3.0–5.0	<3.0	Correlates with rock density and indicates resistance to breakage
RQD$_{50}$	>70%	50–70%	30–50%	<30%	
Alteration of minerals	No alteration	Little alteration	Considerable alteration	Heavy alteration	Alteration inspected in thin sections
Inner bindings of minerals	Excellent	Good	Fairly good	Cleavage visible	Inspections in this sections

The rest of the quarry material was used as core material in the breakwater. The quarry yield in this case turned out to be close to the predicted yield and all the stone material was utilized, thus leading to an economical breakwater structure.

The Icelandic armor stone quality assurance programme, shown in Table 3.19, has been adopted and modified from CIRIA/CUR[171] and Hardarson.[172] Important properties are rock type, density and absorption, strength (point load index), freeze/thaw resistance in cold climates, alteration and inner binding of minerals (in thin section under microscope), and resistance to abrasion in abrasive conditions.

3.12.2. *Requirements of Stone Breaking Strength for Berm Breakwaters*

3.12.2.1. *General*

Although it is advantageous to design a berm breakwater as "nonreshaping, static stable" it may not always be feasible to do so with the quarry yield that can be provided. Since stones on a reshaping berm breakwater are allowed to move, they may be vulnerable to abrasion and breaking. Only limited work has been done until the year 2000 on the issue of the breaking of stones on reshaping berm breakwaters. Frigaard *et al.*[174] considered the stress in one stone hitting another stone based on the method developed by Hertz[175] as reported by Timoshenko and Goodier.[176] However, nothing conclusive was stated as to the possibility of stones breaking. Archetti and Lamberti[177] pursued this concept further by applying Hertz's method to calculate the stress in a stone from impact, and information on the stress a stone can take before breaking, to evaluate if a stone would break or not.

Latham and Gauss[178] discussed drop test results for armor stone integrity. But their objective focused more on establishing a method and criteria that would be accepted for delivery of stones from quarries for normal use of armor stones. The method they described is also referred to in the Construction Industry and Research Information Association/CUR (CIRIA/CUR) Manual[171] and also is considered in the new version, CIRIA CUR CUTMEF.[8]

Tørum and Krogh[46] and Tørum *et al.*[47] developed a method to be used for the assessment of the suitability of quarried rock for berm breakwaters. Their method requires 3–5 days of drop testing of fairly large stones in the selected quarry. Because the method is relatively new and specifically designed for berm breakwaters, the background and the procedures for the method are briefly described here.

There are three fundamentally different ways of reducing the size of rock pieces or stones on a berm breakwater by mechanical action: (1) impact breaking, (2) compression, and (3) abrasion.

The first one could be illustrated by a hammer hitting a stone until it falls to pieces. The second one can be likened to squeezing the stone between plates until it cracks, and the third one would be like rolling the stone back and fourth resulting in chipping off of small fragments. Impact breaking and abrasion are the most dominant mechanisms that break down armoring stones mechanically. The breaking of moving stones when they hit other stones will depend primarily on the impact energy and the ability of the stones to resist this impact energy. The impact energy is equal to $E = 0.5K_i mV_s^2$, where m is mass of stone, V_s is the velocity of the stone and K_i is an impact factor varying between 0 and 1.3 (see below). Statistical information on the velocities of the stones was obtained by measurement of stone velocities on a model berm breakwater by filming the stone motion through a wave flume glass wall. The velocities were made dimensionless by the parameter $(V_s)^2/(gH_s)$. A two-parameter Weibull probability density function was fitted to the data of $(V_s)^2/(gH_s)$

$$f(X) = \frac{\gamma}{X_s^\gamma}(X)^{\gamma-1} \exp\left[-\left(\frac{X}{X_s}\right)^\gamma\right] \qquad (3.107)$$

where $X =$ statistical parameter, in this case $(V_s)^2/(gH_s)$; $\gamma =$ shape factor, in this case 0.877; and $X_s =$ scaling factor, in this case 0.237.

The cumulative two-parameter Weibull distribution function is given by:

$$F(X) = 1 - \exp\left[-\left(\frac{X}{X_s}\right)^\gamma\right] \qquad (3.108)$$

Figure 3.91 shows the probability density of the data and the two-parameter Weibull density function for the dimensionless velocity squared $[(V_s)^2/(gH_s)]$. There are altogether 313 data points behind the probability density of the data shown in Fig. 3.91.

The ability of the stone to resist breaking is dependent on the stresses induced in the stone during impact, the mechanical strength of the solid stone and the number of fissures in the stone. The fissures may be "natural" fissures or fissures imposed during blasting and handling of the stone.

Tørum and Krogh[46] considered the use of Hertz's method and/or the finite element method (FEM) to calculate the stresses in a stone. The primary reason for not using such methods was that stones with internal fissures will experience a different stress distribution than homogenous stones, which is normally assumed for stress calculations. It was decided to make drop test experiments on stones in two quarries at Hasvik and Årviksand,

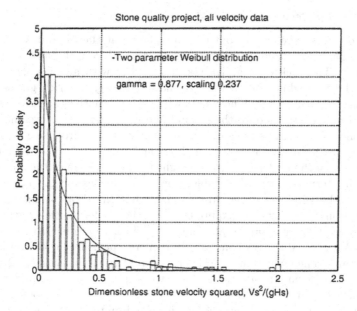

Fig. 3.91. Probability density function for dimensionless stone velocity squared, $(V_s)^2/(gH_s)$.

Norway, following a method developed by Krogh.[179–187] It was thus possible to obtain statistical information on the breaking energies for stones of approximately the same size and for varying sizes. The probability density function for the breaking strength of the stones with respect to impacts, $p_R(P)$, was then obtained. Finally one can calculate the failure probability, P_f, according to the general concept of failure probability when considering two variables, in this case the impact energy, E, and the breaking strength, P, of stones. The probability of failure is then given by (e.g., Melchers[182]):

$$P_f = \int_0^\infty F_R(x) f_s(x) dx \qquad (3.109)$$

where F_R = cumulative distribution function of the resistance (necessary energy to crush the stones), f_s = probability density function of the impact energy available to break the stones, x = parameter, in this case energy.

In addition to the drop test results information was obtained on abrasion, the Young's modulus, uni-axial compressive strength, brittleness, and flakiness of stones from the quarries where the stone investigation was carried out. The results from the latter investigations are difficult to link directly to the conditions on a reshaping berm breakwater due to scaling problems.

3.12.2.2. *Drop tests on quarried stones*

The drop tests were carried out in two quarries, at Hasvik and Årviksand, where armoring stone for the construction of local breakwaters was produced. The rock type at Hasvik is a gabbro with an average density of $3050\,\mathrm{kg/m^3}$, while the rock at Årviksand is a garnet-rich gneiss with a density of $2980\,\mathrm{kg/m^3}$. There are two basic alternatives for the testing: (1) to drop the stones onto the ground and (2) to drop steel weights onto the stones lying on the ground. For alternative 1, the drop height will be the distance from the lower side of the stone to the ground, and for alternative 2 from the underside of the weight to the top edge of the stone.

Both alternatives are acceptable. However, if a constant drop energy — which is the product of the stone weight and the drop height and the acceleration of gravity — is required the drop height has to be set according to the weight of the stone in alternative 1, which will differ within certain limits. Alternative 2 gives constant drop energy providing the drop height is constant. Since constant drop energy is preferable, alternative 2 is the easiest to perform. In addition, it is always difficult to hold and release unevenly shaped stones compared to steel weights easy to hook up. For these reasons alternative 2 was chosen for the drop tests. Figure 3.92 shows a sketch of the test setup.

Because the drop weight has a diameter close to the size of the stone, the weight will always hit the highest point on the stone. Similarly, when two stones hit each other they will usually touch at one point with a force directed according to the movement of the stone.

To manipulate the drop energy for the different stone sizes, three steel weights were used: 51.5, 303.5, and 1709 kg, respectively. Throughout the tests, these weights made it possible to limit the drop height from 0.2 m to 5.0 m. Prior work has shown that the speed of the weight at the moment

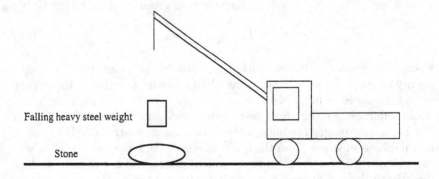

Falling heavy steel weight

Stone

Fig. 3.92. Sketch of drop test set-up.

of touching the stone does not influence on the result of the drop tests as long as the drop energy is kept constant. The weights were hooked up in a release mechanism fastened to a strap on the beam of the excavator. The release mechanism was activated by pulling a rope fastened to it.

A site with solid rock was selected for the drop tests. The surface should be as even as possible where the stones are to be tested. Such a surface will give by far the most reproducible conditions when moving from one site to another. Piling up big stones as a base for the testing, as required for the CIRIA-CUR tests, CIRIA/CUR,[171] cannot be easily reproduced when going from one quarry to another.

The drop test is a statistical procedure, which means that a number of stones have to be tested before having a reliable average value. Three stone classes were tested. The approximate weights of the stones in each class were 20, 200, and 1400 kg. The range of the stone weights in each class was the mean weight $\pm 20\%$. In this procedure, 12 to 25 stones were tested for each set of variables: stone size and drop energy. It could be objected that this was too low a number, but since each stone size was tested at different energy levels, the total number of stones within each size evened out the results statistically. A higher number of stones would have lengthened the test period in the quarry.

During the tests each stone was subjected to only one impact, even though it is recognized that a stone that does not break for one impact may break after a repeated number of impacts. Because of this the test conditions resemble the conditions a stone on a reshaped static stable berm breakwater will experience. The stones move down the slope only once and will frequently experience only one major impact when it hits another stone during maximum velocity. On a reshaped dynamic stable berm breakwater, a stone may be subjected to repeated number of impacts as it moves up and down the breakwater slope. Hence the results obtained with the drop tests method discussed here may be less valid for a reshaped dynamic stable berm breakwater than for a reshaped static stable berm breakwater.

The drop test give values for the applied drop energy (P [Joule]) and the corresponding breaking frequency or probability of being broken, F, for each stone size. These values are plotted in Fig. 3.93 for both the Hasvik and Årviksand quarries. Each curve represents the results of the tests on the three different stone classes from each quarry. Each point in the diagram represents the percentage of broken blocks for a particular drop energy. As mentioned, each point represents 12 to 25 stones. For example, 16 out of 23 stones or 69.5% of the 200 kg stones at Årviksand quarry were broken for a drop energy of 10. 120 J, while only 4 out of 23 stones (or 17.3%) were broken for a drop energy of 2.980 J.

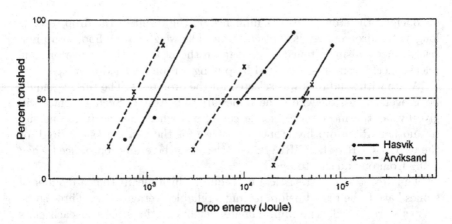

Fig. 3.93.   The probability distributions of strength for different stone sizes (20, 200 and 1400 kg) at the Hasvik and Årviksand quarries.

Krogh[179] found the following relation between the breaking energy, P, the energy to break 50% of the stones, P_{50}, and the probability F of being broken:

$$P/P_{50} = \exp((F - 0.5)/\beta) \qquad (3.110)$$

where β = coefficient obtained from the experiments. The following relations were found for the drop tests at the two quarries in Hasvik and Årviksand:

Hasvik: $\beta = 0.44$, $P/P_{50} = \exp((F - 0.5)/0.44)$

Årviksand: $\beta = 0.50$, $P/P_{50} = \exp((F - 0.5)/0.50)$

The P_{50}-values calculated from the curves in Fig. 3.93 have been plotted in Fig. 3.94 as a function of the stone volume. The volumes are derived from dividing the weights by the stone density. In the log–log diagram the plots define straight lines that can be expressed by the formula:

$$P_{50} = k \cdot V^{\alpha} \qquad (3.111)$$

where

k = coefficient
V = the volume of the block
α = coefficient

From the drop test results the following coefficient values were obtained:

Hasvik: $\alpha = 0.842$, $k = 81070$, $P_{50} = 81070 \cdot V^{0.842}$

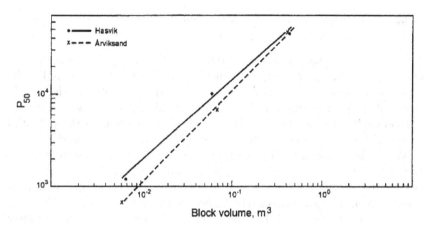

Fig. 3.94. The breaking strength P_{50} (Joule) versus stone volume (m^3) for the Årviksand and Hasvik quarries.

$$(V \text{ in } m^3, P_{50} \text{ in Joule})$$

$$\text{Årviksand: } \alpha = 0.970, \quad k = 98310, \quad P_{50} = 98310 \cdot V^{0.970}$$

The values of α, β, and k were determined for the armor stones at Hasvik and Årviksand from the drop tests performed. These values give the basic breaking properties of the rock in the two quarries. They define the spread in strength for each stone size and the absolute strength as a function of stone volume. By extrapolating from measured values, the energy necessary to break 100% of the stones of a certain size can be determined. The maximum energy that will not break any of the stones of the same size can be further determined. The value P_{50} could be extrapolated to larger stone volumes than those tested. However, all extrapolations should be done with caution.

All such extrapolations are risky, and one must always question the validity of extrapolation. If the testing had stopped at 200 kg stones, one might have objected to the extrapolation beyond that weight. However, stones of 1400 kg were also tested and it was found that one could have accurately extrapolated from the 200 kg and less weights to the 1400 kg values. Therefore it is believed that one can extrapolate significantly beyond the breaking strength for the stone weight of 1400 kg.

The tests performed give the basic strength properties of armoring stone. However, the test procedure requires three to five days of work depending on the type of equipment available and the number of stones to be tested. Therefore, it cannot be considered to be a normal procedure for repeated analysis. However, it might be developed into a standardized method to certify armor stone from different quarries.

3.12.2.3. *Combining results of stone velocity results and drop test results*

When either the stone hits the "ground" or the heavy drop steel weight hits the stone during the drop tests, it will "stop" almost immediately. The impact force is not known, but may be assumed it is related to the kinetic energy of the stone just before it hits the ground (or the heavy steel weight just before it hits the stone). This energy is again equal to the potential energy of stone or the heavy steel weight before they are released.

When the stone rolls on a berm breakwater it will not necessarily lose its kinetic energy completely when it hits another stone. The stone may continue to roll with a new angular and transitional velocity. The impact force between the two stones is not known, but it may be assumed that it is related to the difference of the kinetic energy of the stone before and after the impact. When a stone hits another stone the situation is schematically and simplified as shown in Fig. 3.95.

With reference to Timoshenko and Young,[183] the velocity v just after impact for a sphere with radius r can be calculated. Hence the difference in the kinetic energy of the rolling sphere before and after the inelastic impact can be calculated as follows:

$$E_{\text{diff}} = \frac{7}{5}m\frac{1}{2}(v_c^2 - v^2) = \frac{7}{10}m\left(v_c^2 - \left(\frac{2}{7}\right)^2 v_c^2\left(1 + \frac{5}{2}\left(1 - \frac{h}{r}\right)\right)^2\right)$$

$$= \frac{1}{2}mv_c^2\frac{7}{5}\left(1 - \left(\frac{2}{7}\right)^2\left(1 + \frac{5}{2}\left(1 - \frac{h}{r}\right)\right)^2\right) = \frac{1}{2}mv_c^2 K_i \qquad (3.112)$$

$$K_i = \frac{7}{5}\left(1 - \left(\frac{2}{7}\right)^2\left(1 + \frac{5}{2}\left(1 - \frac{h}{r}\right)\right)^2\right)$$

where h = step height, r = radius of the sphere, K_i = impact factor, v_c = translational velocity of the sphere before the hit.

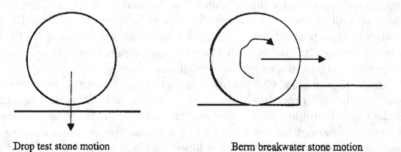

Drop test stone motion   Berm breakwater stone motion

Fig. 3.95. A rolling sphere hits a flat bottom and step.

This difference in kinetic energy can now be considered as available for breaking of the stone. If it exceeds the energy needed to break the stone the stone will break.

In the following the probability of failure of the stones is evaluated by considering only the two variables: (1) the impact energy and (2) the required breaking energy. Although both are dependent on the mass of the stone, the velocity and the strength of the stones are considered to be independent. Hence Eq. (3.107) is used to calculate the probability of failure when only two independent variables are considered.

The impact energy is given by

$$E = \frac{1}{2}mK_iV_s^2 = \frac{1}{2}mK_igH_s\left(\frac{V_s^2}{gH_s}\right) \tag{3.113}$$

The probability function of E for a given significant wave height H_s and a given impact factor K_i is then

$$f(E) = \frac{f\left(\frac{V_s^2}{gH_s}\right)}{\frac{1}{2}mK_igH_s} \tag{3.114}$$

The probability density function $f(V_s^2/gH_s)$ is given previously by a Weibull distribution function with a shape factor $\gamma = 0.877$ and a scaling factor $X_s = 0.237$ (see Sec. 13.2.2).

We also need the cumulative probability distribution function for the breaking strength of the stone and, for curiosity, the probability density function $dF/d(P/P_{50})$ for the breaking energy. These are obtained as follows.

From Eq. (3.108) the cumulative distribution function is obtained:

$$F = \beta \ln\left(\frac{P}{P_{50}}\right) + 0.5 \tag{3.115}$$

and the probability density function

$$f\left(\frac{P}{P_{50}}\right) = \frac{dF}{d\left(\frac{P}{P_{50}}\right)} = \frac{\beta}{\left(\frac{P}{P_{50}}\right)} \tag{3.116}$$

Figure 3.96 shows the cumulative probability distribution and the probability density function for the Årviksand and the Hasvik stone breaking data. In reality, the probability distribution curves have an S-shape. But since no data for the upper (close to $F = 1$) and lower (close to $F = 0$) regions are available no speculation on the real form of the curves in these regions is made. The "true" form of the curves in these regions will not significantly change the probability of failure results.

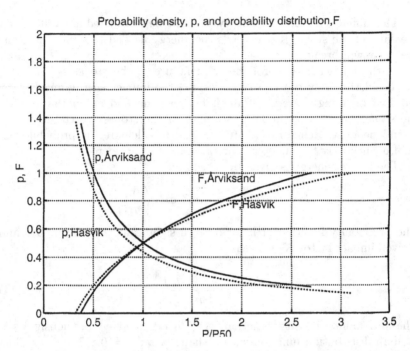

Fig. 3.96. Cumulative probability distribution F and probability density function f for the Hasvik and Årviksand stone breaking data.

In further analysis, the Årviksand stone breaking data [$\beta = 0.504$, $\alpha = 0.966$ and $k = 98306$] are used and extrapolated to larger stone sizes than applied during the field tests.

Figure 3.97 shows an example of the probability density functions for impact energy and strength "energy" of a stone for $Hs = 7.0\,\mathrm{m}$, $Ho = 3.0$, $W_{50} = 8.000\,\mathrm{kg}$, $\rho_s = 2700\,\mathrm{kg/m^3}$ and $K_i = 1.4$. $\mathrm{F}_R * \mathrm{f}_s$ is also shown in the same diagram.

The probability of failure, P_f, for different conditions has been numerically calculated. Figure 3.98 shows as an example the probability of being broken for different stone weights in a stone class for $W_{50} = 2.600\,\mathrm{kg}$, $H_s = 4.0\,\mathrm{m}$, $H_o = 2.5$ and $K_i = 1.4$. The results show that the probability of being broken is virtually the same for all stone sizes in the segment.

It can be concluded from these calculations that the probability of failure is the same for all stones for a given wave height H_s and a given K_i.

The probability of failure for significant wave heights $H_s = 4.0$ and $7.0\,\mathrm{m}$ and for different impact factors K_i have also been calculated. These probabilities are shown in Fig. 3.99. It is seen that both the significant wave

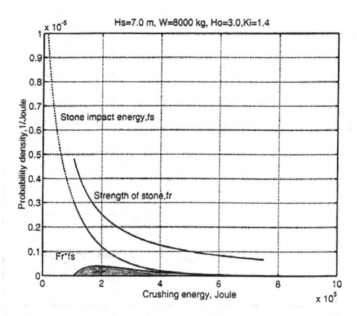

Fig. 3.97. Probability density function for f_s and breaking strength f_r of stone as well as $F_r * f_s$ for $W = 8000$ kg, $H_s = 7.0$ m. $H_o = 3.0$ and $K_I = 1.4$.

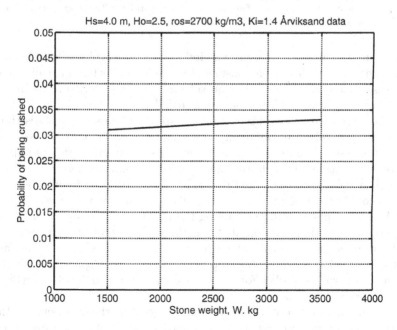

Fig. 3.98. Probability of being broken for different stone size in a stone class. $W_{50} = 2600$ kg.

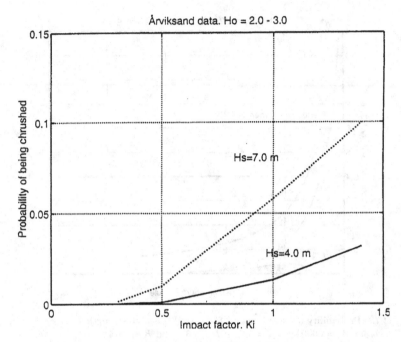

Fig. 3.99. Probability of failure of stones (being) broken) for different wave heights and different impact factors. Range of $H_o = 2.0–3.0$.

height and the impact factor have an influence on the probability of failure. The impact factor has been considered a fixed value, while in reality it is a statistical parameter with a mean value of approximately 0.7. So far the statistical variation of K_i has not been included.

The calculated probabilities of failure are based on statistical distributions of stone velocities obtained from samples from laboratory tests and field drop tests on samples of stones, coupled with an evaluation of impact energies "available" in inelastic impacts when a rolling sphere hits a step. How then should the calculated probabilities of failures be interpreted and what should be considered as an acceptable probability of failure?

As a first approach to interpreting the results the probability of failure is considered as that fraction of stones that will be broken on a reshaped static reshaped berm breakwater where Ho < ~2.7. If an average impact factor $K_i = 0.7$ is considered. Figure 3.99 indicates that approximately 3% of the stones will be broken for $H_s = 7.0$ m and approximately 0.5% will be broken for $H_i = 4.0$ m.

In general, failure criteria should be supported by the experience of the behavior of a structure. No such experience on the failure of stones on a berm breakwater where the breakwater was systematically monitored is available yet.

What are the consequences if small percentage of the stones on a berm breakwater are broken?

Tørum and Krogh[46] observed that a berm breakwater will reshape into a static stable berm breakwater when Ho < ~2.7. The stones will then move primarily downward on the berm slope into a resting position and will seldom move upward again. When Ho > ~2.7 the berm breakwater becomes reshaped dynamically stable, e.g., stones will roll up and down the slope while the profile shape remains stable. In the reshaped static stable conditions the stones will be subjected to only a few impacts, while in the reshaped dynamic stable condition a stone may be subjected to many repeated impacts. Abrasion, in addition to impact breaking, may also then play a more important role in the deterioration of the stones. Tørum and Krogh[46] found, in considering the abrasion tests in a tumbling mill, that there is a scaling problem in relation to converting the tumbling mill abrasion results to full scale use for a reshaped dynamically stable berm breakwater. This scaling problem was also encountered by Mikos and Jaeggi.[184]

During the hydraulic tests it was also observed that there is a redistribution of stone sizes (without any breaking) with increasing wave heights.[61] The larger stones travel down the slope and the smaller stones remain on the upper part of the slope. Hence it seems that one do not need to be afraid if some stones will be broken, even on the upper part of the slope, especially for the reshaped static stable berm breakwater. In fact it seems to be an anachronism to build a reshaping berm breakwater by building a berm that is allowed to reshape. During the reshaping process the larger stones will end at the footing of the berm, while they would have given better protection if they had been on the upper part of the berm. A more stable berm breakwater may have been obtained if the stones with the original mixture had been placed in a profile corresponding to the reshaped profile. This holds probably especially for a reshaped static stable profile. Economical construction methods to build the berm with a profile according to the reshaped profile should be developed.

It has to be born in mind that the breaking tests were carried out for stones with weights up to approximately 1400 kg. Although it is believed that the results can be extrapolated to larger stone sizes, it is recommended that drop tests be followed up with additional breaking tests, including smaller and larger size stones than used in the study of Tørum and Krogh.[46] Similarly it will be useful if a berm breakwater could be monitored after construction to gain experience with regard to breaking and possible abrasion of the stones.

3.12.2.4. *Conclusions of the breaking strength evaluation method*

The main conclusions from the study on the breaking strength of quarried stone for berm breakwaters[46] are as follows:

1. The statistical distribution of the dimensionless velocity $V_s/(gH_s)^{0.5}$ is independent of Ho or HoTo.
2. Previous methods developed for the statistical distribution of the breaking strength of smaller size sand/stone grains can be applied for breakwater stones.
3. The statistical distribution of the dimensionless stone velocities and the statistical distribution of the breaking strength of stones can be combined into a probability of failure (breaking) analysis method.
4. The probability of failure analysis method shows that stones of the quality found in most Norwegian quarries (e.g., the quarries in Hasvik and Årviksand) can be used for reshaped static stable berm breakwaters (Ho < ~2.7) without an excessive number (<~5%) of stones being broken.

3.12.3. *Abrasion*

There has been some concern about stone abrasion of berm breakwater stones, since the stones may be allowed to move (reshaping breakwaters). Tørum and Krogh[46] carried out abrasion tests in an abrasion mill. However, it is not possible to transfer the results of such tests to full scale, Tørum.[185] The abrasion mill results can be used to distinguish between different rock materials, but cannot be used to evaluate long term abrasion on a berm breakwater.

Archetti *et al.*[186] concluded after inspection of several Icelandic berm breakwaters that abrasion was not a problem on these berm breakwaters. It seems that if a berm breakwater can meet the requirement for avoiding excessive crushing, abrasion should not be a problem.

3.13. **Construction of Berm Breakwaters**

The following is partly based on PIANC.[6]

One of the primary benefits of berm breakwaters when compared with conventional rubble mound breakwaters is their greater acceptance tolerances with respect to placement accuracy. The success of the berm breakwater depends to a large extent on the porosity of the structure. It is imperative to try to eliminate material smaller than the minimum required to meet gradation. The contractor should avoid pushing small material onto the berm in order to build pads for the equipment to work on the berm.

Today the construction methods for berm breakwaters take into account these points quite well (see later) similar to what is achieved for conventional rubble-mound breakwaters. A berm breakwater can be constructed using readily available and less specialized construction equipment and labor compared to the construction of a conventional rubble mound breakwater.

The usual equipment for construction of berm breakwaters consist of a drilling rig, two or more backhoe excavators, one or two front end loaders, and several trucks depending on hauling distance and the size of project. In addition stones may be dumped from barges, in most cases split barges.

Backhoe excavators with open buckets or prongs are used to place armor stones. In projects with maximum stone size up to 12–15 tons it is common to use backhoe excavators of 40 to 50 tons. At the Sirevåg berm breakwater, Norway (Figs. 3.2 and 3.22), constructed during the years 2000–2001, a 110-ton backhoe was used to place stones up to 20 tons down to a water depth of −7.0 m water depth and up to 30 tons down to −1.0 m (Fig. 3.100). At Sirevåg the contractor used thick steel plates as a working pad for the excavator.

Recently the Laukvik breakwater in Norway was repaired as a berm breakwater (Fig. 3.101). The design significant wave height is approximately 9.0 m. Several construction strategies were considered by the contractor, but the final choice of machines was the following: A 120 tons backhoe (Fig. 3.102), a 60 tons backhoe, and a 105 tons wheel loader.

Fig. 3.100. Backhoe excavator with a prong used on the Sirevåg berm breakwater for placement of armor stones. The 110-ton excavator is placing Class II stones (10–20 tons). The vertical opening of the prong is 2.5 m and the weight of the stone in the prong is approximately 10 tons. Note that the excavator crawls on the uneven stone layer without any specially prepared work pad, except some steel plates.

Fig. 3.101. The Laukvik breakwater, Norway. Repair work as a berm breakwater. Class I stones 27–36 tons.

Fig. 3.102. Backhoe used on the Laukvik berm breakwater for placement of armor stones. Class I, 27–36 tons.

The construction of the Laukvik berm breakwater was organized as follows, Westeng[187]: All quarry blasting was done before the construction itself started. The armor blocks, 27–36 tons, were transported from the quarry, approximately 0.5 km from the breakwater site, by trucks. The armor stones were stored at the land end of the breakwater and later transported out to the construction site on the breakwater by a wheel loader or by barge for locations below sea level.

Large cranes have been used in some berm breakwater projects, Gilman.[188] But cranes are usually considered more expensive than backhoe excavators. The placing rate with cranes is much lower than with backhoes and the machine cost per hour is higher Sigurdarson *et al.*[189] Cranes need a much finer and more stable work pad than a backhoe, which can crawl on a much more uneven layer (Fig. 3.103).

The tolerances for placement of stones are greater for a berm breakwater than for a conventional rubble mound breakwater and less strict placing techniques are needed. Usually no careful underwater placement is necessary. The front slope is steep and stones can be placed by backhoe excavators. Depending on the stone quality, care has to be taken when placing the stones in such a way that they do not break due to the impact. If the berm breakwater is designed as a reshaped static stable berm breakwater,

Fig. 3.103.   Laukvik berm breakwater. Backhoe crawling on the berm stones.

the stones will take considerable impacts (see Sec. 3.12) and not much care has to be taken from the impact point of view when placing the stones.

Experience from many berm breakwater projects has shown that working with several stone classes and placement of stones only increases the construction costs insignificantly while leading to a better utilization of the quarry material, thus lowering the total costs.

The construction period for larger projects often extends over two years, and possibly more, and experience has shown that partially completed berm breakwaters functions well through winter storms. Thus the risk during construction is also much lower and repairs are also much easier for the berm breakwaters than for the conventional rubble mound structure.

Continuous monitoring and surveying of the structure during its construction phase is a necessity. Nowadays GPS survey technique is used for surveying. Sections have to be measured at the completion of each stone layer. The armor stones have to be weighted, and visual control of the form and structure of the stone matrix has to be carried out. This is necessary to achieve the specified design and to make "as-built" drawings as a reference point for further monitoring. It is also useful to monitor the surrounding area to ensure that every aspect is behaving as expected, such as siltation, scour etc.

3.14. Evaluation of the Probability of Failure of Berm Breakwaters by Monte Carlo Simulations

ISO 21650[95] states the following on uncertainties related to design of coastal structures:

> Actions from waves and currents are classified as variable actions by the definition given in ISO 2394. The related basic variables are the hydraulic parameters characterizing water levels, waves and currents. They are all random variables given by probability distributions, except the astronomical tide level that can be predicted accurately once the time and the location are specified. However, it should be treated as a random variable when the time of wave and current actions is uncertain.

> For design of coastal structures the actions from waves and currents appear in calculations either as hydrodynamic loadings in terms of pressures on the structures, or only indirectly as responses of the structure. The first case applies to loadings on walls and piles, whereas the second case applies to structural response such as rubble mound armor stability and integrity, seabed scour, and to hydraulic responses such as wave run-up and overtopping, wave transmission and reflection. In both cases the methods of estimation of actions and the responses to actions contain uncertainty besides the random variability of the hydraulic basic variables.

Major sources contributing to the uncertainties are as follows:

- Statistical variability of basic variables in nature.
- Errors related to measurement, hindcast, or visual observation of hydraulic basic variables.
- Sample variability due to limited sample size of the available data of basic variables.
- Choice of distribution as a representative of the unknown true long-term distributions for hydraulic basic variables.
- Variability of parameter estimation for a distribution function fitted to basic hydraulic data.
- Accuracy of models for prediction of storm surge water levels.
- Accuracy of models for wave forecasting and/or hindcasting.
- Accuracy of models for transformation of waves and currents, for example from deep to shallow water.
- Accuracy of models for prediction of actions of waves and currents.
- Accuracy of models for prediction of structural and hydraulic responses.
- Reliability of the results of physical model tests for the estimation of loadings and structural and hydraulic responses.
- Variability of structural parameters.

The bias and standard deviation of each source of uncertainty should be investigated and be duly taken into account in evaluation of the actions from waves and currents.

Considerable work has been done on the reliability of traditional rubble-mound breakwaters (e.g., Burcharth[190]). However, little effort has been done on the reliability of berm breakwaters. Tørum et al.[61] studied this aspect, but no final conclusion was obtained.

Burcharth[168] discusses Level II and Level III methods and describe also an advanced partial coefficient system, which takes into account the stochastic properties of the variables and make it possible to design to a specific failure probability level.

Monte Carlo simulations were also mentioned by Burcharth,[190] without further discussions of the method. Later, Burcharth[191] applied Monte Carlo simulations to obtain failure probabilities for a specific conventional rubble mound breakwater with cubic concrete armor blocks. Oumeraci et al.[77] also mention Monte Carlo simulations for vertical breakwaters, but without further discussions, except referring to the method as computational demanding.

Monte Carlo simulations are frequently used within the oil industry.[192,193] It is a computationally efficient method now.

In the following, the Monte Carlo simulation method is applied to assess the probability of failure of a berm breakwater. As a case study and as a demonstration of the method, the Sirevåg berm breakwater (Figs. 3.2

Table 3.20. Design wave conditions for the
Sirevåg berm breakwater.[5,194]

Recurrence period years	Significant wave height H_s (m)	Mean wave period T_z (s)
100	7.0	10.6
1,000	8.2	11.0
10,000	9.3	11.8

and 3.22) is considered. The design wave conditions for this breakwater are shown in Table 3.20. The tidal variation in the area is small and normally in the range ±0.35 m. During storm-surge conditions, the water level has been measured up to 1.12 m above mean water level at Stavanger some 50 km north of Sirevåg.

The mean value of the recession of a berm breakwater is given by Eq. (3.37), which is to a large extent based on model test results for the Sirevåg berm breakwater.

To establish a failure equation some criteria for failure must be determined.

As pointed out previously the berm breakwater should preferably be designed as "nonreshaping" or "reshaping, static" stable. "Reshaping, static stable" is obtained when HoTo < 70. This is assumed to be the ultimate limit state (ULS).

For the accident limit state (ALS) the berm should still be intact (e.g., the recession should be less than the width of the berm (Rec < B, where B = berm width).

Figure 3.104 shows the calculated mean recession for the Sirevåg berm breakwater according to Eq. (3.37). The calculated mean recessions for the 100-year wave conditions is $Rec_{100} \approx 6.0$ m, for the 1,000-year wave conditions ($Rec_{1000} \approx 8.0$ m and for the 10,000 wave conditions $Rec_{10,000} \approx 10.5$ m).

Figure 3.105 shows the mean HoTo versus Hs for the design wave conditions at Sirevåg. The steepness of the design waves, $s_{mo} = 2\pi H_s/(gT_z^2) \approx 0.04$.[5] This steepness is based on wave measurements during large waves hitting the Sirevåg berm breakwater and this steepness has been used when calculating the mean recession, Rec versus H_s (Fig. 3.104), and mean HoTo versus H_s (Fig. 3.105).

The 100-year condition is considered as the ULS condition, while the 10,000-year condition is considered to be the ALS condition. Figure 3.104 shows that the Sirevåg brem breakwater is reshaped static stable for the mean ULS condition (HoTo < 70). For the mean ALS condition, HoTo = 70. If HoTo > 70 Class I stones will be moving up and down the berm breakwater slope.

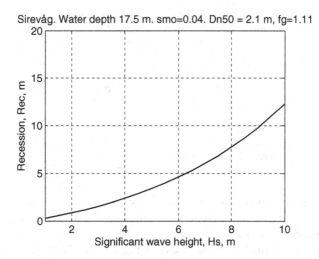

Fig. 3.104. Calculated mean recession of the Sievåg berm breakwater. Wave steepness $s_{mo} = 2\pi H_s/(gT_z^2) = 0.04$.

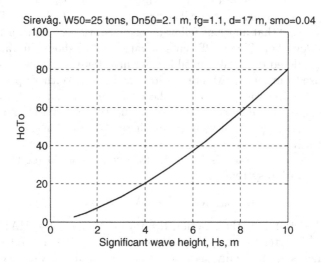

Fig. 3.105. Calculated mean HoTo versus significant wave height for the Sirevåg berm breakwater. Wave steepness $s_{mo} = 2\pi H_s/(gT_z^2) = 0.04$.

Tørum et al.[47] showed that for reshaped, static stable berm breakwater, there is for stones from the two quarries that were investigated small probabilities of crushing of the stones while rolling once down the slope. What effect the repeated impact will have on the crushing of the stones if the

Table 3.21. Standard deviations of the different
parameter for the Sirevåg berm breakwater.

Parameter	Coefficient of variation, $\sigma_x/\mathrm{X_{mean}}$
Wave height, H_s	0.05
Diameter D_{n50}	0.05
Wave period T_z	0.05
Gradation, f_g	0.05
Depth, f_d	0.05
Uncertainty of equation	0.1

breakwater is reshaped into a reshaped, dynamic stable berm breakwater
is not known, but we may assume that this condition may be accepted for
small probabilities of occurrence.

To explore the probabilities of exceeding HoTo = 70 and the recession
exceeding the berm width $B = 20$ m of the Sirevåg berm, a Monte Carlo
simulation was carried out using Eq. (3.37). There is not enough data
on the statistical distributions of the different items that enter into the
equation to know which statistical distributions should be used. Hence it
has been assumed that all items follow a normal statistical distribution with
a standard deviation. The coefficients of variation, as shown in Table 3.21,
which is considered reasonable until further investigations have been made,
has been used during the Monte Carlo simulations. With respect to the
uncertainty on the wave period, it should be mentioned that the given
COV(T_z) applies to the larger waves (e.g., the 100-year waves and above).
In addition to the uncertainty on waves, there is also an uncertainty on the
formula (Eq. (3.37)).

Each item in Eq. (3.37) is given the following value during the simulations in the Matlab system:

$$X = X_{\mathrm{mean}}(1 + \mathrm{COV}(X) * R(N, n)) \qquad (3.117)$$

where the function R is the function R = randn(N, n_i) in the MATLAB®,
MATLAB,[195] system. "Randn" generates random numbers and matrices
whose elements are normally distributed with mean 0 and variance 1. N is
the number of simulations and n_i is the ith number of variables. For all the
simulations, $N = 1,000,000$ has been applied. 1,000,000 simulations take
2–3 s.

As an example, based on the variations given the COV(X) in Table 3.20,
Fig. 3.106 shows the isolated histogram of the significant wave height variation for the 10,000-year wave conditions with the mean value of $H_{s,10,000} =$
9.3 m (Table 3.20).

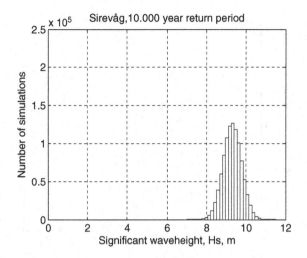

Fig. 3.106. Histogram of the isolated statistical variation of the significant wave height for the 10,000 year wave conditions, $H_{s,10,000} = 9.3$ m. 1,000,000 simulations in total.

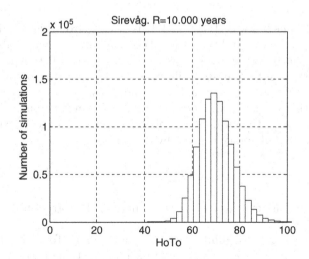

Fig. 3.107. Histogram of the statistical variation of HoTo for the 10,000 year wave conditions, $H_{s,10,000} = 9.3$ m. 1,000,000 simulations in total.

The Monte Carlo simulations has been applied for the 100-year (ULS), 1000-year and 10,000-year (ALS) wave conditions.

Figures 3.107 and 3.108 show as examples the simulated HoTo and recession, Rec, for the 10,000-year wave conditions respectively, including

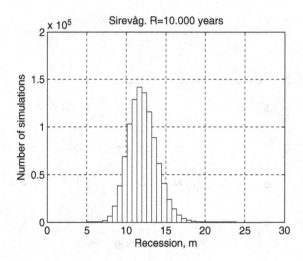

Fig. 3.108.   Histogram of the recession for the 10,000 year wave conditions, $H_{s,10,000} = 9.3\,\mathrm{m}$. 1,000,000 simulations in total.

all the variations listed in Table 3.20. It is obvious from Fig. 3.107 that for the 10,000-year wave conditions that there is a fairly large probability that HoTo $= 70$ will be exceeded, in this case $P_{\mathrm{f,HoTo}>70} = 0.47$. On the other hand the probability that Rec will exceed $B = 20\,\mathrm{m}$ (the berm width) is $P_{\mathrm{f,Rec}>B} \approx 0$ for the 10,000-year wave conditions.

In addition to the probabilities of failure obtained through the Monte Carlo simulations, the encounter probabilities have also to be considered. The encounter probability is given by

$$P_e = 1 - \left(1 - \frac{1}{R}\right)^T \qquad (3.118)$$

where $R =$ return period, $T =$ structural lifetime. The structural life time T is set to 50 years.

The calculated probabilities of exceedance of HoTo > 70, $P_{\mathrm{fHoTo}>70}$ and Rec $> B = 20$, $P_{\mathrm{fRec}>B=20\,\mathrm{m}}$, are shown in Table 3.22. The annual probabilities are also shown as well as the probabilities of encounter.

Table 3.22 shows that the probabilities that the berm of the Sirevåg berm breakwater will be damaged are small with the chosen standard deviations for the parameters governing the recession of the berm. From the model tests on the Sirevåg berm breakwater,[5] it appeared that the rear side might be damaged from the extreme 10,000-year waves. The probability of rear side failure should be further evaluated.

Table 3.22. Probabilities of failure and encounter probability for the Sirevåg berm breakwater.

Wave return period, R, years	Annual probability of exceedance	$P_{f,\mathrm{HoTo}>70}$	Annual probability $P_f A_{\mathrm{HoTo}>70}$	$P_{f,\mathrm{Rec}>B=20\,m}$	Annual probability $P_f A_{\mathrm{Rec}>B=20\,m}$	Encounter probability (P_e)
100	10^{-2}	0	0	0	0	0.40
1000	10^{-3}	0.0162	$0.0162 \cdot 10^{-3}$	≈ 0	≈ 0	0.05
10,000	10^{-4}	0.47	$0.47 \cdot 10^{-4}$	$1.73 \cdot 10^{-4}$	≈ 0	0.006

Anyhow, the probability of failure calculations show that the Sirevåg berm breakwater is extremely strong. This complies with the fact that the breakwater was subjected to two heavy storms of January 28–29, 2002 and January 11–12, 2005. The significant wave height of the January 28–29, 2002 storm was estimated to be somewhere in the range $H_s = 7.1-8.7\,\text{m}$.[3,48] The significant wave height of the January 11–12, 2005 storm was estimated to be somewhere in the range $H_s = 6.3-7.7\,\text{m}$.[49] The reshaping after these two storms were in fair agreement with the model test results,[49] the basis for the probability of failure calculations shown in this section.

3.14.1. *Risk Levels*

Risk levels have to be at the choice of the owner/designer of a breakwater. Oumeraci *et al.*[77] refers to Vrijling *et al.*[196] that two points of view in general can be defined for the acceptable safety level:

1. The individual accepted risk. The probability accepted by an individual person to die in case of failure of the structure. In Western countries this probability is of the order of 10^{-4} per year or smaller.
2. The societal accepted risk. Two approaches are presented, depending on the relative importance of the total number of lives lost in case of failure on the one hand and the total economic damage on the other. If the number of potential casualties is large, the likelihood of failure should be limited accordingly. The accepted probability of occurrence of a certain number of casualties in case of failure of a structure is then restricted proportional to the inverse of the square of this number Vrijling *et al.*[196] If the economic damage is large, an economic optimization equating the marginal investment in the structure with the marginal reduction in risk should be carried out to find the optimum dimensions of the structure.

The two boundary conditions based on the loss of human lives form the upper limits for the acceptable probability of failure of any structure. In case of a breakwater without amenities, the probability of loss of life in case of failure is very small. In that case, the acceptable probability of failure can be determined by economical optimization, weighing the expected value of the capitalized damage in the life of the structure (risk) against the investment of the breakwater.

If, for a specific breakwater, failure would include a number of casualties, the economic optimization should be performed under the constraint of the maximum allowable probability of failure as defined by the two criteria related to the loss of lives.

For the Sirevåg berm breakwater and similar breakwaters, with no amenities, the probability of direct loss of lives during breakwater damage is considered very small. Hence an economical optimization is the direction to go.

Acknowledgments

This chapter is based on a report the author wrote to Statoil, Norway, on breakwaters in arctic areas.[150] The author acknowledges Prof. Ove T. Gudmestad, formerly Statoil, for helping to make this report an open report. The report was also the basis for lecture notes at NTNU, for which several additions were made.

The author also acknowledges Mr. Sigurdur Sigurdarson of the Icelandic Maritime Administration for reviewing and commenting on Sec. 3.13 and for several fruitful and constructive discussions on berm breakwaters.

Finally, the author acknowledges the Norwegian Coastal Administration, which financially supported most of the breakwater research the author has undertaken as well as the research on forces on vertical cylinders upon shoals (light towers), frequently together with students, for letting the reports of this research be open reports and approving publications of the research results.

References

1. R. Iribarren, *Una Formula para el Calcula de los Pisques de Escollera* (Revista de Obras Publicas, 1938).
2. T. Carstens, A. Tørum and A. Trætteberg, The stability of rubble mound breakwaters against irregular waves, *Proc. 10th Int. Conf. on Coastal Engineering*, Tokyo, Japan, ASCE (1966).
3. A. Tørum, S. Bjørdal, M. Mathiesen, Ø. A. Arntsen and A. Jacobsen, Berm breakwaters. Comparison between model scale and prototype stability behavior, *Proc. Conf. Port and Ocean Engineering under Arctic Conditions (POAC)*, NTNU, Trondheim, Norway (June 16–19, 2003).
4. S. Sigurdarson, A. Jakobsen, O. Smarason, S. Bjørdal, C. Urrang and A. Tørum, Sirevåg berm breakwater. Design, construction and experience after design storm, *Proc. Int. Conf. on "CoastalStructures'03*, ASCE, Portland, Oregon, USA, (August 26–30, 2003).
5. A. Tørum, F. Kuhnen and A. Menze, On berm breakwaters. Stability, scour, overtopping, *Coastal Engineering* **49**, 209–238 (2003).
6. PIANC State-of-the-art of designing and constructing berm breakwaters. Report of the MarCom Working Group 40 (Chairman Alf Tørum). International Navigation Association (PIANC), Brussels, Belgium (2003).

7. CEM, *Coastal Engineering Manual 1110-2-1100* (U.S. Army Corps of Engineers. US Army Corps of Engineers, Washington D.C. (in 6 volumes), 2002).
8. CIRIA, CUR, CETMEF, *The Rock Manual. The Use of Rock in Hydraulic Engineering*. 2nd Edn., C683 (CIRIA, London, UK, 2007).
9. H. F. Burcharth, *The Design of Breakwaters* (Department of Civil Engineering, Aalborg University, Denmark, 1993).
10. A. Tørum, Wave induced forces on armor unit on berm breakwater, *Journal of Waterway, Port, Coastal and Ocean Engineering*, ASCE, **120**(3) (May/June 1994).
11. J. R. Morison, M. P. O'Brian, J. W. Johnson and S. A. Schaff, The forces exerted by surface waves on piles, *Petroleum Transactions, AIME* **189** (1950).
12. R. Y. Hudson, Design of quarry stone cover layers for rubble mound-breakwaters; Hydraulic laboratory investigation. Research Report No. 2–2, U.S. Army Engineer Waterways Experiment Station, Vicksburg, MS, USA (1958).
13. R. Y. Hudson, Laboratory investigation of rubble-mound breakwaters, *Journal of the Waterways and Harbours Division, ASCE* 85(WW3), 93–121 (1959).
14. J. W. Van der Meer, Stability of breakwater armour layers. Design formulae, *Coastal Engineering* **11**(3), 219–240 (1987).
15. J. W. Van der Meer, Rock slopes and gravel beaches under wave attack. PhD thesis, Delft University of Technology, Delft, The Netherlands (also Delft Hydraulics Publication No. 396) (1988).
16. H. F. Burcharth and S. Hughes, Fundamentals of design, in *Coastal Engineering Manual, Part VI, Design of Coastal projects, Chapter VI-5, Engineering Manual 1112-2-1100*, ed. S. A. Hughes (US Army Corps of Engineers, Washington DC, USA (CEM(2002), 2002).
17. L. L. Broderick, Riprap stability. A progress report, *Proc. Conf. Coastal Structures '83*, American Society of Civil Engineers (1983), pp. 320–330.
18. M. R. A. Van Gent, A. J. Smale and C. Kuiper, Stability of rock slopes with shallow foreshores, in *Proc. Int. Conf. "Coastal Structures 2003,"* ed. J. F. Melby (Portland, Oregon, USA, August 26–30, 2003); (American Society of Civil Engineers, Washington, DC, 2004).
19. M. Canel and Grauw, Rubble mound breakwater stability with multidirectional waves. Report 5 2149, SOGREAH. EU-MAST I-G6 Coastal Structures (1992).
20. C. J. Galland, Rubble mound breakwater stability under oblique waves, *Proc. 24th International Conference on Coastal Engineering*, Kobe, Japan, ASCE (October 23–28, 1994).
21. T. Hald, A. Tørum and T. Holm-Karlsen, Stability of single layer roubble mound breakwaters, *Proc. 26th Int. Conf. of Coastal Engineering*, ASCE, Copenhagen (June 1998).
22. SPM (*Shore Protection Manual*), 4th Edn. (U.S. Army Engineer Waterway Experiment Station, U.S. Government Printing Office, Washington D.C., USA, 1984).

23. SPM (*Shore Protection Manual*), 3rd Edn. (U.S. Army Engineer Waterway Experiment Station, U.S. Government Printing Office, Washington D.C., USA, 1977).

24. J. A. Melby and P. R. Mlakar, Reliability assessment of breakwaters, Technical Report CHL-97-9, US Army Engineer Watereways Experiment station, Vicksburg, MS, USA.

25. Th. Jensen, H. Andersen, J. Grønbech, E. P. D. Mansard and M. H. Davies, Breakwater stability under regular and irregular wave attack, *Proc. 25th Int. Conf. of Coastal Engineering*, ASCE, Orlando, Florida, USA (September 2–6, 1996).

26. E. P. D. Mansard, M. H. Davies and D. Caron, Model study of reservoir riprap stability, *Proc. 25th Int. Conf. on Coastal Engineering*, Orlando, Florida, USA, ASCE (September 2–6, 1996).

27. T. Meulen, van der, G. J. Schiereck and K. d'Angremond, Toe stability of rubble mound breakwaters, *Proc. 25th Int. Conf. on Coastal Engineering*, Orlando, Florida (September 2–6, 1996).

28. P. Aminti and A. Lamberti, Interaction between main armor and toe berm damage, *Proc. 23rd Int. Conf. on Coastal Engineering*, Orlando, Florida, USA (September 2–6, 1996).

29. H. Oumeraci, Filters in coastal structures, in *Proc. "Geofilters'96"*, eds. J. Jan Lafleur and A. L. Rollin, Ecole Polyteqnique de Montreal, Canada (1996), pp. 337–347.

30. J. Ahrens, Large wave tank tests of rip rap stability. TM-57, US Army Corp of Engineers, CERC, Fort Belvoir (1975).

31. D. M. Thomson and R. M. Shuttler, Design of rip rap slope protection against wind waves, CIRIA Report No. 61 (December 1976).

32. J. H. van Oorshot, Breakwater design and the integration of practical construction techniques, *Proc. Conf. Coastal Structure 83*, ASCE (1983).

33. Engineer Manual — Engineering and design, *Design of Coastal Revetments, Seawalls and Bulk Heads*, Dept. of the Army, US Corps of Engineers, Washington DC, USA (1986).

34. P. R. Rankilor, *Membranes in Ground Engineering* (John Wiley & Sons, New York, 1981).

35. J. L. Sherard et al., Basic properties of sand and gravel filters, *ASCE Journal of Geotechnical Engineering Division* **110**(6), 685–700 (1984).

36. A. F. De Grauw et al., Design criteria for granular filters. Waterloopkundig Laboratorium Delft Hydraulics Laboratory, Publication No. 287 (January 1983).

37. F. Molenkamp et al., Cyclic filter tests in a triaxial cell, *Proc. 7th ECSMFE*, Vol. 2, Brighton, UK (1979).

38. N. N. Belyashevski et al., Behavior and selection of the composition of graded filters in the presence of fluctuating flow, *Gidrotechniskoe Stroiteltvo* **1979**(6) (1972).

39. H. R. Cedergren, Seepage requirements of filter and pervious bases, *Transactions ASCE* **127**(1), 1090–1113 (1962).

40. A. Silveira, An analysis of the problem of washing through in protective filters, *Proc. 6th Int. Conf. on SMFE*, Montreal, Canada (1965).
41. W. Thanikachalam and R. Sakthivadivel, Design of filter thickness based on application of queuing theory, *Journal of Hydrauilic Research* **13**(2), 207–219 (1975).
42. J. Pedersen, Wave forces and overtopping on crown walls of rubble mound breakwaters. PhD thesis, Series paper No. 12, Aalborg University, Department of Civil Engineering, Denmark (August 1996).
43. J. Van der Meer, Design of concrete armour layers, *Proc. Int. Conf. Coastal Structures'99*, Balkema, Rotterdam; Brookfield (2000).
44. J. Van der Meer, Stability of Cubes, Tetrapods and Acropodes, in *ICE, Design of Breakwaters, Proc. Conf. Breakwaters'88*, Thomas Telford, London (1988a).
45. S. Sigurdarson, G. Viggoson, S. Benedictson, S. Einarson and O. B. Smarason, Berm breakwaters, Fifteen years experience, *Proc. 26th Int. Conf. on Coastal Engineering*, ASCE, Copenhagen, Denmark (June 1998).
46. A. Tørum and S. R. Krogh, Berm breakwaters. Stone quality. SINTEF Report STF22 A00207 (July 2000).
47. A. Tørum, S. R. Krogh and S. Fjeld, Berm breakwaters: Stone breaking strength, *Journal of Waterway, Port, Coastal, and Ocean Engineering*, ASCE, **128**(4) (July 1, 2002).
48. A. Tørum, S. Bjørdal, M. Mathiesen, Ø. A. Arntsen and A. Jacobsen, Sirevåg berm breakwater. Comparison between physical model and prototype behavior. Report No. IBAT/MB R1, NTNU Norwegian University of Science and Technology, Depart of Civil and Transportation Engineering (December 16, 2003).
49. A. Tørum, S. Løset and H. Myhra, Sirevåg berm breakwater. Comparison between model and prototype behavior, *Proc. Int. Symp.*, Höfn, Iceland (June 2005).
50. M. N. Moghim, Experimental study of hydraulic stability of reshaping berm breakwaters. PhD thesis, Department of Civil Engineering, School of Engineering, Taribat Modares University, Tehran, Iran (in Persian) (2009).
51. M. N. Moghim, A. Tørum, Ø. Arntsen and B. Brørs, Stability of berm breakwaters for deep water, *Proc. 4th International Short Conference on Applied Coastal Research (SCACR)*, Barcelona, Spain (June 15–17, 2009).
52. J. Van der Meer, Verification of BREAKWAT for berm breakwaters and low crested structures. Delft Hydraulics Report H986. Prepared for CUR C67 (1990).
53. M. R. A. Van Gent, Wave interaction with permeable coastal structures. PhD Thesis, Delft University of Technology, ISBN 90-407-1128-8, Delft University Press, the Netherlands (1995).
54. R. Archetti and A. Lamberti, Parametrizzazione del profilo di frangiflutti berma, *Proc. Congresso AIOM*, Padova (October 3–5, 1996).
55. K. Hall and S. Kao, A study of the stability of dynamically stable breakwaters, *Canadian Journal of Civil Engineering* **18**, 916–925 (1991).

56. A. Tørum, On the stability of berm breakwaters in shallow and deep water, *Proc. 26th Int. Conf. on Coastal Engineering*, ASCE, Copenhagen, Denmark (June 1998).

57. A. Menze, Stability of multi layer berm breakwaters. Diploma thesis, University of Braunschweig, Braunschweig, Germany. Carried out at SINTEF/NTNU (2000).

58. A. Lamberti and G. R. Tomasicchio, Stone mobility and longshore transport at reshaping breakwaters, *Coastal Engineering* **29**, 263–289 (1997).

59. A. Alikhani, On reshaping breakwaters. PhD thesis, Aalborg University, Denmark (September 2000).

60. A. Tørum, S. Petkovic, A. Gürtner, O. T. Gudmestad, S. Løset and Ø. Arntsen, Ice and wave forces on shoulder ice barrier, *Proc. Int. Conf. on "Coastal Structures'07,"* Venice, Italy (June 2007).

61. A. Tørum, S. R. Krogh, S. Bjørdal, S. Fjeld, R. Archetti and A. Jacobsen, Design criteria and design procedures for berm breakwaters, in *Proc. Int. Conf. "Coastal Structures'99"*, ed. I. Losada (A. A. Balkema/Rotterdam/ Brookfield, Santander, Spain).

62. Th. Lykke Andersen, Hydraulic response of rubble mound breakwaters. Scale effects — berm breakwaters. PhD thesis, *Hydraulics & Coastal Engineering*, Department of Civil Engineering, Aalborg University, Denmark (2006).

63. A. Tørum, Some considerations on the probability of failure of rubble mound breakwaters obtained by Monte Carlo simulations. Norwegian University of Science and Technology, Department of Civil and Transport Engineering. Report under preparation (2010).

64. A. Tørum, Berm breakwaters revisited. Internal note, Department of Civil and Transport Engineering, Norwegian University of Science and Technology, Trondheim, Norway (2007).

65. S. Sigurdarson, G. Viggoson, A. Tørum and O. B. Smarason, Stable berm breakwaters, *Proc. Int. Workshop on Advanced Design of Maritime Structures in the 21st Century*, Port and Harbor Reseach Institute, Yokusuka, Japan (March 5–7, 2001).

66. J. Van der Meer and J. J. Veldman, Singular points at berm breakwaters: Scale effects, rear, round head and longshore transport, *Coastal Engineering*, **17**(3, 4) (1992).

67. A. Tørum, Berm breakwaters. EU MAST II Berm breakwater structures. SINTEF Report No. STF22 A97205, Trondheim, Norway (1997).

68. J. Van der Meer and J. P. F. M. Jansen, Wave run-up and overtopping at dikes, in *Wave forces on inclined and vertical wall structures* (ASCE, 1995).

69. J. De Rouck, R. Verdonk, P. Trock, L. van Damme, F. Schlütter and J. de Ronde, Wave run-up and overtopping: Prototype versus scale model, *Proc. 26th Int. Conf. on Coastal Engineering*, Copenhagen, Denmark (June 22–26, 1998).

70. J. van der Meer, Wave run-up and oveertopping, *Dikes and Revetments*, Chapter 8 (A. A. Balkema, 1998), pp. 145–159.

71. J. De Rouck, P. Trock, B. van e Walle, M. van Gent, L. van Damme, J. de Ronde, P. Frigaard and J. Murphy, Wave run-up on sloping coastal structures — prototype measurements versus scale model tests, *Proc. Int. Conf. on "Breakwaters, Coastal Structures and coastlines,"* Institution of Civil Engineers, London, UK (September 26–28, 2001), Tomas Telford, 2002.

72. T. Pullen, N. W. H. Allsop, T. Bruce, A. Kortenhaus, H. Scüttrumpf and J. W. van der Meer, *EuroTop. Wave Overtopping of Sea Defences and Related Structures: Assessment Manual* (Environment Agency, UK, German Coastal Engineering Research Council, Germany and Rijkswaterstaat, Netherlands Expertise Network for Flood Protection, The Netherlands, 2007).

73. Y. Goda, *Random Waves and Design of Maritime Structures*, 2nd Edn. (World Scientific, Singapore, 2000), p. 443.

74. OCDI *Technical Standards and Commentaries for Port and Harbour Facilities in Japan* (The Overseas Coastal Area Development Institute of Japan, Tokyo, 2002), p. 600.

75. S. Takahashi, T. Tanimoto and K. Shimosako, A proposal of impulsive pressure coefficient for the design of composite breakwaters, *Proc. Int. Conf. on Hydro-Technical Engineering for Port and Harbour Construction (Hydro-Port'94)*, Port and Harbour Res. Inst., Yokosuka, Japan (1994a), pp. 489–504.

76. T. Takayama and N. Ikeda, Estimation of sliding failure probability of present breakwaters for probabilistic design, *Report of the Port and Harbour Research Institute*, Yokosuka, Japan, **31**(5), 3–32 (1993).

77. H. Oumeraci, A. Kortenhaus, W. Allsop, M. de Groot, R. Crouch, H. Vrijling and H. Voortman, *Probabilistic Design Tools for Vertical Breakwaters* (A. A. Balkema Publishers/Lisse/Abingdom/Exton (PA)/Tokyo, 2001).

78. S. Takahashi, K. Tanimoto and K. Shimosako, Hydraulic characteristics of a sloping top caisson — Wave forces acting on the sloping caisson, *Proc. Int. Conf. on Hydro-Technical Engineering for Port and Harbor Construction, (Hydro-Port'94)*, Port and Harbour Res. Inst., Yokosuka, Japan (1994b).

79. G. D. Khaskhachich and D. M. Vanchagov, Effect on waves on walls made of circular cells, *ASCE* (WWW), 735–754 (1971).

80. R. Silvester, *Coastal Engineering, I. Generation, Propagation and Influence of Waves* (Elsevier, Amsterdam/London/New York, 1974).

81. A. Jahr, Composite breakwater with circular caissons. Master thesis, Norwegian University of Science and Technology, Department of Civil and Transport Engineering (2010).

82. K. Tanimoto, T. Yagyu and Y. Goda, Irregular wave tests for composite breakwater foundations, *Proc. 18th Conf. on Coastal Engineering*, ASCE, Cape Town, South Africa (1982), pp. 2144–2163.

83. K. Kimura, S. Takahashi and K. Tanimoto, Stability of rubble mound foundations of composite breakwaters under oblique wave attack, *Proc. 24th Int. Conf. on Coastal Engineering*, Kobe, Japan (1994), pp. 1227–1240.

84. C. T. Stansberg, A. Tørum and S. Næss, On a model study of a box type floating breakwater. Wave damping and mooring forces, *Proc. PIANC Conf.*, Osaka, Japan (1990).

85. C. T. Stansberg, Motions of large floating structures moored in irregular waves: Experimental study, *Proc. Int. Workshop on Very Large Floating Structures*, Hayama, Japan (November 25–28, 1996).

86. C. T. Stansberg, J. R. Krokstad and O. H. Slåttelid, Model tests on non-linear slow-drift oscillations compared to numerical and analytical data, *Proc. BOSS'88 (Behavior of Offshore Structures) Conf.*, Trondheim, Norway (1988), pp. 667–686.

87. D. Kriebel, Performance of vertical wave barriers in random seas, *Proc. of the Int. Conf. Coastal Structures '99*, A. A. Balkema/Rotterdam/Brookfield, Sanatander, Spain (June 7–10, 1999).

88. D. Kriebel, Vertical wave barriers: Wave transmission and wave forces, *Proc. 23rd Int. Conf. on Coastal Engineering*, ASCE (1992), pp. 1313–1326.

89. H. Bergmann and H. Oumeraci, Wave pressure distribution on permeable vertical walls, *Proc. 26th Intl. Conf. on Coastal Engineering*, ASCE (1998), pp. 2042–2055.

90. D. Kriebel, C. Sollitt and W. Gerkin, Wave forces on a vertical wave barrier, *Proc. 26th Int. Conf. on Coastal Engineering*, Copenhagen, Denmark, ASCE (1998), pp. 2069–2081.

91. J. Gilman and D. Kriebel, Partial depth pile supported wave barriers: A design procedure, *Proc. Coastal Structures '99 Conference*, ed. I. Losada (A. A. Balkema Publishers, Rotterdam, 1999), pp. 549–558.

92. J. Grune and S. Kohlhause, Wave transmission through slotted walls, *Proc. 14th Int. Conf. on Coastal Engineering*, ASCE (1976).

93. I. Losada, M. Losada and A. Roldan, Propagation of oblique incident waves past rigid vertical thin barriers, *Applied Ocean Research* **14**, 191–199 (1992).

94. I. Losada, M. Losada and R. Losada, Wave spectrum scattering by vertical thin barriers, *Applied Ocean Research* **16**, 123–128 (1994).

95. ISO 21650 Actions from Waves and Currents on Coastal Structures (2007).

96. T. Sarpkaya, In-line and transverse forces in oscillatory flow at high Reynolds numbers, *Proc. 8th Offshore Technology Conf.*, Paper No. OTC 2533, Houston, TX, USA (1976).

97. T. Sarpkaya, Vortex shedding and resistance in harmonic flow about smooth and rough circular cylinders at high Reynolds numbers. Report No. NPS-59SL, Naval Postgraduate Scholl, Monterey, CA, USA (1976).

98. NORSOK Standard N-003. Action and action effects. Norwegian Technology Standards Institution (now Standard Norway), Oslo, Norway (1999).

99. DNV (Det Norske Veritas) Environmental conditions and environmental loads. DNV Classification notes No. 30.5 (1991).

100. ISO 19902 Petroleum and Natural Gas Industries — Fixed Steel Offshore Platforms (2001).

101. API RP 2A WSD, Recommended Practice for Planning, Designing and Constructing Fixed Offshore Platforms — Load and Resistance Factor Design. 1st Edn. (American Petroleum Institute, Washington, DC, USA, 1993).

102. O. T. Gudmestad and G. Moe, Hydrodynamic coefficiens for calculations of hydrodynamic loads on offshore truss structures, *Marine Structures* **9**, 745–758 (1996).

103. G. Moe and O. T. Gudmestad, Prediction of Morison-type forces in irregular waves at high Reynolds numbers, *International Journal of Offshore and Polar Engineering* **8**(4) (1998).

104. R. C. MacCamy and R. A. Fuchs, Wave forces on piles: A diffraction theory, Tech. Memo 69, Beach Erosion Board (1954).

105. D. L. Kriebel, E. P. Berek, S. K. Chakrabarti and J. K. Waters, Wave-current loading on a shallow water caisson. Preprints Offshore Technology Conference, Houston, Texas, USA (May 6–9, 1996). OTC Paper No. 8067.

106. R. G. Dean, R. A. Dalrymple and R. T. Hudspeth, Force coefficients from wave project I and II data, including free surface effects, *Society of Petroleum Engineers Journal* **12**, 779–786 (1981).

107. A. Tørum, Wave forces on pile in the surface zone, *Journal of Waterway, Port, Coastal and Ocean Engineering* **115**(4) (July 1989).

108. R. G. Dean and R. A. Dalrymple, *Water Wave Mechanics for Engineers and Scientists* (World Scientific, Singapore, 1993).

109. O. M. Faltinsen, *Sea Loads on Ships and Offshore Structures*. (Cambridge Ocean Technology Series, Cambridge University Press, Cambridge, 1990).

110. H. Gjøsund, S. Løset and A. Tørum, Wave loads on an offshore oil terminal in shallow water, *Proc. Int. Conf. on Port and Ocean Engineering under Arctic Conditions (POAC)*, Trondheim, Norway (June 16–19, 2003).

111. J. Wienke, Druckschlagbelastung auf Schlanke zylindrische Bauwerke durch brechender Wellen. — theoretishe und grossmasstäblishe Labountersuchungen (Slamming forces from breaking waves on slender cylinders — theoretical and large scale laboratory investigations). Dr. Ing. Dissertation, Technischen Universtät Carolo-Wilhelmina zu Braunschweig (2001) (in German).

112. J. Wienke and H. Oumeraci, Breaking wave impact force on a vertical and inclined slender pile, *Coastal Engineering* **52**, 435–462 (2004).

113. V. Lie and A. Tørum, Waves over shoals, *Coastal Engineering* **15**(5, 6) (1991).

114. A. Grønsund Hanssen, Bølgekrefter på trefoting. (Wave forces on a tripod). Master thesis, NTH, (1994).

115. A. Grønsund Hanssen and A. Tørum, Breaking wave forces on tripod concrete structure on shoal, *Journal of Waterway, Port, Coastal and Ocean Engineering*, **125**(6) (1999).

116. Y. Goda, Wave forces on circular cylinders erected upon reefs. Coastal Engineering in Japan, 16, Tokyo (1973).

117. S. I. Hovden, Bølgekrefter på ein vertikal sylinder på ei grunne. (Wave forces on a vertical cylinder upon a shoal) Master thesis NTH, Norway (1990).

118. S. I. Hovden and A. Tørum, Wave forces on a vertical cylinder on a reef, *Proc. 3rd International Conference on Port and Coastal Engineering for Developing Countries*, Mombasa, Kenya (1991).

119. A. Kyte and A. Tøum, Wave forces on vertical cylinders upon shoals, *Coastal Engineering* **27**, 263–286 (1996).

120. N. Janbu, *Slope Stability Computations*. Embankment-dam Engineering Casagrande volume (John Wiley and Sons, New York, 1972), pp. 377–393.

121. J. De Rouck and P. Troch, Pore water pressure response due to tides and waves based on prototype measurements. PIANC Bulletin No. 110 (April 2002), pp. 9–31.

122. Ø. Arntsen, I. J. Malmedal, B. Brørs and A. Tørum, Numerical and experimental modeling of pore pressure variation within a rubble mound breakwater, *Proc. 17th Int. Conf. on Port and Ocean Engineering under Arctic Conditions (POAC'03)*, Trondheim, Norway (June 16–19, 2003).

123. R. Sandven, E. Husby, J. E. Husby, K. O. Jønland, Roksvåg, F. Stæhli and R. Tellugen, Development of a sampler for measurement of gas content in soils, *Journal of Waterway, Port, Coastal, and Ocean Engineering ASCE.* **133**(1) (2007).

124. J. De Rouck, De stabiliteit van storsteengolfbrekers. (Stabilty of rubble mound breakwaters. Slope stability analysis; a new armor unit). Dr thesis Katholieke Univesiteit Leuven, Faculteit Toegepaste Wetenschappen, Departemente Burgerlijke Bouwkunde, Labratorium voor hydraulica, Leuven, Belgium (1991).

125. J. De Rouck and L. van Damme, Overall stability analysis of rubble mound breakwaters, *Proc. 25th Int. Conf. on Coastal Engineering*, Orlando, Florida, USA, ASCE (1996).

126. C. C. Mei and M. A. Foda, Wave-induced responses in a fluid-filled poro-elastic solid with a free surface — a boundary layer theory, *Geophysical Journal of the Royal Astronomical Society* **66**, 597–637 (1981).

127. J. H. Prevost, Mechanics of continuous porous media, *Journal of Engineering Science* **18**, 787–800 (1980).

128. C. C. Mei, *The Applied Dynamics of Ocean Surface Waves*. Advanced Series on Ocean Engineering, Volume 1 (World Scientific, Singapore, 1989).

129. D. Bonjean, P. Foray, I. Piedra-Cueva, H. Michaller, P. Breul, Y. Haddani, M. Mory and S. Abadie, Monitoring of the foundations of a coastal structure submitted to breaking waves: Occurrence of momentary liquefaction, *Proc. 14th Int. Offshore and Polar Engineering Conf.*, Toulon, France (May 23–28, 2004).

130. M. Hovland, D. Orange and O. T. Gudmestad, Gas hydrates and seeps — Effects on slope stability: The "hydraulic model". *Proc. 11th Int. Conf. on Offshore and Polar Engineering Conference*, Stavanger, Norway (June 17–22, 2001).

131. Wm. Sackinger, Ice action against rock mound structure slopes, in *Design and Construction of Mounds for Breakwaters and Coastal Protection*, ed. Per Bruun (Elsevier, Amsterdam, 1985).

132. A. T. Chen and C. B. Leidersdorf (eds.), *Arctic Coastal Processes and Slope Protection Design* (Technical Council on Cold Regions Engineering Monograph. ASCE, 1988).

133. G. W. Timco, D. H. Willis and B. D. Wright, Ice action on armor rocks with application to an artificial island concept, *Proc. 2nd Int. Conf. on Development of Russian Arctic Offshore, RAO'95*, St. Petersburg, Russia (1995).

134. M. D. Coon, Scale factors for distorted ice model tests, *Proc. 7th Int. Conf. on Port and Ocean Engineering under Arctic Conditions (POAC)*, Helsinki, Finland, ESPOO (1983).

135. S. H. Iyer, Size effects in ice and their influence on the structural design of offshore structures, *Proc. 7th Int. Conf. on Port and Ocean Engineering under Arctic Conditions (POAC)*, Helsinki, Finland, ESPOO (1983).

136. S. K. Løseth, O. T. Shkhinek, P. Gudmestad, E. Strass, R. Michalenko, Frederking and T. Kärnä, Comparison of the physical environment of some Arctic seas, *Cold Regions Science and Technology* **3**(29), 204–214 (1999).

137. API RP 2N, Recommended Practice for Planning, Designing and Constructing Offshore Structures in Ice Environment, 1st Edn. (American Petroleum Institute, Washington DC, USA, 1988).

138. T. Vinje, The physical environment, Western Barents Sea. Drift, composition, morphology distribution of the sea ice fields in the Barents Sea. In Norsk Polarinstitutt Skrifter, Norway, Nr. 179 (1985), p. 26.

139. A. Egorov, V. Spichin and V. S. Mironov, Estimation of the average and maximal level ice thickness in the Barents and Kara Seas, *Proc. 2nd International Conference on Development of the Russian Arctic Offshore (RAO-95)*, St. Petersburg (1995).

140. T. Sanderson, *Ice Mechanics: Risk to Offshore Structures* (London, 1988).

141. K. Croasdale, Ice engineering for offashore petroleum exploration in Canada, *The 4th Int. Conf. on Port and Ocean Engineering under Arctic Conditions (POAC)*, New Foundland, (September 26–30, 1977), pp. 1–32.

142. E. Mironov, V. A. Spinchkin and A. Egorov, Season variability and their variations in the region of mastering of the Barents and Kara seas offshore, *Proc. 1st Int. Conf. on Development of Russian Arctic Offshore, RAO'93*, St. Petersburg (September 21–24, 1993), pp. 439–449.

143. K. Riska, Ice conditions along the North-East passage in view of ship traffic ability studies, *Proc. 5th Int. Offshore and Polar Engineering Conference*, The Hague, Vol. II, pp. 420–427 (1995).

144. R. Ettema, M. G. Horton and J. F. Kennedy, Ice study for the port development at Nome, Alaska, *Proc. 7th Int. Conf. on Port and Ocean Engineering under Arctic Conditions (POAC)* — VTT Symp. 28, Technical Research Centre of Finland, Espoo, Helsinki, Finland (April 5–9, 1983).

145. P. Perdichinni, M. G. Horton and R. Ettema, Planning and design of a medium draft port in Nome, Alaska, *Proc. 7th Int. Conf. on Port and Ocean Engineering under Arctic Conditions (POAC)* — VTT Symp. 28, Technical Research Centre of Finland, Espoo, Helsinki, Finland (April 5–9, 1983).

146. K. J. MacIntosh, G. W. Timco and D. H. Willis, Canadian experience with ice and armour stone, *Proc. 1995 Canadian Coastal Conf.*, Vol. 2, Nova Scotia (October, 1995), pp. 597–606.

147. K. J. MacIntosh, Ice interaction with rubble mound breakwater, W. F. Baired & Associates Coastal Engineerings Limited, Report to Small Craft Harbours Directorate of Fisheries and Oceans, and Master theis, Carleton University, 121 pp., Ottawa, Ont., Canada.

148. D. S. Sodhi, S. L. Borland and J. M. Stanley, Ice action on riprap. Small-scale tests. US Army Corps of Engineers, Cold Regions Research & Engineering Laboratory, CRREL Report 96-12 (1996).

149. K. R. Croasdale, N. Allyn and W. Roggensack, Arctic slope protection: Considerations for ice, In *Arctic Coastal Processes and Slope Protection Design*, eds. A. T. Chen and C. D. Leidersdorf (Technical Council on Cold Regions Engineering Monograph, ASCE, 1988).
CUR, *Manual of the Use of Rock in Hydraulic Engineering* (A. A. Balkema, Rotterdam/Brookfield, 1995).

150. A. Tørum, Breakwaters in Arctic areas. Review and research needs. Norwegian University of Science and Technology, Department of Civil and Transport Engineering. Open report to Statoil, Report no. BAT/MB-R2/2004 (December 14, 2004).

151. A. Gürtner, Experimental and Numerical Investigations on Ice-Structure Interaction. PhD thesis, Norwegian University of Science and Technology, Department of Civil and Transport Engineering (January 2009).

152. S. F. Daly, J. Zufelt, L. Zabiinski, D. Sodhi and K. Bjella, Estimation of ice impacts on armor stone revetments at Barrow, Alaska, *Proc. 19th IAHR Int. Symp. on Ice*, Vancouver, British Columbia, Canada (July 6–11, 2008).

153. T. Kitamura, M. Sato, M. Kato, Y. Watanabe and H. Saeki, The effect of drift ice on the stability of armor stones covering a doubly placed submerged breakwater in a wave field, *Proc. 15th Int. Symp. on Ice*, IAHR. Gdansk, Poland (August 28 to September 1, 2000).

154. S. Kunimatsu, Study of size of drift ice at the Okhotsk Sea of Hokkaido, *Proc. Civil Engineering in the Ocean* **9**, 95–100, (1993).

155. H. J. Lengkeek, K. R. Croasdale and M. Metge, Design of ice protection barrier in Caspian Sea, *Proc. 22nd Int. Conf. on Offshore Mechanics and Arctic Engineering (OMAE03)*, Cancun, Mexico (June 8–13, 2003).

156. A. Barker, G. Timco and M. Sayed, Numerical simulations of the broken ice zone around Molikpaq: Implication for safe evacuations, *Proc. 16th Int. Conf. on Port and Ocean Engineering under Arctic Conditions (POAC)*, Ottawa, Canada (2001), pp. 505–515.

157. J. L. Wuebben, Ice effects on riprap, in *River and Shoreline Protection: Erosion Control Using Riprap and Armourstone*, John Wiley & Sons. Ltd. (1995).

158. A. Gürtner, Field development in the Northern Caspian Sea — establishment of ice loads on offshore structures and ice mitigation measures in this area. Master's thesis at the University of Stavanger (2005).

159. A. Gürtner, O. T. Gudmestad, A. Tørum and S. Løset, Innovative ice protection for shallow water drilling. Part I: Presentation of the Concept, *Proc. of the 25th Int. Conf. on Offshore Mechanics and Arctic Engineering*, OMAE 2006 (2005).

160. K. U. Evers and P. Jochmann, An advanced technique to improve the mechanical properties of model ice developed at the HSVA Ice Tank, *Proc. 12th Int. Conf. on Port and Ocean Engineering under Arctic Conditions (POAC)*, Hamburg (August 17–20, 1993), pp. 877–888.

161. J. Schwarz, New development in modeling ice problems, *Proc. 4th Conf. on Port and Ocean Engineering under Arctic Conditions (POAC)*, St. Johns, Canada (1977), pp. 73–82.

162. M. Bjerkås, Review of measured full scale ice forces to fixed structures, *Proc. 26th Int. Conf. on Offshore Mechanics and Arctic Engineering*, San Diego, California, USA, OMAE2007–29048 (2007).

163. A. Gürtner and O. T. Gudmestad, Innovative ice protection for shallow water drilling. Part II: SIB model testing in ice, *Proc. 27th Int. Conf. on Offshore Mechanics and Arctic Engineering, OMAE2008*, Estoril, Portugal (June 15–20, 2008).

164. S. Petkovic, Wave forces on the Shoulder Ice Barrier (SIB). Master thesis, Norwegian University of Science and Technology, Department of Civil and Transport Engineering (June 26, 2006).

165. A. Tørum, S. Petkovic, A. Gürtner, O. T. Gudmestad, S. Løset and Ø. Arntsen, Ice and wave forces on shoulder ice barrier, *Proc. Int. Conf. "Coastal Structures'07"*, Venice, Italy (June 2007).

166. A. A. Malyutin, A. R. Gintovt, Y. Y. Toporov and V. A. Chernetsov, Off-shore platforms for oil and gas production on the Russian arctic shelf, *Proc. 17th Int. Conf. on Port and Ocean Engineering under Arctic Conditions, POAC'03*, Trondheim, Norway (June 16–19, 2003).

167. I. V. Lavrenov, V. N. Bokov, T. A. Pasechnik, V. I. Dymov, N. P. Yakovleva and I. N. Davdan, Estimation of the extreme heights of wind waves in the Arctic Sea, *Proc. OMEA'99, 18th Int. Conf. on Offshore Mechanics and Arctic Engineering*, St. Johns, New Foundland, Canada (June 11–16, 1999). OMEA99/P&A-1105.

168. Sandwell Engineering Inc. Comparison of international codes for ice loads on offshore structures. Report to National Research Council of Canada, Ottawa, Ontario. PERD/CNC Report 11–20. Prepared by Sandwell Engineering Inc., Vancouver, Canada, and Central Marine Research & Design Institute, St. Petersburg, Russia. Project 142211 (May 31, 1998).

169. O. Smarason, S. Sigurdarson and G. Viggoson, Quarry yield prediction as a tool in breakwater design. Key note lectures NMG-2000 and 4th GIGS, Finish Geotechnical Society, Helsinki, Finland (2000).

170. H. Leeuwestein, A. Franken, V. Homebergen and J. K. Vrijling, Quarry-based design of rock structures, in *River and Shoreline Protection: Erosion Control using Riprap and Armour Stone*, eds. C. R. Thorne, S. R. Abt,

F. B. J. Barends, S. T. Maynord, and K. W. Pilarczyk (John Wiley & Sons Ltd, New York, 1995).

171. CIRIA-CUR, Manual on the use of rock in coastal and shoreline engineering. CIRIA Special Publication 83 — CUR Report 154 (CIRIA = Construction Industry Research and Information Association, UK. CUR = Centre for Civil Engineering Research and Codes, The Netherlands, 1991). This manual was revised in 2004.

172. B. Hardarson, Study of Icelandic Basalt for the use in Breakwaters. B.Sc. thesis, University of Iceland (1979).

173. O. Smarason, Laukvik in Lofoten. Quarry for extra large armor stone. Report to the Norwegian Coastal Administration from STAPI, Consulting Geologists (August 2005).

174. P. Frigaard, T. Hald, H. F. Burcharth and S. Sigurdarson, Stability of reshaping breakwaters with special reference to stone durability, *Proc. 22nd Int. Conf. on Coastal Engineering, ASCE* (1996), pp. 1640–1651.

175. H. Hertz, *J. Math.* (Crelle's J), 92 (1881). (See Timoshenko and Goodier 1970).

176. S. P. Timoshenko and J. N. Goodier, *Theory of Elasticity*, 3rd Edn. (McGraw-Hill, New York, 1970).

177. R. Archetti and A. Lamberti, Stone movement and stresses during reshaping of berm breakwaters, *Proc. COPEDEC'99*, ed. G. P. Mocke (Creda Communications, Cape Town, South Africa), pp. 1550–1561.

178. J.-P. Latham and G. A. Gauss, The drop test for armor stone integrity, in *Proc. Conf. "River, Coastal and Shoreline Protection,"* eds. E. R. Thorne, F. B. J. Bahrends, S. T. Maynord and K. Pilarczyk (Wiley, Chichester, 1995), pp. 481–499.

179. S. Krogh, Determination of crushing and grinding characteristics based on testing of single particles, *Trans. AIME* **266**, 1957–1962 (1980).

180. S. Krogh, Crushing characteristic. *Powder Technology* **27**, 71–181 (1980).

181. S. Krogh, Energy model for crushing single particles, *Proc. Mineral Technology Conf.*, Swedish Mineral Processing Research Association, Stockholm, Sweden (1999).

182. R. E. Melchers, *Structural Reliability — Analysis and Prediction* (Ellis_Horwood, Chichester, 1987).
    G. Moe and O. T. Gudmestad, Prediction of Morison-type forces in irregular waves at high Reynolds numbers, *International Journal of Offshore and Polar Engineering* **8**(4), (1998).

183. S. P. Timoshenko and D. H. Young, *Engineering Mechanics* (McGraw-Hill, New York, 1951).

184. M. Mikos and M. N. R. Jaeggi, Experiments on motion of sediment mixtures in a tumbling mill to study fluvial abrasion, *Journal of Hydraulic Research* **33** (1995).

185. A. Tørum, Abrasion mill test results in relation to stone movement of berm breakwaters. Report No. IBAT/MB R2/2003. Norwegian University

of Science and Technology, Department of Civil and Transport Engineering, Trondheim, Norway (2003).

186. R. Archetti, A. Lamberti, G. R. Tomassicchio, M. Sorci, S. Sigurdarson, S. Erlingson and O. B. Smarason, On the application of a conceptual abrasion model on six Icelandic breakwaters, *Proc. 28th Int. Conf. on Coastal Engineering*, ASCE (Cardiff, UK, June 2002).

187. K. Westeng, Rubble mound and berm breakwater. Emphasis on Sirevåg and Laukvik. Project work, Norwegian University of Science and Technology, Department of Civil and Transport Engineering (Fall Semester 2009).

188. J. Gilman, St. George harbor — berm breakwater construction issues, Paper presented at the Fifth Regional Symposium on Pacific Congress on Marine Science and Technology (PACON); San Francisco, California, USA (July 8–11, 2001).

189. S. Sigurdarson, B. J. Bjørnson, J. Skulason, G. Viggoson and K. Helgason, A berm breakwater on weak soil, Extension of the port of Hafnarfjördur, *Proc. COPEDEC*, Cape Topwn, South Africa, pp. 1395–1406.

190. H. F. Burcharth, Reliability evaluation of structures at sea, *Proc. Int. Workshop on "Wave barriers in deep waters,"* Port and Harbor Research Institute, Yokosuka, Japan (1994).

191. H. F. Burcharth, Verification of overall safety factors in deterministic design of model tested breakwaters, *Proc. Int. Workshop on Advanced design of Maritime Structures in the 21st century*, Port and Harbor Research Institute, Yokosuka, Japan (5–7 March 2001).

192. A. Næss, O. Gaidai and S. Haver, Efficient estimation of extreme response of dragdominated offshore structures by Monte Carlo simulations, *Ocean Engineering* **34**, 2188–2197 (2007).

193. A. Næss, B. Leira and O. Batsevuch, System reliability analysis by Monte Carlo simulations, *Structural Safety* **31**, 349–355 (2009).

194. M. Mathiesen, Sirevåg harbor. Wave analysis. SINTEF Report No. STF60 F95029 (in Norwegian), SINTEF, Trondheim, Norway (1999).

195. MATLAB *MATLAB Reference Guide* (MathWorks, Inc., Natick, MA, USA, 1992).

196. J. K. Vrijling, W. van Hengel and R. J. Houben, A frame work for risk evaluation, *Journal of Hazardous Materials*, Amsterdam, The Netherlands, Elsvier Science B. V. Vol. 43, pp. 245–261 (1995).

Chapter 4

Kaumālapa'u Harbor: Design and Construction Challenges of an Exposed Deepwater Breakwater

Scott P. Sullivan

Sea Engineering, Inc., Makai Research Pier, Waimānalo,
Hawai'i, USA 96795
ssullivan@seaengineering.com

Kaumālapa'u Harbor is a small barge harbor located on the southwest coast of the island of Lāna'i in the state of Hawai'i, USA. The harbor was protected by an old, deteriorated, rock rubblemound breakwater, originally constructed in 1925. The breakwater was recently completely rebuilt by the US Army Corps of Engineers, Honolulu District, in association with the state of Hawai'i. Oceanographic design considerations included a water depth of about 21 m (70 ft) and direct exposure to possible 10.7 m (35 ft) hurricane strength storm waves. The design process involved both numerical models and 2-D and 3-D physical models. The repairs utilized an armor layer of 31.8 tonne (35 ton) Core-Loc^{TM} concrete armor units, the largest units of this type used to date. The use of Core-Loc units, a relatively new armor unit type, presented both design and construction challenges to insure that their performance capability was maximized. Particular attention was placed on developing a Core-Loc placement plan to achieve the required unit packing density, and utilizing rigorous construction techniques to achieve the specified armor layer placement. This project contributed significantly toward furthering design and construction guidelines for Core-Loc use. Construction was completed in 2007, and the project won the ASCE (Hawai'i Section) 2008 Outstanding Civil Engineering Achievement Award.

4.1. Introduction and General Description

Kaumālapa'u Harbor is a small barge harbor located on the southwest coast of the island of Lāna'i in the state of Hawai'i, USA (see Fig. 4.1). Lāna'i is the sixth largest island in the state, covering about 140 square miles with about 2% of the state's land area. The island has about 2,500 full-time residents, most of them living in or around the island's only town, Lāna'i City. Almost the entire island is privately owned by Castle & Cooke, Inc., the third largest private landowner in the state.

Kaumālapa'u Harbor was constructed in 1925 by the Hawaiian Pineapple Company (Dole Company) for the export of pineapple, the island's primary product and business up until the early 1990s. The harbor is located in a small embayment providing a 4-hectare (10-acre) berthing area with water depths of 6.1 to 18.3 m (20 to 60 ft). A breakwater was constructed extending toward the south from the northwest headland of the embayment. The original length of the breakwater was about 107 m (350 ft), with a crest elevation of +3 m (+10 ft) mean lower low water (MLLW) or less. The breakwater was constructed of quarried stone and field stone from the island. Shoreside facilities on the north side of the embayment in the lee of the breakwater include a 122-m (400 ft) long wharf, sheds, and barge loading and unloading equipment.

In the early 1990s, the growing of pineapple was terminated, two luxury resort hotels were built, and the island changed from an agricultural economy to an economy based on tourism. Primary use of the harbor changed from the export of pineapple to the import of fuel and goods to support the new economic base industry. The State Department of Transportation, Harbors Division, now operates the harbor as part of the statewide harbor system.

It is not known whether the original breakwater structure was an "engineered" design, or simply a rubblemound constructed of a random mix of available stone. Even if it was an engineered structure, it would not have met the criteria required by modern breakwater design practice, nor would sufficient information regarding oceanographic design conditions have been available to the designer. The breakwater suffered extensive damage over the years, and has been repaired numerous times. Maintenance of the breakwater appears to have been done using rock, concrete rubble, concrete-filled pineapple wagons, and even several concrete-filled school buses. A severe storm resulting from a low-pressure front southwest of Hawai'i severely damaged the breakwater in January 1980, and Hurricane Iwa caused further damage in 1982. Repairs were made using dolos concrete armor units, however the armor unit size using existing forms available in Hawai'i was apparently too small for the design conditions as the units were quickly

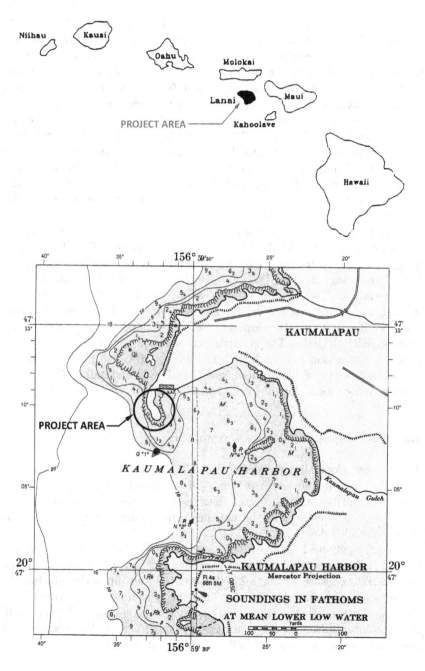

Fig. 4.1. Location map.

broken and turned into concrete rubble. In 1992, Hurricane 'Iniki badly damaged the breakwater, and only a portion of the structure remained above water level. Seventy-five years of existence, storm events, and repairs resulted in a large, broad rubblemound, with a side slope of about 1 vertical on 1.5 horizontal (1V:1.5H). By the 1990s, the deteriorated condition of the breakwater permitted significant wave energy to enter the harbor and reach the wharf, resulting in berthed vessel motion that rendered cargo handling and fuel off-loading difficult, and at times hazardous or impossible. The purpose of the Kaumālapa'u Harbor breakwater project was to repair the existing structure in order to reduce wave energy in the harbor and increase harbor safety and usability. The breakwater was rebuilt on the footprint of the old rubblemound structure, and used the existing material for the core of the new breakwater.

The project was designed and constructed by the US Army Corps of Engineers, Honolulu District, for the state of Hawai'i. Detailed project design and construction plans and specifications were completed by Sea Engineering, Inc. (SEI) under contract with the Honolulu District administration. The repairs utilized an armor layer of CORE-LOC^{TM} (Core-Loc) concrete armor units, recently developed by the US Army Engineer Research and Development Center, Coastal and Hydraulics Laboratory (ERDC/CHL). The construction was completed in June 2007 at a final construction cost of US\$21,229,000. In 2008 the project won the American Society of Civil Engineers (ASCE)'s Outstanding Civil Engineering Achievement Award in the Hawai'i section.

4.2. Oceanographic Design Parameters

The bottom slope directly seaward of the breakwater is about 1V:10H, with the toe of the old rubblemound breakwater in 21 m (70 ft) of water. A multibeam sonar system was used to accurately map the ocean bottom, the shoreline, and the existing rubblemound topography at the start of detailed design engineering.

The project site has direct exposure to long period southern swell and westerly waves generated by low pressure systems, passing through Hawai'i, and partial exposure to north swell refracting and diffracting during the winter season around the islands. The site is completely sheltered from the prevailing northeast tradewind waves by the island of Lāna'i itself. Strong winter storms in the southern hemisphere create large waves that propagate across the central Pacific. Traveling distances of up to 8,000 km

(5,000 miles), these waves arrive with relatively low deepwater wave heights of 0.3 to 1.2 m (1 to 4 ft) and periods of 14 to 20 s. On occasion, deepwater wave heights reach 2 m (6 ft) with peak periods of 25 s. Westerly storm waves are associated with winter low-pressure systems passing through the islands, with strong southwesterly winds and deepwater heights up to about 4.6 m (15 ft) and periods of 6 to 10 s. During the winter months in the northern hemisphere strong storms in the North Pacific and Aleutian Islands generate large swell approaching from the northwest to northeast, and which arrives in Hawai'i with little attenuation of wave energy. These are the waves that have made Hawai'i's north shore surfing beaches world famous. Deepwater wave heights often reach 4.6 m (15 ft) and occasionally 9 m (30 ft), with periods between 12 and 20 s. Kaumālapa'u Harbor is sheltered from the direct approach of north swell by the islands to the northwest, but some of their energy refracts and diffracts around the islands to reach the site.

Although hurricanes occur very infrequently in Hawaiian waters, Kaumālapa'u is potentially directly exposed to hurricane strength storm waves approaching from the southeast to southwest. Hurricane Iwa (1982) and hurricane Iniki (1992) are the only two hurricanes to strike the Hawaiian Islands in the past 50 years, and both passed west of Kaumālapa'u and struck the island of Kaua'i. Wave hindcasts for these two storms showed deepwater wave heights offshore of Kaumālapa'u of 6.7 and 6.1 m (22 and 20 ft) for hurricanes Iwa and Iniki, respectively. Predicted scenario hurricane parameters were used to determine extreme design waves for Kaumālapa'u. Numerical model generated deepwater wave heights of 10.7 m (35 ft) plus are estimated to result from a direct hurricane strike on the project site.

In general, wave conditions at the project site are moderate, and an incident wave height of about 4.6 m (15 ft) was considered reasonable as an upper limit for harbor operating conditions, after which surge conditions at the dock would preclude vessel berthing and cargo handling. A hurricane event, with deepwater wave heights of 10.7 m (35 ft), was selected for structural design of the new breakwater.

The tide range in Hawai'i is low, 0.7 m (2.2 ft) between mean lower low and mean higher high water at the site. (Note: all elevations are referenced to the mean lower low water datum.) A stillwater level rise of +1.5 m (+5 ft) is estimated during hurricane conditions, including a mean high water tide, a rise due a drop in atmospheric pressure and storm surge due to wind stress. In Hawai'i, with deepwater extending very close to shore, the effects of wind stress surge are almost negligible. And, because the seaward toe of the breakwater is in deepwater and the waves are non-breaking, there is no appreciable wave setup.

4.3. Breakwater Improvement Design

4.3.1. *Initial Design*

Initial planning and design for the Kaumālapa'u project was accomplished
by the Honolulu District administration, and included numerical and
physical modeling to select the plan of improvement.[1] Numerical modeling was done using the software HARBD, developed by the US Army
Corps of Engineers, Engineer Research and Development Center, Coastal
and Hydraulics Laboratory (ERDC/CHL). This is a steady-state hybrid
element model based on a linearized mild slope equation, used to calculate linear wave response in harbors of varying sizes and depths. This
model incorporates many of the important processes affecting harbor wave
response, including the effects of wave energy refraction and diffraction,
bottom friction, and boundary reflection. Nonlinear processes such as wave
breaking and transmission over or through protective structures are not
modeled. The numerical model was used for initial evaluation of the performance of various improvement plans, and based on the results a preliminary
breakwater improvement, layout was selected. (Note: HARBD has been
replaced by CGWAVE and BOUSS2D, also developed by ERDC/CHL.)

An undistorted, three-dimensional (3-D) physical model of the harbor
was used to evaluate both operational conditions and breakwater stability
during severe storm wave conditions. The model was constructed at a scale
of 1:49, model to prototype. Early in the planning and design process Core-
Loc concrete armor units were selected by the Honolulu District administration for use in the breakwater repair, and the model scale was partly
selected based on available model Core-Loc unit sizes. The wave generators
were computer-controlled and generated irregular waves, i.e., waves having
varying heights and periods, in order to provide a more realistic representation of sea conditions. 3-D breakwater stability tests were conducted
using the physical model. The proposed new breakwater was constructed
on top of the existing rubblemound, which was reshaped as necessary for
the new structure, and consisted of a Core-Loc armor layer over a stone
underlayer. The initial proposed crest elevation was +6.1 m (+20 ft), and
the crest width was 6.1 m (20 ft). The Core-Loc armor extended to a depth
of 10.7 m (35 ft) on the ocean side, and into the existing rubblemound on
the harbor side to a depth of one Core-Loc height. Core-Loc units weighing
18.1 tonnes (20 tons) and 31.8 tonnes (35 tons) were tested, placed in a
single layer using a selective random placement, and with packing densities
of 0.58 to 0.63. The first two rows at the toe were placed uniformly to
achieve maximum interlocking and stability. A concrete rib cap was used
on the crest as a means of anchoring and buttressing the top armor units,

however it was anchored in the model and thus not dynamically similar to a prototype situation. Therefore, while it provided proper transmission, reflection and dissipation of wave energy in the model, the actual stability of the rib cap was not tested. For the Core-Loc stability tests, deepwater wave heights ($H_{1/3}$ or H_s) of 10.8 m (35 ft) and 8.5 m (28 ft) were used. The stability tests showed the following:

i. A toe trench was necessary to obtain armor stability in the shallow nearshore water at the ocean side landward terminus of the breakwater.
ii. Densely placed (packing density of 0.63) 18.1 tonne (20 ton) Core-Locs were stable for wave heights up to 8.5 m (28 feet), but failed for higher wave heights, particularly around the breakwater head.
iii. The 31.8 tonne (35 ton) units were found to be stable everywhere on the structure for all wave heights tested.

4.3.2. Detailed Design

Detailed design of the Kaumālapa'u Harbor breakwater repair project was initiated in 1998, and went through several iterations, including preparation of 35% construction plans and specifications, value engineering, and then final design. Preparation of the 35% design construction plans and specifications was accomplished based primarily on the project formulation analysis contained in the Honolulu District's 1996 report *Kaumālapa'u Harbor Special Design Report*,[2] which summarized the results of engineering to date, including the numerical and physical model studies. The initial 35% design included using 25.4 tonne (28 ton) Core-Loc units placed to a depth of −10.7 m (−35 ft) on the ocean side and with a packing density of 0.62, placed over an underlayer of 3.6 to 6.4 tonne (4 to 7 ton) stone. The crest elevation was +5.6 m (+18.5 ft) with a cast-in-place concrete rib cap. A Value Engineering (VE) study was conducted by the Honolulu District administration following completion of the 35% design. VE recommendations included three possible design revisions:

1. Reduce the breakwater crest elevation to +3.7 m (+12 ft), decrease the Core-Loc toe depth to −9.1 m (−30 ft), and reduce the packing density from 0.62 to a more achievable 0.58.
2. Reduce the underlayer stone size to 1/10th of the armor unit weight.
3. Reevaluate the concrete rib cap design to reduce its size in order to reduce cost.

It was decided to conduct additional two-dimensional (2-D) wave flume tests in order to evaluate the VE recommendations, as well as to confirm the required stable Core-Loc weight.

4.3.2.1. *Breakwater stability study*

Two-dimensional breakwater stability studies were conducted by ERDC/
CHL to investigate the effect of VE recommendations on the structure
stability — reducing the crest elevation, decreasing the armor layer toe
depth, and decreasing the Core-Loc packing density.[3] In addition, the design
wave heights and the required Core-Loc unit weight were revisited.

The model was constructed at a geometrically undistorted linear scale
of 1:54.3, model to prototype, for the proposed breakwater cross-section.
A 1V:10H slope represented the prototype nearshore bathymetry seaward
of the structure. Waves were generated by a piston-type, electro-hydraulic
system, controlled by a computer-generated irregular wave command signal.
Core-Loc armor units were tested with prototype equivalent weights of 18.1
and 31.8 tonnes (20 and 35 tons), with a packing density of 0.58 to 0.59.
Underlayer stone had a prototype equivalent weight of 2.2 to 3.4 tonnes (2.4
to 3.8 tons). The rib cap was geometrically similar to the prototype, however
it was anchored to the sidewalls in the model to ensure proper transmission,
reflection, and dissipation of wave energy, and thus its stability was not
tested. Design wave heights from 4.6 to 12.2 m (15 to 40 ft) were tested
in 1.5 m (5 ft) wave height increments, with periods of 12 and 16 seconds.
Lower wave heights representative of more prevailing conditions when the
harbor would be expected to be operational were also tested.

Results and conclusions from the 2-D testing are summarized as follows:

1. The Core-Loc size of 18.1 tonnes (20 tons), used on the sea side in the
   previous 3-D physical model study, was not stable for 12-s, 8.5 m (28 ft)
   waves.
2. The heavier 31.8 tonne (35 ton) Core-Loc appeared to be stable with
   no damage for waves up to 10.7 m (35 ft) if the depth of the armor
   toe was placed in sufficiently deep water (−19.8 m [−65 ft]) on the sea
   side, and −6.1 m (−20 ft) on the lee side. On the sea side, toe stones
   were observed to rock in place and displace for incident waves in which
   the wave drawdown approached the depth of the toe. Displacement of
   toe stones led to subsidence of the armor layer and loss of interlocking
   between Core-Locs, which led to substantial damage to the breakwater.
   On the lee side, energy from overtopping waves displaced toe stones
   placed high in the water column, which led to leeside toe failure.
3. If 31.8 tonne (35 ton) sea-side units were placed to a depth of −13.7 m
   (−45 ft), with a stone toe buttress at least 3 m (10 ft) wide, the sea
   side of the breakwater suffered minor damage with 12-s, 12.2 m (40 ft)
   waves, and moderate to major damage with 16-s, 12.2 m (40 ft) waves.

No displacement was observed with 10.7 m (35 ft) waves. Minor damage occurred on the lee side units with 10.7 m (35 ft) waves if the toe was placed at −6.1 m (−20 ft).

4. The 31.8 tonne (35 ton) Core-Locs showed significant rocking in place during waves 7.6 (25 ft) and higher. This could possibly result in armor unit damage.

5. Observations of wave overtopping during the study indicated that if the crest elevation was +4.6 m (+15 feet) overtopping was minor for incident waves less than 4.6 m (15 ft), moderate for 4.6 m (15 ft) waves, and major for 6.1 m (20 ft) waves and higher with a stillwater level of +0.7 m (+2.2 ft). For experiments with a crest elevation of +3.7 m (+12 ft), moderate overtopping occurred for 3 m (10 ft) waves, and major overtopping occurred for 4.6 m (15 ft) waves and higher.

Results from the 2-D study indicated the most stable plan consisted of 31.8 tonne (35 ton) Core-Locs placed from a crest elevation of +4.6 m (+15 ft) to a sea-side toe depth of 13.7 m (45 ft), and a leeside toe depth of 6.1 m (20 ft). The armor toe was protected by a toe berm constructed of underlayer stone placed 3 m (10 ft) wide on the sea side, and 9.1 m (30 ft) wide on the lee side.

The 2-D tests also showed that the VE suggested crest elevation of +3.7 m (+12 ft) was too low for the design operating incident wave conditions of up to 4.6 m (15 ft). However, it also showed that the crest could be lowered to about +4.6 m (+15 ft) without compromising harbor use during the maximum expected operating incident wave conditions. The studies indicated the likelihood of significant toe instability with a Core-Loc toe shallower than about −13.7 m (−45 ft), thus negating the VE suggestion of reducing the toe depth to −9.1 m (−30 ft). The 2-D tests did show that the 31.8 tonne (35 ton) Core-Loc's would be stable with a packing density of 0.58 to 0.59, in lieu of 0.62, as suggested by the VE study.

4.3.2.2. Core-Loc strength investigation

Kaumālapa'u Harbor breakwater repair utilized the largest Core-Loc armor units constructed to date (31.8 tonne), and the deepest Core-Loc toe (−13.7 m) and longest overall slope length (approximately 34 m) of any Core-Loc armored structure. The combination of large unit size and long slope length introduces a greater potential for structural failure of the Core-Locs due to static loads than has existed on previous projects. At the request of the Honolulu District administration, ERDC/CHL conducted a study of the structural response of the Core-Locs to be used for the

Kaumālapa'u Harbor project.[4] Laboratory model tests were conducted with scale model instrumented Core-Locs to determine maximum stresses in the toe region of the proposed armor layer, and estimate maximum design level stresses. Finite element method model tests were also accomplished to define static stress levels in the casting yard and to study the effect of varying chamfer and fillet shapes on tensile strength. Study results indicated a maximum flexural stress in the toe row of Core-Locs of 3,450 kPa (500 psi). Based on these results, a Core-Loc concrete flexural strength of 4,825 kPa (700 psi) was recommended by ERDC/CHL and was required in the construction specifications.

4.3.3. Final Design Elements

4.3.3.1. General breakwater configuration

The project site was topographically complicated, with a shallow nearshore rock bench at the seaward toe of the breakwater, and then a very steep drop to the sea floor. Thus the design required a very three-dimensional approach. The breakwater alignment was designed for a best-fit position on the existing rubblemound, with the head of the breakwater in the approximate location of the configuration tested in the 3-D physical model study. The existing rubblemound was excavated and shaped to form the core of the new breakwater. Excavated material was used as core material for the breakwater head, and as stone for a harbor side toe berm. The selected crest cap elevation is +4.4 m (+14.5 ft), and provides for no overtopping during all prevailing wave conditions when the harbor can reasonably be expected to be operating, and only minor overtopping during typically occurring storm wave conditions. Portions of the Core-Loc units randomly extend about 1.5 m (5 ft) above the crest cap. The crest width is 12.2 m (40 ft) at the top of the underlayer stone (+2.9 m [+9.5 ft] elevation), which would permit use of a ringer crane for construction. A horizontal row of Core-Loc units was placed on the ocean side of the crest to improve armor stability and energy dissipation during wave overtopping conditions. It was not considered practical to construct a nonovertopping structure for the very infrequent extreme storm wave conditions to which the breakwater could possibly be subject to. The sheer size of such a structure would not fit the physical confines of the project site, and the cost would have been prohibitive. In addition, there are limited harbor facilities that would be damaged by wave overtopping. Instead, the breakwater has been designed to be stable under possible significant wave overtopping conditions. The ocean side and harbor side breakwater slope is 1V to 1.5H.

4.3.3.2. *Toe depth and configuration*

Ocean Side: The landward end of the breakwater toe was placed in a 1.2 m (4 ft) deep toe trench excavated into hard rock, and then secured by filling the trench with tremie concrete. This extends for approximately 30 m (100 ft), at which point the toe transitions down the existing bottom slope to a toe trench excavated into the existing rubblemound at the −13.7 m (−45 ft) depth. The Core-Loc was placed over underlayer stone in the toe trench, and buttressed by a 4.6 m (15 ft) wide stone berm, with an elevation of −12.6 m (−41.5 ft). The toe berm also aids in stabilizing the structure during possible seismic (earthquake) loading conditions.

Harbor Side: The harbor side toe extends to a toe trench excavated into the existing rubblemound at a depth of −6.1 m (−20 ft), and is buttressed by a 9.1 m (30 ft) wide stone berm, with a crest elevation at −4.6 m (−15 ft).

4.3.3.3. *Armor layer*

The armor layer consists of 31.8-tonne (35-ton) Core-Loc concrete armor units, placed in a random matrix and oriented to achieve maximum inter-locking. A detailed Core-Loc placement grid pattern was developed and specified in order to ensure that the desired packing density was maintained. A 0.58 packing density was initially considered to be the tightest placement that could realistically be achieved with such large units. This was revised to 0.62 during construction after a change was made in the Core-Loc shape and additional placement testing was accomplished. The placement grid specifies a precise x,y,z-coordinate (position) for each Core-Loc unit. The placement pattern also considers the physical reality of placing 31.8 tonne (35 ton), nearly 4 m (13 ft) long armor units, and makes minor adjustments to the toe elevation and toe trench widths to permit proper unit placement. The placement grid is based both on the hydraulic considerations of stable toe depth (as determined in the 2-D model tests), and achieving a uniform top surface and Core-Loc termination at the crest to maximize stability during design wave overtopping conditions.

4.3.3.4. *Concrete crest cap*

The initial breakwater design utilized a concrete "rib" cap to buttress the top row of Core-Loc units. The rib cap concept has been developed and utilized on several breakwater projects in Hawai'i involving large concrete armor units. The rib cap is cast in place, and fastened in place by dowelling them into the large cap stones underneath, with the ribs formed around the armor units where they randomly contact them. The primary purpose for utilizing a rib cap design is its porosity, and thus ability to vent and relieve uplift forces during storm wave attack. The rib cap design also reduces the volume of concrete required.

The rib cap design, however, was not accurately modeled in either the 3-D or 2-D model tests such that its performance could be evaluated. In the 2-D stability tests, the model rib cap was actually fastened to the sides of the flume to hold it in place. It was noted in the 2-D tests the settlement and consolidation of armor units on the breakwater slope would result in exposure and loss of underlayer stone and core material from beneath the rib cap. In addition, the breakwater repair design did not utilize large capstones on the crest on which the rib cap could be constructed and fastened to. In fact, the underlayer stone was approximately the same size as the rib spacing. Given that there would be significant overtopping during design storm wave conditions, there was concern about loss of stone from around and under the rib cap, and possible movement of the cap itself. Should the Core-Loc units shift away from the crest rib cap, exposing underlayer stone that could be removed by wave action, the stability of the rib cap and crest Core-Loc units could be in jeopardy.

To improve the durability of the crest, the final design utilized a solid mass-nonreinforced concrete crest cap to better contact and buttress the Core-Locs, and to contain the underlayer stone on the crest. On the ocean side, the crest is positioned such that the front edge has minimal exposure to wave uprush and pressure within the stone underlayer, and a horizontal row of Core-Loc units is placed on the underlayer stone crest fronting the concrete crest cap to help further dissipate overtopping wave energy. The crest cap is then completely formed around the Core-Loc crest units. The 1.5 m (5 ft) thick, nominal 7.6 m (25 ft) wide, crest cap was cast in 4.6 m (15 ft) sections, with shear keys between each section and a continuous key into the breakwater crest along the axis of the breakwater. The cap weighs about 8.2 tonnes (9 tons) per linear foot and 122 tonnes (135 tons) per section.

4.3.3.5. *Breakwater plan and section*

Figures 4.2 and 4.3 show the final breakwater repair plan and typical sections. A summary of final design parameters and dimensions is shown on Table 4.1.

4.4. Core-Loc Design Considerations

During initial planning and design studies for the project, a decision was made by the Honolulu District administration to use the newly developed

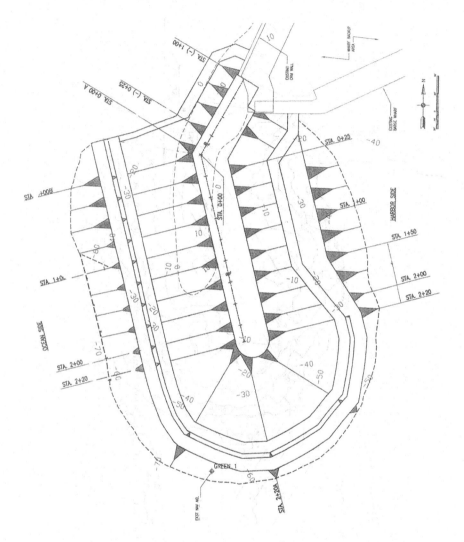

Fig. 4.2.   Kaumālapa'u Harbor breakwater repair project plan.

Core-Loc concrete armor unit, created by ERDC/CHL. The Core-Loc shape is patented by the US Army Corps of Engineers, and there are licensing requirements for both Core-Loc design and use. There are, however, no restrictions or royalties required for Core-Loc use by other Corps of Engineers Districts or Divisions on their projects. The Core-Loc unit has very good interlocking and thus a relatively high design stability coefficient

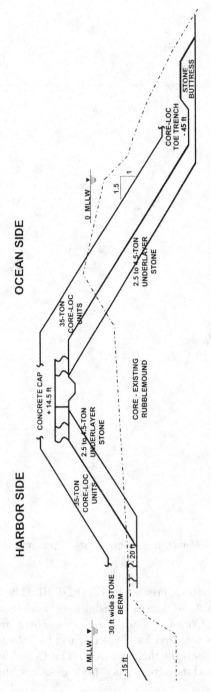

Fig. 4.3.   Kaumālapa'u Harbor breakwater repair typical breakwater section.

Table 4.1. Final breakwater design parameters and dimensions.

Crest	Elevation:	+4.4 m (+14.5 ft) MLLW (top of concrete cap)
	Width:	Approx. 7.6 m (25 ft) clear width across concrete cap between ocean and harbor side Core-Locs. 12.2 m (40 ft) wide across top of underlayer and core stone at the +2.9 m (+9.5 ft) elevation.
	Type:	Solid mass concrete cap, 1.5 m (5 ft) thick, unreinforced, formed around top row of Core-Loc units.
Armor Layer Toe	Ocean Side: Sta. 0+00 to (−)1+00:	Elevation −1.2 m (4 ft) below existing hard bottom
		Buttress Toe Trench with Tremie Concrete
	Sta. 0+00 to 2+20:	Elevation −13.7 m (−45 ft)
		Buttress 4.6 m (15-ft) wide stone berm
	Harbor Side:	Elevation −6.1 m (−20 ft)
		Buttress 9.1 m (30 ft) wide stone berm
Armor Layer	Type	Concrete Core-Loc
	Weight	31.8 tonnes (35 tons)
	Placement	Single layer, random orientation, individual unit position specified by a x,y,z-coordinate placement grid to achieve a 0.62 packing density
Underlayer	Material	Stone (Quarried and Salvaged)
	Weight	2.3 to 4.1 tonnes (2.5 to 4.5 tons)
	Placement	Layer thickness of 2.1 m (7 ft)
Core	Material	Salvaged stone and concrete rubble

(recommended $K_D = 13$ and 16 for breakwater head and trunk, respectively), and a robust shape which reduces the possibility of breakage of the unreinforced units. The units are also designed to be placed in a single layer, which reduces the volume of concrete required over typical two layer systems, and thus reduces cost. The units were so new that at the time of their selection for use at Kaumālapa'u Harbor there had not yet been a completed project built with Core-Loc units anywhere.

4.4.1. *Core-Loc Design Guidelines*

When detailed breakwater repair design began in 1998, the only available Core-Loc design guidelines were those contained in the paper *CORE-LOC*[TM] *Concrete Armor Units: Technical Guidelines* (1997) published by ERDC/CHL.[5] This paper presents details of the Core-Loc geometry, design guidelines for stable weight estimation, armor layer thickness, and packing density, as well as recommendations for fabrication, handling, and placement. Core-Loc design aspects for Kaumālapa'u Harbor were initially developed using these guidelines. However, Core-Loc design practice was rapidly evolving as prototype structures were being built, and as a result of the lessons being learned significant design changes related to the use of Core-Loc armor were considered necessary. ERDC/CHL personnel recommended that the repair design be reviewed in light of recent prototype experience with Core-Loc structures. Discussion was initiated with W.F. Baird & Associates, Ltd., which has licensed Core-Loc design engineers with considerable experience with erecting Core-Loc structures in the Middle East and elsewhere. Changes to the repair plan based on the new Core-Loc design information were accomplished by Sea Engineering with the assistance of Baird & Associates. Thus, the Core-Loc design for Kaumālapa'u changed and evolved through the course of the project.

4.4.2. *Core-Loc Shape*

The Core-Loc armor unit shape was initially based on guidance provided in Ref. 5. However, the Kaumālapa'u project would be using the largest Core-Loc units constructed to date, and, because of the deep water depth, the armor slope would be about 34 m (110 ft) long. The combination of large Core-Loc size and long slope length, which can result in unit movement as the armor layer consolidates over time and was considered to introduce a greater potential for failure of the Core-Locs due to static loads than exists on previous Core-Loc projects. The Honolulu District administration tasked ERDC/CHL to do further study of the Core-Loc strength in order to define the structural response of the large Kaumālapa'u units. The study included physical model and numerical model studies. The maximum stress in the units was determined to be 3,450 kPa (500 psi), and thus below the specified concrete flexural strength of 4,830 kPa (700 psi). Small radius fillets were suggested to help relieve stress magnification in the chamfer areas.

As design for Kaumālapa'u Harbor proceeded, another project to repair a breakwater protecting a US Navy air base in the Azores Archipelago (Portugal) was initiated, and which used almost exactly the same size Core-Loc units. Construction of this project began in 2003, and following the 2003–2004 winter storm season, a significant number of the large Core-Loc units were found to have broken. No obvious design problems were found, and the Core-Loc concrete generally exceeded the specified strength requirements. Some Core-Loc form and fabrication issues were determined, and the placement may not have resulted in the specified packing density being met. One result of the breakage and subsequent investigations was a recommendation to modify the shape of the Core-Loc unit in an effort to further reduce stress concentration areas and improve fabrication techniques to maximize concrete strength. This led to a change in the Kaumālapa'u Core-Loc shape shortly after award of the Kaumālapa'u construction contract. The shape changes were principally made to add larger radiused fillets in the chamfer regions at the intersection of the Core-Loc legs, and thicken the center section to accommodate the fillet. The new Core-Loc shape is more "compact," and thus heavier for the same given characteristic length ("C" dimension, or controlling height of the unit). There were no hydraulic model tests done to determine the performance of the new unit shape, and the assumption was made that performance would be similar to the old shape. Thus, the hydraulic stability would be similar at the same total weight (31.8 tonnes), and the therefore the "C" dimension was adjusted to obtain the same weight with the new unit. The C dimension changed from 3.93 m (12.9 ft) to 3.84 m (12.6 ft) for the new unit. The final Core-Loc dimensional shape schematic is shown on Fig. 4.4.

4.4.3. *Packing Density and Placement*

In order to provide adequate and uniform coverage on the breakwater slope and maintain unit-to-unit interlocking and contact, a sufficient packing density must be achieved. This is particularly critical for concrete armor units in a single layer configuration. Packing density is defined as the number of individual units required to cover a given area of slope, and is defined as a coefficient related to the volume of the unit. In general, the larger the unit becomes the lower the practicable achievable packing density. Larger units become more difficult to handle, which limits the ability to pack them together, and the concrete crushes where they contact, increasing friction and making them "sticky" so they do not slip together as easily as lighter units. Initial testing of Core-Loc by ERDC/CHL showed stable structures built with packing densities ranging from 0.54 up to 0.64, with 0.60 generally recommended. The 3-D hydraulic model studies conducted for the Kaumālapa'u project used a packing density of 0.62. Subsequent 2-D flume tests of breakwater stability used a packing density coefficient of 0.58–0.59. In the initial Kaumālapa'u breakwater design a packing coefficient of 0.58 was used to prepare the placement plan. Following the Core-Loc shape change, and additional packing density experiments conducted by ERDC/CHL, Honolulu District administration, and Baird & Associates, the packing density coefficient was revised to 0.62 for preparation of the prototype placement plan. This packing density was estimated based on limited model scale test placement in a dry test box with various test grids and a string and quick release system used to simulate placement by crane. This

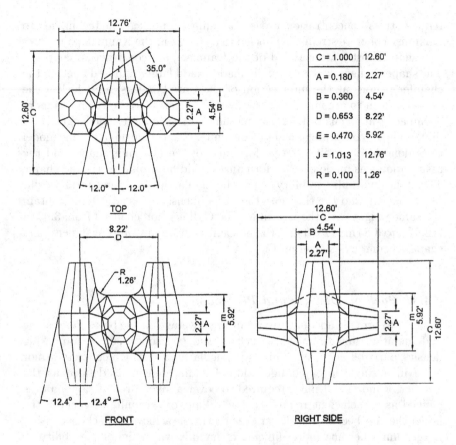

Fig. 4.4.   Core-Loc shape and dimensions.

is a high coefficient for units weighing 31.8 tonnes (35 tons), however the contractor was able to achieve it during construction of the Kaumālapa'u project.

Core-Loc was designed to be placed in a single-layer thickness in a random orientation with a fixed packing density. The specified packing density must be strictly maintained during construction to assure proper interlocking between units in order to achieve the desired hydraulic stability. For the Kaumālapa'u project, a Core-Loc placement plan was specified, with each unit being assigned a specific x, y, z-coordinate for its location on the structure. Although Core-Loc units can be placed randomly along the toe, a uniform pattern placement has been shown to be more stable if it can be achieved in the field. The Kaumālapa'u construction plans specified that the Core-Loc toe units be placed in the toe trench in a "cannon style" pattern placement. The first course was set with the central fluke pointing seaward at a 45-degree angle like a cannon barrel, and were placed side-by-side with a minimal space between adjacent units. The second course was placed such that they straddled the first course toe units. Subsequent armor units moving up the slope to the crest were placed in a random matrix and oriented to achieve maximum interlocking. The typically very clear water at the project site greatly facilitated placement of the toe units in a uniform pattern, and permitted camera and diver placement verification prior to the placement of subsequent units.

4.4.4. *Construction Specifications*

Construction specifications included "Formwork for Concrete Core-Loc," "Concrete for Core-Loc," and "Placing Core-Loc Armor Units."

4.4.4.1. *Formwork for concrete Core-Loc units*

The formwork specification was essentially a generic precast concrete armor unit form specification, written to insure the adequacy and suitability of the forms for their intended purpose of casting 31.8 tonnes (35 ton) Core-Loc units. The specification stated that forms shall not be removed from the Core-Loc within 24 and 72 hours after casting for nonsupporting and supporting forms, respectively. A provision was made to permit earlier removal of the forms provided that a structural analysis of concrete strength was made to show that the concrete in the forms had compressive and flexural strength sufficiently higher than the minimum required for form removal. The Kaumālapa'u contractor did not elect to remove any forms early.

4.4.4.2. *Concrete for Core-Loc units*

The specified flexural strength of the Core-Loc concrete was 4,830 kPa (700 psi) at 28 days. The maximum allowable water/cement ratio was 0.40 by weight, and the total air content could not exceed 5%. Particular attention was given to the concrete temperature during curing. At no time could the temperature of the concrete exceed 74°C (165°F), and the maximum temperature differential between the interior and exterior concrete could not exceed 20°C (36°F). Portland Type II low alkali cement was specified. Pozzolan, if used, had to be fly ash conforming to the requirements of ASTM C 618, Class F. (Pozzolan is not commonly used in Hawai'i, and was not used for this project.) Fiber-reinforced concrete was specified, in accordance with ASTM C 1116, Type III, synthetic reinforced concrete, with 7.5 pounds of structural fibers per cubic yard of concrete used (forta fiber or equal). The concrete-placing temperature was not permitted to exceed 30°C (85°F), which basically necessitated that the Contractor cast the Core-Loc at night. In addition, in order to meet the concrete curing temperature requirements, the contractor used a significant amount of water in the form of ice in the mix. The concrete mix resulted in a concrete unit weight of $2,400 \, \text{kg/m}^3$ (150 pounds/cubic foot).

4.4.4.3. *Placing Core-Loc armor units*

This specification contained the requirements for placement of Core-Loc toe and slope units, and positioning requirements. The toe units (first and second rows) were placed in the "cannon" style of uniform placement. Placement of Core-Loc units on the slope was based on the following parameters.

1. Maintain the specified packing density by insuring that each unit is placed at its specified location according to the placement plan, with the centroid of each unit being within $0.38 \, \text{m}$ (15 inches) of the specified location. Care must be taken to avoid small error "creep" in the placement of individual units that may be additive, eventually resulting in being unable to place the units at their specified location.

2. Moving up the slope, successively higher units should be "keyed" into and between two units below (i.e., fit the higher Core-Loc into the "pocket" between two adjacent lower units). Keying into the lower units should result in contact between at least one, and usually both, of the lower units. The keying into and between the lower units also results in units on the same horizontal row not being in contact.

3. Every Core-Loc unit must rest on and contact the underlayer stone.

4. The units shall be placed randomly, with different attitudes so as to interlock with and contact adjacent units to the maximum extent practicable. Every reasonable effort should be made to rotate and adjust the individual unit orientation so as to achieve the best interlocking and contact possible. However, it is recognized that interlocking and contact will vary, and that every unit may not have direct contact between all adjacent units.

5. Effort should be made to not place units with an H-member parallel to the slope, and less than one-third of the units shall be oriented this way, and they should be scattered throughout the structure and not placed in groups.

The specification for Core-Loc placement not only included the placement requirements, but also the requirements for a physical scale model of the breakwater repair plan to be built by the contractor, for building a test section of the Core-Loc placement on land, and for the contractor to retain a Core-Loc specialist to assist with construction. The purpose of the scale model was to give the contractor an opportunity to see how the various component parts fit together, and thus aid in working out solutions to construction questions before actual construction takes place. Constructing with 31.8 tonne (35 ton) armor units is not easy, and errors in prototype construction can be costly and time consuming. This is also why a test section of the Core-Loc placement on land was required prior to starting work on the actual breakwater. This gave the contractor a chance to practice slinging units and an understanding of unit attitude variation to facilitate interlocking, and practice adhering to a precise placement plan prior to placing units underwater. The requirement for the contractor to retain a Core-Loc specialist (CLS) was to insure that the contractor had access to an experienced Core-Loc design and construction specialist. The CLS advised on the preparation of the contractor's required Construction Execution Proposal, provided Core-Loc construction training, supervised test section practice, prepared modifications to the placement plan, and assisted with quality control. This turned out to be an important requirement, and greatly facilitated construction of the breakwater. The CLS for Kaumālapa'u Harbor was W.F. Baird & Associates.

4.5. Core-Loc Related Construction Highlights

Traylor Brothers (Pacific), Inc. (Traylor), headquartered in Irvine, California, was the successful construction bidder, and performed the site preparation and all the on-site breakwater construction work. GPRM Prestress, LLC was the precast fabricator for the Core-Loc armor units, which were made at their Campbell Industrial Park, Oah'u, precast facility and shipped by barge to Lāna'i. The Kaumālapa'u Core-Loc units won the 2008 Precast/Prestressed Concrete Institute award for "Best Custom Solution." American Marine Corporation were subcontractors for tug & barge and diving services. The CLS was W.F. Baird & Associates Ltd. Construction management and inspection was accomplished by the Honolulu District administration. Design-related construction assistance was provided to the district by Sea Engineering, Inc.

4.5.1. *Contractor's General Methodology*

On-site construction was accomplished primarily by one piece of heavy equipment, a Manitowoc 2250 crane that was used to shape the existing rubblemound to form the core, place all the stone, and place the Core-Loc units.

Fig. 4.5.   Kaumālapa'u breakwater construction site.

Ancillary equipment included a smaller Manitowac 999 crane for handling
of the Core-Loc units on the barge, and low boy trailers used for moving
Core-Locs to and from the inland stockpile area. Underlayer stone was
obtained from a quarry site on the island of Moloka'i and barged to Lāna'i.
Suitable stone salvaged from on-site, as approved by the Honolulu district,
was also utilized in the construction. Positioning for survey of the lines and
grades of the core and underlayer and precise positioning of the Core-Loc
units was accomplished by using a Novatel Propak L1L2 RTK DGPS Base
Station OEM4, and a Rover OEM4 Receiver with an antenna mounted on
the crane boom head, outputting to Winops Positioning Software. The construction site and nearly completed breakwater are shown on Fig. 4.5.

4.5.2. Core-Loc Handling and Placement

The Core-Loc units were transported to the construction site on low boy
trailers and could be placed on the trailers in two different orientations.
A sling from the crane was placed around the unit for lifting, and the sling
had three ways of wrapping around the unit. Thus six different Core-Loc orientations could be achieved during the lifting operation. Traylor developed

a lifting frame for the Core-Loc units to which the sling attaches, and for which the rotation can be controlled with tugger winches from the crane. The combination of different unit sling orientations and the ability to rotate the unit permitted a good range of Core-Loc orientations to be achieved and maintained underwater. For the cannon-style toe unit placement, two slings were used to maintain the proper uniform Core-Loc orientation. The slings were connected around the unit using a quick release hook.

It is extremely important that the Core-Loc units are placed at the correct grid locations. The use of a predetermined placement plan with each unit assigned an x, y, z-coordinate location assures that the units are placed at the right packing density on the slope, and that there are no voids or gaps between units. The placement grid is tied to the project coordinate system so that each unit can be presented to the slope at the correct x, y location. Correct underlayer stone lines and grades insures that the z (elevation) coordinate will be met at the respective x, y (horizontal) locations. The control is provided by the global positioning system (GPS) that reports the precise location of the headworks of the crane. The GPS x, y location is fed into a computer at the crane controls, and with positioning software is over-layed onto the placement plan. Using a targeting feature in the software, the crane operator can locate the headworks of the crane, and thus the center of mass of the Core-Loc unit hanging directly below it, to the correct x, y location. It should be noted that this system was not perfect, as wind, waves (particularly long period swell), and currents will cause the Core-Loc to move slightly around the position reported by the GPS antenna. In general, however, the system worked very well and resulted in very accurate placement with regard to the "ideal" predetermined placement plan.

Visual observation for Core-Loc placement was obtained three ways:

1. The Core-Loc lifting frame had provision for up to four cameras to be mounted in protected locations at the outside corners of the frame. Typically two, and sometimes three, frame cameras were used.
2. In shallow water, less than about 6 m (20 ft), a camera was mounted on an extendable pole and deployed by personnel from a small work boat. This was used primarily in shallow water when it was too rough for divers to work.
3. A diver with helmet camera was used for final verification of correct Core-Loc placement before it was released from the sling. The diver was particularly important for placement of the cannon-style toe units and the second row above them. Traylor also used diver placement and verification assistance for almost all the below water Core-Loc placement. The remote cameras worked well and reduced the reliance on divers as

well as providing visual observation as the units were being moved into position when it would have been too dangerous for a diver to be close. However, the mobility of a diver and the ability to see close-up was very important to be sure the unit interlocking was optimized and that the required contact between adjacent units and the underlayer stone was maintained for every unit placed.

The camera views were all fed into a split screen monitor for the crane operator to see what was going on from several angles simultaneously. This greatly aided him in achieving correct and optimal unit placement. It should be noted that the very clear water and excellent underwater visibility greatly facilitated the visual observations. In turbid water with poor visibility correct Core-Loc placement may be difficult to achieve.

A summary of the general Core-Loc placement procedure is as follows:

1. Following placement of a unit, the lifting frame was moved to the next unit location and the general layout and orientation of the units around the new location were observed by the crane operator. He then gave instructions as to which of the two trailer unit orientations and which of the three sling arrangements for the next unit should be used. (With practice the crane operator became increasingly adept at looking at the spot the unit was going and estimating the optimal lifting orientation of the unit.)

2. Two trailers, one for each of the trailer unit orientations, were staged at the work site. The selected orientation was backed up to be off-loaded, the crane swung back to shore, and the sling was placed around the unit in the selected configuration (see Fig. 4.6).

3. The unit was then moved to the correct x, y location, and slowly lowered into the water (see Fig. 4.7). Once in the water, the crane operator watched both a computer screen for the targeting information (Fig. 4.8, lower left) and the camera monitor for visual observation to guide the unit into the proper position. The split screen monitor in Fig. 4.8 shows two frame cameras in the top two quadrants, the diver camera safely off to the side in the lower left quadrant, and a frame camera angled back toward the underlayer slope in the lower right quadrant. The crane operator was also in voice contact with the diver, who helped fine tune the placement.

4. Once the unit was in position and the crane operator was satisfied that it had been placed, and moved around in position so as to achieve the best orientation, a final verification of its x, y coordinate was made, the sling was slackened, and the diver made a visual check of the contact with the underlayer and adjacent units. After a verification report by the diver,

Fig. 4.6.   Unit rigged for offloading from trailer.

Fig. 4.7.   Unit being lowered into water at selected location.

Fig. 4.8. Positioning system in crane cab.

Note: TV monitor shows camera views, top left and right and bottom right are lifting frame cameras, bottom left is diving helmet camera. The computer screen on the left displays the placement plan overlain by the GPS positioning/targeting software. Also note small model units at bottom center of the figure, used to help the crane operator visualize how the units are fitting together.

and with an ok from the crane operator (and the CLS, who was on-site for much of the underwater Core-Loc placement), the sling was released and the placement process started again for the next unit. On occasion, when either the diver or the crane operator was not satisfied that the placement requirements, or even the best placement for a particular location, had been achieved, the unit was reslung and adjusted as necessary.

It should be noted that the use of divers greatly facilitated underwater Core-Loc placement. Divers were used to (1) verify contact between Core-Locs and the underlayer stone and between adjacent Core-Locs, (2) to release the lifting sling once the correct placement was verified, and (3) to resling Core-Locs if they had to be repositioned.

Core-Loc placement above the water line was much easier and faster, than placement underwater. Figure 4.9 illustrates Core-Loc placement on the underlayer above water. Ultimately, a total of 819 Core-Loc units were used for construction of the breakwater. Of these, one was broken during construction of the test section on land, one broke during placement on

Fig. 4.9. Core-Loc placed on underlayer stone.

the head above water, and four broke underwater after placement when subsequent units were placed above them. No obvious explanation for any of the breakage was evident.

4.5.3. *Concrete Crest Cap Construction*

The concrete crest cap was constructed in segments, beginning from the breakwater head and proceeding landward. The segments were initially to be 3 m (10 ft) long, however Traylor requested approval to pour them in 4.6 m (15 ft) lengths in order to optimize use of the batch plant production capability. Figure 4.10 shows a crest cap segment being prepared. Note the shear key female halves cast in the end of the previous segment, and the 1 m (3 ft) deep key trench excavated into the crest core material. Heavy wire mesh fencing material lined with geotextile filter fabric was used as the flexible formwork around the Core-Loc crest units, between the crest units and the cap. This worked very well, it could easily be made to conform to the varying contours of the Core-Locs, and it could be made to project into gaps between units. The mesh was held in place by a steel cable system until the concrete cured, and then any mesh and cables projecting above the concrete were cut off. The completed cap is illustrated on Fig. 4.11.

Fig. 4.10.   Crest preparation for pouring concrete cap.

Fig. 4.11.   Completed crest cap.

4.5.4. Toe Trench Construction

The Core-Loc toe trench at a depth of -13.7 m (-45 ft) on the ocean side was constructed of 2.3 to 4.1 tonne (2.5 to 4.5 ton) underlayer stone. The required cannon style Core-Loc toe unit placement necessitates careful and uniform placement of the units in order that they provide a solid foundation for the armor layer. Gaps between the large underlayer stone can be as big as a Core-Loc fluke, and it was noted during construction of the test section on land that unit placement could be affected by gaps between stones. There was also concern that even if breakwater toe units were placed solidly on stone, but with a fluke near a gap between stones, the unit could shift slightly during loading as subsequent units were placed up the slope, resulting in fluke slipping into a gap and a greater movement and possible loss of contact occurring than was necessary. To eliminate this problem Traylor and the CLS recommended chinking of the bottom of the trench between the large stones with 45 to 450 kg (100 to 1,000 pound) stone. This would provide a more secure and stable platform for the Core-Loc toe unit to rest on, and help insure that a fluke was not inadvertently placed near a large void between trench stones. This was approved provided that (1) the stone was well graded between the upper and lower size range so as to form a tight matrix, (2) no chinking stone was placed above the level of the trench stone elevation neat line so that the Core-Locs rested primarily on the large trench stone, and (3) no chinking stone was to be placed seaward of the Core-Locs so as to not effect interlocking of the second layer of 2.3 to 4.1 tonne (2.5 to 4.5 ton) stone placed seaward of the Core-Locs to form the seaward side of the trench and the toe buttress.

It should also be noted that the construction plans called for the toe trench to be backfilled with 225 to 2,265 (500 to 5,000 pound) stone following Core-Loc placement. Thus the chinking stone would be covered with larger stone. The purpose of the stone backfill was to help prevent movement of the Core-Loc toe units.

4.5.5. Core-Loc Placement Modifications

During construction, Core-Loc placement requirements were revised slightly for two areas of the breakwater — at the head above water and the toe coming up slopes.

As units approached the waterline at the head, the curvature becomes more pronounced. The spacing of the placement plan grid pattern is based on the centroid location of the units, and not the slope contact point. As the curvature becomes more severe, the lower edges of the units touching the slope begin to interfere with each other, causing difficulties in placement.

The placing pattern grid becomes slightly rotated around the head, and although the density is maintained, placement difficulty increases because gravity is not necessarily working to advantage in assisting with interlocking of the units. In some areas, the tightest unit to unit interlock may be horizontal, and the crane has limited ability to interlock a unit horizontally, especially while gravity is trying to force the unit downhill. In addition, once out of the water, the cushioning effects of the water in slowing the swing of the unit, and the weight reduction due to buoyancy and resultant friction reduction, is lost. One Core-Loc unit on the head was broken during placement, through no apparent fault of Traylor. Because of these factors, the CLS recommended that the Core-Loc placement around the head above the waterline be revised to require that each unit be carefully placed in an optimal manner relative to the previously placed adjacent units following a random placement with random orientation protocol. Each unit was placed and oriented for its best fit into the structure, and its position recorded. Care was taken to achieve the greatest possible packing density, while minimizing the risk of breaking units during placement.

The cannon style placement pattern is not possible on slopes, i.e., on the ocean side from Sta. 0+00 to Sta. (−) 1+00 and on the harbor side between Sta. 2+20 and Sat. 1+50. In these areas random toe placement was used, similar to the placement recommended for the tight radius head section above the waterline as described above. Instead of a uniform cannon placed in the first and second rows, each toe unit was carefully placed in an optimal manner relative to the previously placed adjacent units following a random placement with random orientation to achieve maximum interlocking between adjacent units and the toe trench or existing bottom substrate.

4.6. Lessons Learned

The Kaumālapa'u Harbor project was unique in several ways, primarily related to its use of Core-Loc concrete armor units. It was the first use of Core-Loc in the United States to construct a new breakwater, it used the largest Core-Loc units ever cast and placed anywhere to date, and the design parameters were very rigorous — deepwater at the structure toe and a 10.7 m (35 ft) design wave height. There were lessons learned during construction of this project, the more important of which include the following Core-Loc related items.

Use of Concrete Armor Units: The large design wave heights necessitate the use of concrete armor units with high stability coefficients. Core-Loc units have some advantages, e.g., single layer system to reduce cost and good stability (if placed properly), however these advantages come at some design and construction effort. Considerable care and control must be placed on the armor unit concrete mix design and the casting process, and their placement on the structure, in order to realize their advantages and reduce the possibility of failure. The use of the Core-Loc units at Kaumālapa'u resulted in the design and construction becoming something of an on-going and evolving project, requiring revision and adjustment as the project proceeded through construction.

Experience: Core-Loc armor units are unique, and design and construction utilizing them is best accomplished by persons and firms having Core-Loc specific design and construction experience. The US Army Corps of Engineers, who have developed and hold a patent for the Core-Loc design, recognizes the unique design aspects of the unit and has licensed specific firms as Core-Loc designers. While US government projects may design and construct using Core-Loc without constraint or royalty, any private or foreign government project utilizing Core-Loc must involve a licensed Core-Loc design firm. The use of an experienced Core-Loc designer will contribute significantly to the success of a project.

Complete and Detailed Pre-Bid Information: The use of Core-Loc units requires the use of rigorous construction requirements specific to Core-Loc — from fabrication of the units to precision unit placement on the structure. Construction firms without prior Core-Loc experience may not fully appreciate the construction requirements associated with the use of the units. It is important to ensure that the construction documents are complete and very clear with regard to Core-Loc requirements, and to carefully explain them to potential contractors at a pre-bid conference.

Core-Loc Placement Plan: The use of Core-Loc units requires a uniform packing density in order to achieve their design stability, which in turn requires relatively precise placement to achieve. The Core-Loc Placement Plan prepared for the Kaumālapa'u project, and included in the construction drawings, worked very well to ensure that the construction contractor was achieving the correct packing density, and also aided the construction inspector in verifying that the plan requirements were being met.

On-Land Test Section: The Kaumālapa'u specifications required the contractor to build a test section on land representative of the lines and grades of the structure to be built, on which the construction crew could practice placement of the Core-Loc units — slinging and lifting the units, how to follow the placement plan and achieve the correct x, y, z-coordinate for each unit, and how to vary the unit orientation in order to maximize interlocking and the required contact between the underlayer stone and adjacent units. This test section worked very well to aid in avoiding errors and placement problems when the contractor began work in the water, and which would have been difficult and costly to fix after the fact. The specifications also required the contractor to construct a physical scale model of the project on-site, to be used in training and verification of Core-Loc placement. The contractor noted the usefulness of the model to provide a three-dimensional visualization of what was being constructed.

Visual Observation During Core-Loc Placement: Core-Loc units are designed to be randomly oriented on the structure slope, however careful orientation of the units to achieve maximum interlocking is necessary in order to achieve the design stability. In addition, the placement specifications require specific contact between the Core-Loc and the underlayer stone and between adjacent units. The Kaumālapa'u specifications required that divers verify the placement of all units prior to their being released into position. The contractor also used an underwater camera system mounted in the lifting frame to aid in positioning and orienting the units, and while this worked reasonably well, it could not always reliably replace the diver for final verification of the required contact between units. Construction of the Kaumālapa'u project was greatly aided by the generally crystal clear water at the site, which made camera and diver visibility excellent. Given the need for correct Core-Loc placement in order to achieve the design stability, proper construction without divers or in turbid water would likely be more challenging. It should be noted that techniques for Core-Loc placement in low visibility conditions are being developed, such as the POSIBLOC System, which permits real time 3-D positioning and orientation of Core-Loc units underwater.

4.7. Post-Construction Monitoring

The Kaumalapau Harbor project was selected for postconstruction monitoring by the US Army Corps of Engineers (USACE) Monitoring of Completed Navigation Projects (MCNP) Program, which conducts data collection and analysis on federal navigation projects in an effort to improve the methods and technologies used to construct and maintain safe and efficient navigation features.[6] One component of the Kaumālapa'u Harbor monitoring is the use of Tripod (or Terrestrial) Light Detection and Ranging (T-LiDAR) to map the breakwater surface and provide information about how Core-Loc armor units consolidate following construction, and in turn how consolidation, nesting, and settlement affects armor unit stability, concrete crest cap performance, and possible armor unit breakage. T-LiDAR data was collected by USACE, the US Geological Survey, and the University of Hawai'i, shortly before the breakwater construction was completed in June 2007, and again one year later in July 2008, to measure movement of the above-water breakwater using ultra dense sub-centimeter three-dimensional point cloud imagery.

Nesting and downslope consolidation of the armor units over time due to gravitational settling is expected, and increased movement is expected during high wave events. The head of the breakwater with its greater and more variable wave exposure, and where it was more difficult to place the fixed size armor units around a decreasing radius curve with elevation above water, is particularly susceptible to potential armor unit movement. Preliminary assessment of the T-LiDAR data obtained one-year postconstruction indicates a range of armor layer movement from a few centimeters to greater than a meter along the ocean side of the breakwater, with significantly less movement observed along the harbor side. No unusually high wave events occurred during the 2007–2008 period between measurements. As expected, the larger changes in Core-Loc position occurred at the ocean side breakwater head, where both settling and rotational movement of some units was indicated. In general, most of the measured armor unit changes are characterized by a landward rotation and downward translation of the Core-Loc units close to the waterline, where wave action is greatest, and where armor unit placement was more difficult due to bends in the breakwater alignment and around the curving head. No significant change in the elevation of the concrete crest cap was noted, and no Core-Loc unit breakage was observed. The overall stability of the structure in its first year has been good, and the packing density and interlocking of the armor units achieved during construction has been adequate to limit movement that could open up gaps in the armor layer cover or result in armor unit breakage.

References

1. U.S. Army Corps of Engineers, Waterways Experiment Station *Wave Response of Kaumalapau Harbor, Island of Lanai, Hawaii*, Technical Report CHL-98-11 (1998).
2. U.S. Army Corps of Engineers, Pacific Ocean Division Kaumalapau Harbor Special Design Report, Island of Lanai, Hawaii (1996).
3. Ernest R. Smith, U.S. Army Engineer Research and Development Center, Coastal and Hydraulics Laboratory Kaumalapau Harbor, Lanai, Hawaii, Two-Dimensional Breakwater Stability Study, ERDC/CHL TR-01-3 (2001).
4. J. A. Melby, PhD, U.S. Army Engineer Research and Development Center, Coastal and Hydraulics Laboratory *Kaumalapau, Hawaii, Breakwater — Core-Loc Strength Investigation* (2002).
5. G. F. Turk and J. A. Melby, U.S. Army Corps of Engineers, Waterways Experiment Station, *CORE-LOC*^{TM} *Concrete Armor Units: Technical Guidelines*, Miscellaneous Paper CHL-97-6 (1997).

6. J. H. Podoski, G. W. Bawden, D. Bond, T. D. Smith and J. Foster, (In Publication), *Post-Construction Monitoring of a Core-Loc*^{TM} *Breakwater Using Tripod-Based LiDAR, Proc. Coasts, Marine Structures and Breakwaters*, Institution of Civil Engineers/Thomas Telford Ltd., Scotland, UK (2009).

Chapter 5

Waterfront Developments in Harmony with Nature

Karsten Mangor[*], Ida Brøker[†], Peter Rand[‡] and Dan Hasløv

DHI, Agern Allé 5, DK-2970 Hørsholm, Denmark
[*] *km@dhigroup.com*
[†] *ibh@dhigroup.com*
[‡] *prd@dhigroup.com*
[‡] *Hasløv & Kjærsgaard I/S, Architects and Planners,*
Copenhagen, Denmark
dbh@hogk.dk

The main development theme in many coastal countries is to utilize the attractiveness of water in a broad context. Nowadays emphasis has shifted from coastal protection to shoreline management, which includes waterfront developments.

Waterfront development can be considered as an artificial piece of new nature. The artificial beaches and lagoons, so to speak, still follow a natural trajectory of growth. Consequently, such landscape elements will follow the natural marine and coastal processes in the construction area. These parameters cannot be changed. It is therefore important to understand the prevailing natural processes responsible for creating attractive beach and lagoon environments as the basis for the design of well-functioning artificial coastal and marine elements.

The focus of this chapter is to provide the reader with a basic understanding of how to create a well functioning waterfront development with regard to the important hydraulic elements — beaches and lagoons.

Design guidelines for *artificial beaches* and *lagoons* are presented as well as guidelines for *landscape elements of waterfront developments*. Examples include a popular new beach park in Copenhagen and a new type of offshore development scheme. Finally, investigation methodology for waterfront developments is presented.

5.1. Introduction

The art of developing waterfront projects involves utilizing the possibilities provided at a specific site to the benefit of the project, i.e., to integrate the possibilities provided by the marine environment with the demands of the society. This includes perceiving the marine forces such as waves and tides, as external opportunities, which shall be used to maintain high-quality artificial beaches and lagoons, contrary to the traditional approach of perceiving these external forces as problem generators, against which protection is required. The chapter is divided into the following sections:

— The characteristics of natural landscape elements
— Design guidelines for artificial beaches
— Design guidelines for artificial lagoons
— Landscape elements of coastal and offshore developments and their hydraulic design
— Example of beach park development
— Presentation of a new concept for an offshore development scheme.

5.2. The Characteristics of Natural Landscape Elements

5.2.1. *Characteristics of Natural Beaches*

Attractive and safe recreational beaches are always characterized as being exposed to moderate wave action, the tide is micro to moderate (tidal range $< \sim 1.5\,\text{m}$), clean and transparent water, no rock outcrops, well-sorted medium sand and minimal amounts of natural and artificial debris.

Examples of attractive natural beaches are presented in Fig. 5.1.

These beaches are all characterized by being exposed to waves, the sand is clean beach sand, and the water is clean. The type and color of the sand is different, but all types are natural beach sand of great beauty and recreational value. The exposed beaches have a sandy and clean appearance due to the wave action that prevents settlement of fine sediments and organic matter. However, there are also many examples of good quality beaches along coasts where the water contains high amounts of suspended sediments, at least during the rainy season and/or during rough weather conditions. This is, for example, the situation along Malaysia's E-coast and along Sri Lanka's coasts. The reason for the clean sandy beaches in these environments is that these beaches are exposed to waves.

Natural beaches appear differently when they are lacking wave exposure. This is clearly seen in the examples presented in Figs. 5.2 and 5.3.

Fig. 5.1. Examples of attractive beaches. Upper: The Skaw Spit in Denmark. Lower left: NW-Mediterranean coast in Egypt. Lower right: Sunset Beach in Dubai.

Fig. 5.2. Natural beaches lacking wave exposure. Left: Beach in a natural Lagoon in the UAE, which suffers from algae and deposition of fine sediments. Right: Muddy beaches in creek in the UAE.

It is evident from the above examples of exposed and protected natural beaches that the wave exposure is of paramount importance for the type of natural beach that develops in an area. Lack of wave exposure on an artificial beach will allow settlement of suspended matter on the seabed

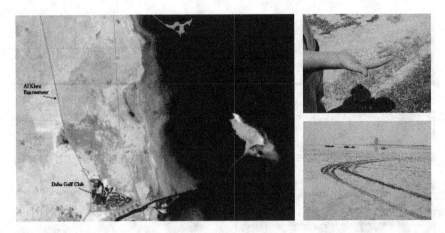

Fig. 5.3. Example of correlation between type of beach and wave exposure. Left: Location "Map," North Beach in Doha, Qatar. Note that the southern part is protected by an island and associated reefs and has a muddy tidal flat type of beach (photo lower right), whereas the northern part is exposed and has a sandy beach (photo upper right).

and on the beach, also in cases where the beaches have been built of clean sand. This will with time lead to the seabed being covered with a layer of soft sediments. Such beaches feel muddy when walking on them, which is unattractive for recreational beaches.

5.2.2. Characteristics of Natural Lagoons

Natural lagoons are attractive from a recreational point of view due mainly to the open water body they offer and not due to their beaches, which are normally of poor quality. Coastal lagoons are characterized by the following elements:

— One or more so-called tidal inlets that connect the lagoon and the sea
— Tidal exchange of water between the lagoon and the sea known as the tidal volume
— Rich flora such as sea grass beds, mangroves, and meadows
— Rich fauna such as mussel banks, nursery areas for many fish species, and rich bird life
— The openings are sometimes stable and sometimes suffering from sedimentation; this is dependent of the balance between the tidal range and the wave exposure
— The lagoon environment is protected and is therefore often characterized by settlement of fines, which in many cases leads to the formation of mud flats.

These natural conditions offer the following attractions:

— Protected water environment, which is traditionally used by coastal societies as a natural location for settlements based on natural harbor facilities
— Possibilities for a great number of commercial activities such as fishing, hunting, aquaculture, location for water intakes/outlets of different kinds, salt production, etc.
— Recreational activities such as water sports, navigation in protected waters, fishing, bird watching, etc.

However, settlements and the many associated activities in the lagoons imply on the other hand also the risk of many impacts on the lagoons such as:

— Pollution leading to degradation of the water quality and associated degradation of flora and fauna
— Installation of sluices leading to changes in the salinity, etc.
— Reclamations leading to changes in the tidal volume
— Regulation of inlets and dredging of navigation channels leading to changes in the tidal volume, which may lead to local erosion or general siltation
— Navigation that may lead to pollution and erosion, etc.

An example of an attractive lagoon environment is the semi-open Marsa Matrouh lagoon located at the NW-Mediterranean coast of Egypt (see Fig. 5.4). This lagoon offers attractive sandy beaches and good water quality due to the wide opening, which allows some wave penetration and good flushing.

5.3. Design Guidelines for Artificial Beaches

The most important landscape elements in many waterfront developments are attractive sandy beaches. An artificial beach is the construction of a new beach by supply of sand, the so-called beach fill. The design requirements to a good-quality recreational beach are outlined in the following.

5.3.1. *Exposure to Waves*

A beach shall be exposed to waves in order to obtain a good-quality beach. However, a recreational beach shall not be too exposed, as this endangers bathing safety. This means that there are two opposing requirements:

— There shall be a certain exposure to secure a self-cleaning beach.

Fig. 5.4. Upper: Marsa Matrouh Lagoon, NW-Mediterranean coast of Egypt. Lower photos: Western lagoon beach. Lower left: Semi-exposed sandy beach towards NW. Lower right: Semi-exposed beach towards SE.

— The exposure shall not be too large in order to secure safe bathing conditions.

The shoreface (the active part of the coastal profile) shall extend to a water depth of $d_l \sim 2\,\mathrm{m}$ in order to secure a clean and attractive sandy seabed for the bathers, where d_l is the so-called Closure Depth. It is the consistent movement of the sand by the wave action during rough conditions that maintains a nice sandy beach and shore face through preventing settlement of the content of fines, which are often present in seawater. Furthermore, the wave exposure prevents sea grasses from growing on the shore face.

Fig. 5.5. Left: Exposed artificial beach in the UAE, nice clean beach due to the wave exposure. Right: Artificial Beach in artificial lagoon in the UAE, note the muddy seabed, which is due to lack of wave exposure. The lack of freshness is clearly seen.

The difference in attractiveness between two artificial beaches in a beach park in the UAE area, where the first beach is exposed to waves and the second beach is a protected lagoon beach, is clearly seen in Fig. 5.5.

There exist no internationally agreed criteria related to wave height relative to safe bathing conditions. Bathing safety is mainly related to the occurrence and type of breaking waves and the wave-generated currents in the breaking zone. These conditions are discussed in the following.

Spilling breakers are often associated with the formation of bars and rip currents, which can carry both adults and children out into deep water. This situation is typical for strong wind and storm conditions at sandy (ocean) coasts. Plunging waves are dangerous because the violent breaking can hit a bathing person. Plunging breakers typically occur on ocean coasts with moderate wave conditions, such as under monsoon and trade wind conditions on coasts with relatively coarse sand. It is evident from the above that ocean coasts are the most dangerous. An upper limit for wave heights in relation to safe bathing conditions is estimated at: $H_s < 0.8$–1.2 m during the bathing season. The low limit is valid for long-period waves (swell) and the high limit for steep waves (wind-waves).

This means that protective measures are required if a site is more exposed during the bathing season than given in the above rough criteria, e.g., in the form of specially designed coastal structures.

Rip currents generated as a result of the presence of coastal structures can also be very dangerous. The condition that makes this situation dangerous is that during a storm situation there will be partial shelter for the waves behind the coastal structure, but in the same area there will be strong currents due to eddy formation, which can carry poor swimmers out into deep water. In conclusion, areas with partial shelter provided by coastal structures near the coast generate a false feeling of safety and such arrangements should consequently be avoided.

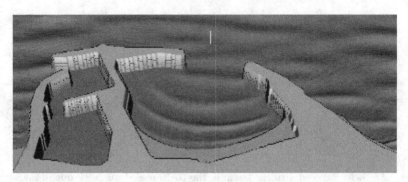

Fig. 5.6. Upper: Artificial beach at Alexandria's exposed coast. Designed by DHI, Hasløv & Kjærsgaard and ECMA. Upper: Modelled wave conditions. Lower: The newly constructed beach.

The principle of providing safe partial shelter and shoreline stability at an exposed coast is presented in Fig. 5.6, which shows an optimized layout for an artificial beach and a marina in Alexandria, Egypt incorporating

— Partial shelter provided by the large breakwaters
— A width of the opening to the open sea adjusted to provide suitable wave exposure to fulfill the requirements of moderate exposure for securing a good beach quality and semisheltered conditions for providing safe swimming conditions
— A stable beach under the resulting wave conditions
— Exclusion of dangerous rip currents due to the long distance between the breakwaters and the beach and because of the equilibrium shape of the beach.

5.3.2. *Minimum Wave Exposure*

A recreational beach shall have an active profile out to a water depth of about 2.0 m relative to low tide. This recommendation of a depth of 2.0 m relative to low tide is caused by the requirement that bathers walking on the seabed shall experience an attractive clean sandy seabed without deposition of fines, which results in muddy seabed. The clean and active seabed is secured by the requirement that the active coastal profile shall extend out to a water depth of 2.0 m, i.e., $d_l \geq 2.0$ m or in terms of wave height: $H_{s,12h/y} \geq 1.0$ m. This means that the seabed out to this water depth will be exposed regularly to waves preventing fine sediments to settle on the shoreface. Furthermore, the growth of sea grasses will also be prevented.

If the natural shore face does not allow these requirements to be fulfilled, e.g., if the shore face is shallower than the equilibrium profile, then there are two possibilities to fulfill the requirements:

— The beach is shifted seaward.
— The existing coastal profile is excavated to accommodate the equilibrium profile.

5.3.3. *Exposure in Relation to Tidal Range*

A certain tidal range and storm surge activity will cause a wide beach. However, a tidal flat may develop if the tidal range is much larger than the yearly average wave height. Furthermore, a high tidal range constitutes a major challenge for the layout of an artificial beach. Thus, a good quality recreational beach is normally characterized by a micro to moderate tidal regime, which means a tidal range smaller than approximately 1.5 m.

5.3.4. *Beach Plan Form*

The beach should be stable in plan form (horizontally) in order to secure minimum maintenance. This means that the alignment of the beach shall be perpendicular to the direction of the prevailing waves, or in other words, the orientation of the beach shall be in the equilibrium orientation, which is the orientation providing net zero littoral transport. This often leads to the requirement of supporting coastal structures for stabilizing the beach in an orientation, which is different from the natural orientation of the coastline in the area of interest.

At an exposed beach with oblique wave attack, the supportive coastal structures for an artificial beach shall be designed to fulfill the following conditions:

— Build the beach in the equilibrium orientation and prevent loss of sand out of the artificial beach area by constructing terminal structures

— The terminal structures shall have a streamlined form and an open angle
with the shoreline in order to minimize trapping of floating debris and
in order to avoid dangerous currents along the beach adjacent to the
structures
— Provide partial protection against waves if the wave exposure is too high
related to bathing safety
— All coastal structures should also have recreational functions.

5.3.5. *Beach Profile Form*

The beach profile shall be stable, which means that a beach shall be built in
the form of an equilibrium profile. A beach adjusts to the equilibrium profile
in the active littoral zone. The shape of the profile is mainly dependent of
the grain size characteristics of the sand. The equilibrium shape, Dean's
equilibrium profile, follows the shape $d = A x^{2/3}$ where d is the depth in
the distance x from the shoreline, both d and x in meters. A is Dean's
constant, which is dependent on the grain size of the sand according to the
directions given in Table 5.1.

Table 5.1.   Correlation between mean grain size d_{50}
in mm and the constant A in Dean's Equilibrium
profile equation. For typical beach sand.

d_{50}	0.20	0.25	0.30	0.50
A	0.080	0.092	0.103	0.132

The equilibrium profile concept is valid only for the active littoral zone,
i.e., out to the closure depth, d_l. As a rule of thumb $d_l \sim 2H_{s,12h/y}$ can be
used for normal wind waves.

5.3.6. *Design Level for Coastal Areas*

The required design levels for various kinds of coastal land areas are
dependent mainly of the following conditions:

— Land use
— Safety level
— Lifetime and sea level rise (due to climate changes) at end of lifetime
— Design water level.

The land use and the recommended safety level are interdependent;
the recommended criteria for development projects in the coastal area are
presented in Table 5.2.

Table 5.2. Relation between type of land use and recommended return period for flooding (only for guidance).

Type of land use		Criteria	Recommended recurrence period in years
	Boat piers	Low vulnerability, access to boats important	0.5–1
Waterfront Developments	Park areas	Low vulnerability, nearness to water important	1–10
	Inhabited and roads	Vulnerable to flooding, risk of damage to houses	50
Industrial/public utilities		Very vulnerable to flooding, risk of interruption in supplies	100
Low-lying inhabited areas		Extremely vulnerable to flooding, risk of losing lives	200 or higher dependent of size of area and density of population

The design level is dependent on the following conditions:

— Extreme still water level, which includes the effect of tides and storm surges
— Allowance for the influence of future sea level rise due to the Greenhouse effect
— Wave run-up, which is the sum of the wave set-up and the wave swash. The wave set-up is the local elevation in the mean water level on the foreshore caused by the reduction in the wave height through the breaker zone. The wave swash is the propagation of the waves on the beach slope
— Special phenomena such as the risk of tsunamis should be taken into account when relevant however, surges caused by tsunamis are normally only included in the design for very critical installations, such as nuclear plants.
— Safety margin.

The difference in the recommended recurrence periods for the different types of land use categories opens for terraced layouts of the water edges as proposed and discussed in Section 5.5.

In the planning of waterfront schemes there are sometimes several conditions to be taken into consideration such as improvement of the beaches and the need for sea defense. An example of a scheme where both these conditions have been built into a beach park project is the Køge Bay Beach Park, south of Copenhagen, where new exposed beaches were established and the dunes were designed as a sea defense structure safeguarding the low-lying build up area behind the scheme (see Fig. 5.7).

Fig. 5.7. Køge Bay Beach Park south of Copenhagen. A dike is integrated in the dunes and runs behind the marinas thereby protecting the entire hinterland against flooding.

5.3.7. Beach Fill Material

The fill material for artificial beaches shall fulfill the following criteria in order to provide a high quality recreational beach:

— The characteristics of the fill sand shall be similar to that of the natural sand in the area if the new artificial beach is connected to an existing beach, however, as a rule slightly coarser
— The sand shall be medium, i.e., $0.25\,\mathrm{mm} < d_{50} < 0.5\,\mathrm{mm}$, preferably coarser than 0.3 mm, which minimizes wind loss
— Minimum content of fines, i.e., silt content less than 1–2%
— Gravel and shell content less than 3%
— Well sorted sand, $u = d_{60}/d_{10}$ less than 2.0
— Color shall be white, light gray, or yellow/golden
— No content of organic matter
— Thickness of sand layer shall be minimum 1 m, preferably thicker.

The reason for the requirement for the beach sand to be medium, well-sorted sand with minimum content of fines is discussed in the following. If this requirement is not fulfilled, i.e., if the sand is graded (the opposite of well-sorted) with some content of fines, the permeability will be low, which means that the beach will drain very slowly at falling tide. This implies that the beach will be wet at all times and it will have a tendency to be swampy, and thereby unpleasant to walk on. This criterion is especially important for artificial beaches built at protected locations as there will not be enough wave action to wash the fine sediments out of the beach. Furthermore, algae may start to grow on the beach, which makes it greenish in color and unaesthetic. Finally in such environments the beach will be occupied by beach crabs and their pellets. It is especially important at protected sites to use clean sand with zero content of organic matter because the combination of lack of wave exposure and content of organic matter may lead to anoxic conditions, resulting in formation of hydrogen sulfide, which causes a bad smell and dark coloring of the sand (see examples in Fig. 5.8).

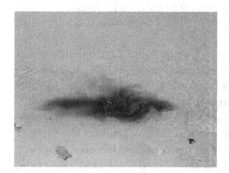 

Fig. 5.8. Examples of anoxic conditions and the formation of hydrogen sulfide at protected artificial beaches. Left: Dark substance suspended in the water when seabed at shallow water is disturbed, artificial beach in lagoon environment at the Egyptian NW coast. Right: Dark substance at beach in artificial lagoon in the Red Sea.

The requirement to a small content of gravel and coarse fractions is important for the quality of the beach surface as the action of the waves will wash away the fine fractions, leaving the beach armored with the coarse fractions. Such a beach surface is unpleasant to walk on (see examples in Fig. 5.9).

Fig. 5.9. Artificial beaches with too much coarse material. Left: Beach Park in Copenhagen. Section with too high content of gravel and pebble. Right: Artificial beach in the UAE. Too high content of shells and coral debris.

It is evident from the above examples that it is of great importance to use very good quality fill sand for the construction of artificial recreational beaches.

5.4. Design Guidelines for Artificial Lagoons

The most important landscape elements in many coastal development schemes are attractive tidal lagoons. Such lagoons provide an attractive protected marine environment, in addition to posing major technical challenges. Design guidelines for elements of artificial lagoons are discussed in the following.

Lagoon mouth and channel sections: The stability of tidal inlets is a science in itself, and so it would not be discussed in detail here. It shall just be pointed out that the stability of tidal inlets is a major issue at littoral transport coasts. This means that careful studies are needed to secure a stable tidal inlet, which is not blocked by sedimentation. The required cross-sections of the inlet will be dependent on the tidal volume in the lagoon. No specific criteria can be given. However, the cross-section area of mouth and channel sections of artificial lagoons shall be so large that peak tidal current velocities are less than ~0.8 m/s. The width and depth shall also be designed according to guidelines for safe navigation if the lagoon is to accommodate navigation by motor and sailing yachts.

Open water body: The main purpose of introducing artificial lagoon and channel elements in a coastal development scheme is to add attractive landscape elements adjacent to an urban development area. The lagoon may be designed to accommodate water sports, navigation, sport fishing, and bathing, but bathing will never be as attractive in a lagoon as in the sea. The most import function is to provide the inherent attraction of water to an area that does not have this in its native condition. The lagoon shall therefore be properly designed as an important landscape element.

A set of water quality objectives for the lagoon area has to be established in order to secure an appropriate water quality in the lagoon elements. This could, for example, be in the form of international bathing water quality standards, e.g., the EU Standard, Ref. 2. The water quality objectives reflect the designated use of the planned water body elements and specify sets of water quality criteria (transparency of the water, concentration of nutrients and pathological bacteria, etc.) that have to be met in order to secure the planned recreational uses of the water body.

The water quality in the lagoon has consequently to be planned and regulated to meet the established water quality objectives based on factors such as:

— Control of local sources of pollution such as point sources (rivers and sewage discharges) and diffuse sources (surface run-off and drainage waters)
— Degree of flushing of the lagoon with water from the sea and/or from rivers considering also the water quality in these.

The retention time of the water in the lagoon and channels is often relatively long because the extension of these elements is maximized in order to facilitate the recreational use of the lagoon. Consequently, it is imperative to avoid or seriously reduce all point sources of pollution as well as to avoid or reduce the use of fertilizers or treated sewage water for irrigation of green areas in the proximity of the lagoon elements. There are two reasons for these requirements. (1) Primarily to avoid seepage of nutrients to the lagoon waters; (2) Fallen leaves and cut grass from these recreational areas are often mineralized locally to improve the quality of the soil, whereby all phosphorous and most of the nitrogen contained in the fertilizers or the treated sewage waters used for irrigation are eventually washed to the groundwater and further to the lagoon water. These supplies of nutrients to the lagoon waters may cause eutrophication unless controlled by increased flushing. As brackish waters are generally less diverse than freshwater or marine ecosystems, it is advisable to avoid bigger discharges from rivers in the lagoon.

The need for flushing with sea water is dependent on a number of factors such as:

— Flow and quality of river water discharged to the lagoon (control of salinity and water quality)
— Local discharges of nutrients and bacteria from point and diffuse sources as such discharges may cause eutrophication and reduced bathing water quality
— Water quality in the ambient seawater used for flushing (a high flushing with polluted water does not solve the problem).

The flushing can be expressed in terms of a characteristic "flushing time" T_{50}, which is the time it takes before 50% of the water in the lagoon system has been exchanged with clean water from the sea outside the lagoon during a design scenario. The design scenario shall be a calm and warm period, as this is most critical for flushing and water quality. There are no specific requirements for the flushing time. An acceptable flushing time for a natural lagoon with minimal local sources or freshwater supply, pathogenic bacteria, or nutrients will normally be 5–7 days, but the flushing time for artificial recreational lagoons should preferably be shorter.

It will under many conditions be recommendable with more than one opening to accommodate sufficient flushing, but two openings do not necessarily provide sufficient flushing as this depends on the local tidal conditions. Consequently, it will sometimes be required to establish forced flushing by tidal gates or by additional pumping. Other rules of thumb are:

— Water depths shall not be larger than 3–4 m
— There must be no local depressions in the seabed
— There must be no discharge of pollutants to the lagoon such as sewage, storm water, brine, cooling water, pesticides and nutrients.

Perimeters: Normally it will be difficult to obtain a good-quality beach inside a lagoon for the reasons discussed in the previous section. The following guidelines should be followed to obtain the best possible lagoon beaches if lagoon beaches are embarked on despite the above "warnings":

— Use only high-standard beach sand as explained under the design guidelines for beaches
— Construct the wanted beach profile from the beginning
— Build only beaches at exposed locations in "large" lagoons with dimensions preferably >2–5 km and water depth not less than 2 m or select a beach site near the opening to the sea where some wave exposure is secured
— Build only beaches if the amount of suspended substances in the lagoon water is very small, say in the order of less than 5–10 mg/ℓ.

It is strongly advised to consider other alternatives than beaches inside lagoons.

5.5. Landscape Elements of Waterfront Developments

The recreational and landscaping requirements in relation to design of the marine elements of waterfront developments are discussed in the following. The characteristic elements are the following:

— Beaches
— Other types of water edge treatments
— Landscape behind the shoreline perimeter
— Lagoon areas.

Far too many coastal projects are developed without a clear understanding and respect for the natural hydraulic and coastal processes that are decisive for the overall layout of the marine elements of a development. Consequently, many projects are developed without a clear idea of which layouts are feasible and of how the different elements can support each other. This may lead to unsuitable shoreline perimeter solutions of a poor quality and consequently major expenses for maintenance, often resulting in poor results.

The planning process for a waterfront development shall consequently secure a balance between the objectives of the developer and the possibilities that the specific development site can offer in terms of artificial marine elements, and taking environmental impacts into consideration. The main objectives of the developer of a waterfront development, being a public authority or a private company are typically the following:

— Enhancement of economic development possibilities in an area
— Expansion of the length of water perimeter through establishment of artificial water bodies inland or by reclaiming land in the sea
— Development of attractive marine landscape elements combined with recreational and service facilities
— Provide balanced public and private access
— Provision of attractive sandy beaches and other water edge treatments
— Establishing coastal structures combining technical requirements with recreational requirements
— Establishing optimum internal functionality and minimum environmental impacts.

These objectives shall be balanced against the possibilities and constraints offered by the natural conditions in the development area — where can beaches be located and how shall they be orientated, how can good flushing and water quality be secured in artificial lagoons, etc. These conditions are already discussed in the previous sections where design guidelines for artificial beaches and artificial lagoons are presented.

In the design and planning process for marine landscape elements and coastal structures, it is especially important to secure that technical requirements are combined with recreational requirements, which means that most marine elements shall be designed as multifunctional facilities. Examples of such layouts are listed below and examples are presented in the following figures:

— A terminal structure designed as a viewing headland (see Fig. 5.10)
— A groin structure designed as a recreational pier (see Fig. 5.11)
— A marina designed as a terminal structure for a beach section, note the detached breakwater providing a soft transition (see Fig. 5.12)

Fig. 5.10.   Amager Beach Park, Copenhagen with terminal structure as viewing headland.

Fig. 5.11.   Amager Beach Park, Copenhagen, groyne designed as recreational pier.

Fig. 5.12. Kalø Vig Marina, Denmark. Marina with detached breakwater providing soft transition between beach and marina.

— Smooth transition between structures and beaches provided by open angle between beach and terminal structure (see Fig. 5.10)
— Artificial lagoons, canals, marinas and sea baths designed to provide a combination of the following assets: Attractive landscape elements, areas for water sports, navigation and mooring, viewing, promenading, bathing and provision of open space (see Figs. 5.13 and 5.14)
— A lagoon opening designed as a marina (see Fig. 5.19 in next subsection)
— Water edge perimeter structures designed to combine technical requirements with recreational requirements. Examples of structures could be: (A) Water edge treatments with low-lying stepped promenades/platforms providing close connection to the water and different

Fig. 5.13. Amager Beach Park, Copenhagen. Lagoon with water sport activity.
Note: Nonexposed beach of low quality.

Fig. 5.14. Port of Copenhagen, bathing activity.

safety levels against flooding (see Figs. 5.15 and 5.16). (B) Revetments equipped with steps for access to the water etc. (see Fig. 5.17).

— A seawall designed as an integrated part of the beach furniture (see Fig. 5.18).

Fig. 5.15. Terraced water edge treatment providing close connection to the water with different safety levels against flooding.

Fig. 5.16. Low platform providing possibility for water sport activity.

Fig. 5.17.   Malmø West Harbor, revetment equipped with steps and platforms for access to the water.

Fig. 5.18.   Amager Beach Park, Copenhagen. Upper: Promenade separated from beach by a seawall.

Landscaping principles for layout of artificial beaches should focus on as long and uninterrupted beach sections as possible and minimization of the number of structures as these principles will enhance the natural appearance of a beach section.

An integrated planning based on workable marine elements will make it possible for developers and planners to create unique and sustainable coastal developments where the possibilities and requirements of nature are combined with the economic and recreational requirements of the developer.

5.6. An Example of a Successful Beach Park Development

A new beach park was recently built in Copenhagen using the principle of making the new beaches exposed by moving them out to deep water, thereby avoiding the shelter provided by the existing shallow shoreface. An aerial photo of the new beach park just after finalization of the civil works is presented in Fig. 5.19.

Fig. 5.19. Aerial photo of Amager Beach Park, which consists of the following main elements: Island with terminal structures north and south and a separating headland between northern and southern beaches and a lagoon. Designed by DHI, Hasløv, and Kjærsgaard and NIRAS.

The main wave directions at the site are NE and SE, which have been utilized to make two sections of beaches separated by a headland, one facing towards NE and one facing towards SE. Note the Y-shape of the headland structure providing a smooth transition between the structure and the beaches, which secures minimum trapping of floating seaweed and debris in the transition areas. The new beaches have been constructed on an island and a new lagoon (excavated) has been built between the island and the old shoreline. There is always a good current just off the beach park as it is located in the Sound, the strait between Denmark and Sweden, where the water exchange between the Baltic and the North Sea takes place. This results in a good flushing of the lagoon.

The beach park has been very well received by the inhabitants of Copenhagen and it has in an opinion poll been nominated as the best beach in the Copenhagen area. It has also received a reward from the "Society for the Beautification of the Capital", and people are enjoying all the facilities in the park (see Fig. 5.20).

Fig. 5.20. Amager Beach Park, Copenhagen. Kite surfing on the Lagoon, the beach island in the background.

5.7. New Concept for an Offshore Development Scheme

A new concept for an offshore development scheme utilizing the principles presented in this article has been developed by DHI and Hasløv &

Kjærsgaard, Planners and Architects, Copenhagen. The scheme is universal in the way that it can be implemented at any location where there is a development need and where there is wave exposure. The scheme has been developed under the motto: "Work with Nature" understood in the way that the wave exposure at the site is considered as a valuable gift from nature, which is utilized for developing a superb recreational facility in the scheme, namely a high-quality exposed beach.

The jewel of the scheme (Fig. 5.21) is the unique half moon shape ocean bay providing a superb sandy beach. The scheme thus offers the possibility for developing an extremely high-quality urban beach, which can be equipped with an attractive cornice along which most of the important recreational and leisure functions of the new city can be developed such as promenades, retail downtown areas, advanced apartment schemes, hotels, marine sport facilities, entertainment facilities, parks, etc.

Fig. 5.21. "The Universe" concept for an offshore development scheme, Artist's impression of the beach in "The Universe" and a possible concept for a marina. © DHI and Hasløv & Kjærsgaard. All rights reserved.

5.8. Investigation Methodology

5.8.1. *General Requirements*

The building of a waterfront development scheme in a coastal area is on the one hand imposing an impact on the marine and coastal conditions at the site and on the other hand it is exposed to the marine and coastal processes at the site. The art of developing a successful and sustainable waterfront development is to minimize the negative impacts and at the same time to utilize the possibilities given by the natural conditions at the site such as flushing and wave exposure, for a successful development of the scheme. What is characterizing a successful scheme is often that it solves existing problems at a site such as coastal erosion and flooding, at the same time as it introduces new coastal and marine facilities to the area such as artificial beaches and lagoons, marinas, viewing points, and sea promenades, etc.

It is relevant to obtain information on the natural conditions in the area. The types of relevant data to be collected are listed in the following:

— Geological data
— Topographic and bathymetric surveys and maps
— Historical and recent aerial photos and satellite images
— Shoreline development maps
— Sediment sources and sinks in the area
— History of coastal structures and nourishment
— Sedimentation and maintenance dredging in ports, navigation channels, and tidal inlets
— Land use maps
— Meteo-marine data (winds, tides and storm surges, sea level rise, waves, currents and ice).

Studying these data will most probably provide a good impression of the important processes in the area, the impact caused by the existing structures and the possible problems.

5.8.2. *Hydraulic Studies*

The reliability of numerical models representing coastal processes has increased dramatically in recent years. This development is driven by the ever-increasing computer power, combined with research and increased understanding of the physics (hydrodynamics, sediment transport, and ecology), which is reflected in the development of numerical models.

The models can provide detailed information of wave and flushing conditions and on littoral transport and shoreline development, which can provide insight into the cause of present problems and the consequences of different schemes for remedial action. The numerical models can also be used in functional analysis of waterfront development schemes and for layout optimization as well as hydraulic design. Finally the models can be used for analysis and quantification of environmental impact. The relative ease of using such models, however, can be deceptive. The choice of the modeling methodology and the interpretation of the simulation results require a thorough understanding of phenomena treated, both the large-scale coastal morphology and the physical processes involved at a smaller scale. The numerical models are only useful when applied in combination with qualified engineering judgments. It can thus be concluded that predictions by numerical models can be extremely useful provided that the following conditions are fulfilled:

— The input data shall be of good quality.
— Data shall be available for calibration, validation, and production runs.
— The models shall be of high scientific standard and cover the relevant subjects such as waves, hydrodynamics, sediment transport, and ecology. It is important to realize which assumptions have been introduced, what the model can represent and with which accuracy.
— The scheme to be tested shall be defined on basis of an understanding of the physical processes. It is important to realize that even the most up-to-date numerical hydraulic model only provides documentation of the functional characteristics of the scheme having been defined for testing; the model does not in itself provide the optimal scheme. It is up to the hydraulic engineer and the master planners to define a layout, which has near optimal functional characteristics. The model can then be used to document the actual functional characteristics of the tested scheme, on which basis further optimization of the scheme can be introduced and tested. It is therefore of utmost importance that the definition of the hydraulic layout of a scheme is made on basis of a profound hydraulic knowledge and experience in order to secure an optimal outcome of its testing using numerical models. The important aspect for good functionality of coastal schemes is the subject of the previous sections of this chapter.

This chapter summarizes the capabilities and limitations of different modeling concepts, and provides some indication of the predictions that can be obtained by proper application of a given type of model. The emphasis is placed on models describing sandy coasts and on models primarily based on a description of physical processes rather than on data-driven or rule-based models.

Specific modeling techniques such as single grid, multiple grids, or flexible mesh techniques will not be addressed. A summary of the following characteristics for numerical models typically used for layout analysis and hydraulic design of coastal development schemes are presented in Table 5.3:

— Phenomena to be simulated
— Type of numerical model
— Model size

Table 5.3. Summary of characteristics and types of numerical models applicable of layout analysis and hydraulic optimization of coastal development schemes.

Phenomenon to simulate	Type of model	Model size in m	Input data	Typical results
Wave transf. from offshore to nearshore	Spectral wind-wave Propagation model (WP)	$>10^3$	Bathymetry Offshore waves Wind over modeling area	Wave patterns Statistical distribution of waves Design wave conditions Input to HD, ST and LT models
Wave disturbance in ports and local bays and the like	Short Wave model (SW)	$10^2 - 10^4$	Bathymetry Nearshore wave data	Wave height/direction distribution in port/bay area Wave reduction factor distribution
Tides, storm surges and currents	Hydrodynamic model (HD), 2D or 3D	$>10^2$	Bathymetry Tide Wind Radiation stress from wave model	Tidal analysis Storm surge design data Current patterns Input to Flushing model Input to ST models
Flushing, spreading and sedimentation of substances	Advection-Dispersion model (AD)	10^3-10^5	Hydrodynamics from HD model Source of substances	Flushing pattern and time Transport, dispersion, decay and sedimentation of dissolved and suspended substances
Water Quality	Ecological process model	10^3-10^5	Hydrodynamics from 3D HD model Source of substances WQ data	Distribution of: ammonia, nitrite and nitrate, total nitrogen, phosphate, total phosphorous, chlorophyll a and Secchi disk depths, and others

(*Continued*)

Table 5.3. (Continued)

Phenomenon to simulate	Type of model	Model size in m	Input data	Typical results
Sediment Transport over an area	2D Sediment Transport model (ST) for sand/(MT) for mud	10^3–10^5	Bathymetry Sediment charac. Input from HD and WP models	Transport fields for selected storms and "annual" situation Initial and annual sedimentation and erosion patterns
Morphological Development over an area	Morphological evolution model for sandy environments	10^3–10^4	Bathymetry Sediment characteristics Boundary conditions on waves and hydrodynamics	Evolution of seabed during selected storm events
Littoral Transport	Annual Littoral Transport in a coastal profile (LT)	10^2–10^3	Bathymetry Sediment characteristics Wave statistics or time series	Annual littoral transport in a profile Transport distribution in the profile Equilibrium orientation Closure depth
Shoreline Evolution	One-line shoreline evolution model	10^2–10^5	Initial shoreline and coastal structures Annual transport data from LT model	Present shoreline evolution Future shoreline evolution (without and with stabilization scheme) Shoreline variability
Profile Evolution	Surf zone coastal profile model	10^2–10^3	Coastal profile Sediment characteristics Waves from WP	Profile evolution during storms

— Input data
— Typical results of the modeling.

   Physical scale models are generally mainly used in the detailed design phase for stability of armor layers, overtopping of structures, and wave agitation in port basins and semiclosed bays.

5.9. Conclusions and Recommendations

It is concluded that it is crucial for a successful design of beach and lagoon elements in waterfront developments that the hydraulic, coastal, and environmental aspects are included in the planning from the early planning stage. The reason for this is that the design of these elements has to follow the "rules of nature", which impose certain restrictions on the design.

The main issues to observe are the following:

Artificial beaches:

— Good-quality recreational beaches shall be moderately exposed to waves, they shall be orientated towards the direction of the prevailing waves to be stable and terminal structures shall be constructed to prevent loss.
— Artificial beaches shall be constructed by good quality beach sand: medium, i.e., $0.25\,mm < d_{50} < 0.5\,mm$, well sorted, attractive color, minimum content of fines and minimum content of coarse fractions and no content of organic matter.
— Coastal structures adjacent to beaches shall be designed so that no dangerous currents are generated.

Artificial lagoons:

— High water quality standards shall be secured in recreational lagoons. The "flushing time" T_{50} shall be better than 5 to 7 days, which may require special precautions.
— The lagoon mouths shall be stable and free of sedimentation.
— Water depths shall be between 2 and 4 m.
— There must be no local depressions in the seabed.
— There must be no discharge of pollutants to the lagoon such as sewage, storm water, brine, cooling water, pesticides and nutrients.

Waterfront developments in general:

— A thoroughly planned location and layout of the urban elements integrating recreational demands with the natural dynamics of artificial beaches and lagoons are important. Coastal structures shall also have recreational functions.

References

1. K. Mangor, *Shoreline Management Guidelines*. Published by DHI Water & Environment, ISBN 87-981950-5-0 (2004).
2. Directive 2006/7/EC of the European Parliament and of the Council of February 15, 2006 concerning the management of bathing water and repealing Directive 76/160/EEC.
3. Amager Beach Park, Copenhagen, Technical reports (Coastal hydraulics, landscaping, detailed design and EIA) for Amager Beach Park I/S by DHI, Hasløv & Kjærsgaard, Planners and Architects, and NIRAS, Consulting Engineers and Planners (2003–2005).
4. K. Mangor, *Successful Waterfront Developments — Work with Nature*. Key Note Speech at ARABIANCOAST 2005, Dubai (2005).
5. K. Mangor, I. Brøker and D. Hasløv, *Waterfront Developments in Harmony with Nature*. Article in *Terra et Aqua*, No. 111, June (2008).

Chapter 6

Risk-Based Channel Depth Design Using CADET

Michael J. Briggs

Coastal and Hydraulics Laboratory,
U.S. Army Engineer Research and Development Center,
Vicksburg, MS, USA
Michael J.Briggs@usace.army.mil

Andrew L. Silver and Paul J. Kopp

Naval Surface Warfare Center, Carderock Division,
West Bethesda, MD, USA

Channel analysis and design evaluation tool (CADET) is a coupled set of computer programs to determine the "optimum" dredge depth for entrance channels. CADET is summarized and two validation examples are presented and compared. In the first example, field and laboratory data for a bulk carrier from Barbers Point, HI, are analyzed. In the second example, CADET predictions are compared with differential global positioning system (DGPS) field measurements for a tanker and containership in Ambrose Channel, New York. The CADET predictions showed good agreement with the measurements in both cases, especially when considering wave heights were very small in both examples.

6.1. Introduction

Historically, port designers relied on deterministic approaches with large safety factors for channel design.[1-3] Recent trends are to incorporate a risk-based methodology in the design process to minimize dredging and define a useful design life with an acceptable level of risk of accidents or groundings.[4,5] Channel analysis and design evaluation tool (CADET) is a program to determine the "optimum" dredge depth for entrance channels. This optimum dredge depth is defined as the depth that provides the greatest accessibility for the least amount of dredging and is determined by predicting ship underkeel clearance (UKC) for different wave, ship, and channel combinations. It is based on probabilistic risk analysis techniques to evaluate the accessibility of a series of channel reaches for multiple vessel geometries, loading, and wave conditions.

The CADET technology was developed from the environmental monitoring and guidance system (EMOGS), which provides real-time operational guidance on the expected UKC for deep draft naval vessels.[6,36] Although EMOGS has been used primarily in the naval arena, this technology is just now being applied to commercial ships. Previously, the CADET model was compared with a probability assessment model composed of Poisson/Bernoulli modules for the Barbers Point data with very good results.[7,8] Thus, this chapter summarizes the CADET theory and its operational features, and compares the CADET predictions of vertical ship motions in two validation examples. In the first example, field and laboratory measurements for the *World Utility* bulk carrier are presented and compared.[9,10] In the second example, field DGPS measurements for the *Bear G* tanker and *Frankfort Express* containership are compared during transits of Ambrose Channel, New York.[11]

In the second section of this chapter, the CADET theory including background, vertical UKC calculation, and uncertainty and risk analysis is summarized. In the third section, CADET ship, project, analysis, and results features are discussed. The first validation example for Barbers Point is presented in the fourth section and contains a summary of the project site, field measurements, physical model, and wave-induced vertical motion comparisons. The fifth section contains a description of the Ambrose Channel validation example. It contains the channel description, field measurements, waves, DGPS calculations, and vertical motion allowance comparisons. The last section contains a summary and conclusions.

6.2. CADET Theory

6.2.1. *Background*

CADET was developed by the Naval Surface Warfare Center, Carderock Division (NSWCCD) under contract to the U.S. Army Engineer Research and Development Center (ERDC), Coastal and Hydraulics Laboratory (CHL).[12] CADET is an expansion of the technology developed to determine the depth of entrance channels to new homeports for Nimitz-class Aircraft Carriers (CVN 68). The CADET technology was initially developed for EMOGS.[6,36] EMOGS provides operational guidance on the expected UKC of a vessel, given real-time wave and water level measurements or observed conditions at a particular port. For each UKC prediction, it also calculates the uncertainty and risk of touching the channel bottom for those conditions. EMOGS evaluates clearance and risk for a single specified ship at one channel depth, using a single wave spectrum, for a transit in one direction at a specific date and time. Astronomical tide effects on the water level are included and the duration of transit for a given ship speed and entrance channel configuration are taken into account. Meteorological effects on water level due to barometric pressure are also included. EMOGS is installed at two naval stations in the United States and has been in operation for over 20 years. During this time, no known incident of bottom touching or grounding has occurred, and the users have not complained that the results are too restrictive.

CADET differs from EMOGS in several respects. It is more of a design tool as it evaluates clearance and risk for a range of possible water depths. In addition, it evaluates the entrance channel depths for any channel cross-section. Annual local wave statistics are used to determine the accessibility of the transit channels, expressed in days per year. Astronomical and meteorological tide effects are not explicitly included since, for design purposes, a transit could occur in either direction at any time. Water level changes can be included by varying the project depth relative to ship draft. A tide calculator is a postprocessor option that can be used to indicate additional days of accessibility due to hindcast of 20-year tidal cycles for a particular location.

6.2.2. *Vertical UKC Calculation*

CADET calculates the vertical UKC of a specific ship, commercial or naval, at a specified channel location, and provides information to aid in determining the optimum dredge depth. This optimum depth is defined as the shallowest depth that allows the maximum days of access for any given year at that location. The accessibility of the channel is determined by calculating the vertical UKC and the risk of the vessel touching the channel bottom under all wave conditions that are present. The general rule is that if the risk α of the ship touching a flat channel bottom is less than 1 in 100 (i.e., $\alpha = 0.01$) for each wave condition in a climatology during a given transit, then the channel is considered accessible for that depth. The Navy is comfortable with this level of risk and corresponding accessibility. The number of days per year the channel is accessible is dependent on the persistence of the local wave conditions obtained from the local wave climatology.

The dynamic UKC of the vessel is influenced by five major parameters that include:

- Static draft and trim of the ship at rest
- Underway sinkage and trim
- Wave-induced vertical motions
- Hydrologic factors of channel depth at mean lower low water (MLLW) project depth
- Change in water level due to the astronomical tides.

Because CADET is primarily a channel-depth design tool, ephemeral parameters such as meteorological tides are not factored into the calculation. As mentioned previously, CADET does have a postprocessing option for tidal effects. Otherwise, the user can input equivalent tides in the range of water depths used for the predictions. Also, CADET does not explicitly include channel width or bank effects for ship motions, but they are taken into account when calculating sinkage and trim.

Figure 6.1 shows the major parameters considered when calculating the vertical clearance of the ship in a channel. The static UKC is the difference between the nominal channel depth and the static at-rest draft of the vessel. Static trim must also be taken into account. As the ship travels at speed along the channel, the ship both sinks and trims (i.e., squat or midship

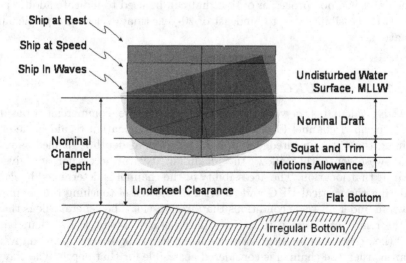

Fig. 6.1.　Cross-section of a ship in a channel.

sinkage and trim by the bow or stern) due to a pressure field between the hull of the vessel and the channel bottom. The net effective UKC_j at location j (i.e., jth control point on hull surface usually corresponding to bow, stern, rudder, or bilges) is given by

$$UKC_j = D_c + E_t - (T_j + S_j + A_j) \tag{6.1}$$

where D_c is the nominal channel depth at MLLW, E_t is water level due to tide relative to MLLW, T_j is static draft, S_j is ship squat, and A_j is the vertical motions allowance.

A_j is determined from the vertical wave-induced ship motions of heave, pitch, and roll. The magnitude of the vertical displacement at a point on the ship is dependent upon the height and period of the waves in the channel, the ship speed, the relative ship heading to the waves, and the channel depth. For the coupled heave and pitch motions at the bow and stern, a vertical displacement complex transfer function $H_j(f, \theta)$ is calculated as

$$H_j(f, \theta) = z(f, \theta) + X_j \Theta(f, \theta) \tag{6.2}$$

where j corresponds to either the bow or stern locations, $z(f, \theta)$ is the vertical heave motion transfer function, X_j is the longitudinal distance from the ship's center of gravity to the jth control point (in this case, the bow or stern), and $\Theta(f, \theta)$ is the pitch transfer function. A similar transfer function is calculated for the coupled heave and roll motions to determine the vertical displacement on the sides of the ship. These transfer functions are then used in the calculation of the σ_j RMS (root mean square) displacement at the bow or stern given by

$$\sigma_j = \sqrt{\sigma_j^2} = \sqrt{\sum_f \sum_\theta S(f, \theta) |H_j(f, \theta)|^2 \Delta f \Delta \theta} \tag{6.3}$$

where $S(f, \theta)$ is the directional wave spectrum, $|H_j(f, \theta)|^2$ is the square of the modulus of the complex transfer function, known as the response amplitude operator (RAO), and Δf and $\Delta \theta$ are the increments in frequency and direction.

Because of phase differences, the σ_j calculated from individual wave conditions may not provide the largest vertical excursion the ship can experience during a transit. Therefore, higher order extremal statistics are used to define an expected extreme motion allowance A_j during a given transit as

$$A_j = \sigma_j \sqrt{2 \ln \left[\frac{T_d \sigma_{vj}}{2\pi \alpha \sigma_j} \right]} \tag{6.4}$$

where T_d is exposure time in the channel (i.e., reach length/ship speed), σ_{vj} is the standard deviation (RMS) of the vertical motion velocity (i.e., time derivative of σ_j) at location j, and α is the risk parameter, normally taken to be 0.01 (i.e., 1/100) in CADET.[13] If $\alpha = 0.01$, then the ship has a risk of 1 in 100 that the predicted motions allowance A_j will be exceeded for the given set of wave conditions over the given exposure time.

6.2.3. *Uncertainty and Risk Analysis*

Each of the parameters in Eq. (6.1) has inherent uncertainties. As an example, Table 6.1 lists the uncertainty (i.e., bias and variability) in each of the major parameters that make up UKC_j for a large naval vessel. First, the channel depth for CADET has no bias or variability because it is a deterministic parameter. Second, the uncertainty in the static and dynamic drafts comes from the estimation of the draft at the pier, from the draft marks, and the method that calculates the sinkage and trim. The error band in the static draft is assumed to be known within a range of ±1%. The critical points of the bow, stern, and bilge (i.e., port and starboard amidships on the keel) therefore have an error band of ±0.5 m (1.8 ft) or within 4.5% of the actual value at the bow and stern and ±0.2 m (0.6 ft) or within 1.5% of the actual value at the bilge. The sinkage estimate is usually based on an analytical method or model test results. The uncertainty in the sinkage stems from the scatter of the data from model tests and how well the calculated results fit the model test results. This gives a variability of the sinkage parameter of ±0.1 m (0.4 ft) with no bias, as shown in Table 6.1.

The final two parameters in Table 6.1 are the wave statistics and the wave-induced motions. The wave statistics are usually generated from hindcast wind models, and then transformed from offshore to the channel with a shallow water wave model. Most models are validated with buoy

Table 6.1. Uncertainty in major CADET parameters (all values represent ±2σ range).

Parameter	Bias	Variability
Channel depth	None	None
Static Draft		
Bow and stern	None	0.5 m (1.8 ft) 4.5%
Bilge	None	0.2 m (0.6 ft) 1.5%
Squat (sinkage & trim)	None	0.1 m (0.4 ft) 1.0%
Transformed wave spectra	0.2 m (0.6 ft)	80%
Wave-induced motions based on measured wave data	20% over predicted	34%

data. The bias and variability shown in Table 6.1 for the transformed wave spectra assumes a measurement bias and large variability in the wave input. The usual models for global ocean wave prediction using wind hindcast data introduce a bias of ±0.2 m (0.6 ft), and an average RMS error of 0.8 m (2.6 ft), that translates to an error band of about 80%. The variability of the wave input, especially wave height, is the main driver of the variability in the motion estimates. There are also uncertainties in the calculation method of the response amplitude operators.

The shallow water motions calculation for Navy ships was based on a software program that was a hybrid of the Navy Standard Ship Motion Program (SMP). The motions of the commercial ships used in CADET were computed through the shallow-water version of SCORES.[14,15] The bias and uncertainty shown in Table 6.1 reflect those of a large Navy ship. As more experience using CADET is attained for commercial ships, the bias and variability could change. However, a large component of the uncertainty and bias in the motions calculation comes from the uncertainty in the wave measurement. Therefore, the difference in the uncertainty of the motions between commercial and Navy ships may be small.

The primary objective for calculating uncertainty is to provide a measure of risk of the vessel touching the various project depths being considered. Risk is defined as that proportion of all possible transits under statistically constant conditions in which the minimum channel clearance would be negative. The risk model accounts for the uncertainty in each of the parameters by assuming a Gaussian distribution for static ship draft and underway sinkage and trim, and a Rayleigh distribution for the vertical motion and velocity variances. The Rayleigh distribution reflects the most likely probability distribution of the waves. Using these distributions, the probability density of the largest motion excursion or the minimum UKC is determined and its area up to a minimum clearance of zero is calculated.

Under this definition of risk, it is necessary to compute the probability density of the net effective clearance and determine the area up to zero net effective clearance. The net effective clearance, therefore, is defined as the difference between the random variables that make up the effective channel depth and the effective vertical displacement of the ship. These random variables are a function of the uncertainty in each of the major parameters that make up the net effective clearance.

Thus, a risk analysis is performed to determine the probability of any one of the critical points of the deep draft vessel touching the channel bottom for inbound and outbound transits. The critical locations on the vessel usually are the bow at the keel, the rudder(s), and the port and starboard bilges at amidships. The risk analysis is performed for each of the wave conditions in a wave climatology for the port. The significant wave height, the peak or modal period, and the primary direction define the wave condition. The result of the risk analysis provides a probability of the vessel touching the channel bottom under each of the wave conditions for a specified project depth. It is assumed that if the risk is greater than some threshold value (normally 1 in 100), then the channel is inaccessible by the vessel. The days of accessibility of the channel are calculated by determining the persistence of the wave condition that produces the risk of 1 in 100 or greater. The risk calculation is performed for each wave condition and a range of project depths. When complete, the optimum channel depth is the one with the greatest number of days of accessibility per year and the least amount of dredging.

6.3. CADET Organization

CADET is a set of computer programs that calculates effective UKC and bottom touching risk probability for any number of ships and loading conditions over a range of multiple project depths. It is organized into four basic modules in FORTRAN and C++ for defining and performing calculations relative to:

- Ship
- Project
- Analyses
- Results

English units (i.e., ft, ft^2/Hz, etc.) are the standard for the program, although metric equivalents can be converted for some inputs.

6.3.1. *Ship Parameters*

The first module contains all of the ship parameters to define a ship relative to geometry and loading. Figure 6.2 shows the nine categories for defining the ship that include (a) static draft and trim, (b) ship speeds, (c) loading parameters, (d) motion risk parameter, (e) water depths, (f) wave frequencies, (g) sinkage and trim, (h) critical point locations, and (i) ship motion transfer functions.

The most critical input is the ship geometry file that is represented by the "ship lines" drawing in Fig. 6.2. Ships are defined by a hull geometry file that is independent of loading condition. The geometry file represents the ship in terms of hull offsets, from the keel to the deck-at-edge, at 21 equally

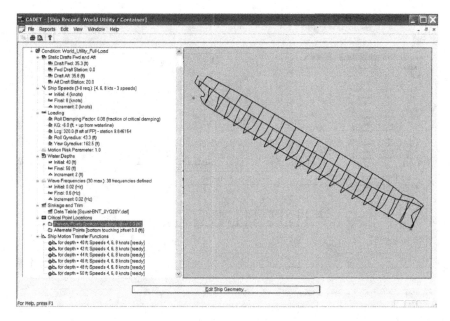

Fig. 6.2. Example CADET ship record with red dots denoting the critical point locations.

spaced stations between the forward and aft perpendicular. These geometry data files can be prepared externally (manually or by conversion from other representations) and imported into CADET, or they can be created using a built-in graphical geometry editor. The spacing between these 21 stations is determined by the ship's waterline length or the length between the forward and aft perpendiculars, L_{pp}. Figure 6.3 shows an example of the station cross-sections. The ship's beam B is defined by these cross-sections. Points defining the x, y, and z coordinates along the ship lines have specific "types" to properly identify the ship lines at beginning and ending points and discontinuities. The origin of the ship geometry is defined in CADET as follows. The x-location identifies the station number starting at zero at the forward perpendicular and increasing to the aft perpendicular with each station equidistant from each other. The y-location is the transverse distance from the centerline of the ship amidships. The z-location is positive upward from the baseline of the ship.

6.3.1.1. *Draft and ship speed*

The static draft and trim are defined at either (a) the forward (T_{FP}) and aft (T_{AP}) perpendiculars or (b) amidships draft and a trim angle in degrees.

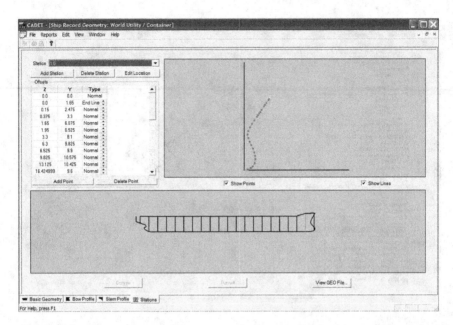

Fig. 6.3. Example CADET ship record geometry for stations. The green highlighted station ship line cross-section is shown in the upper right plot.

Measurement errors in forward and aft drafts can be specified to account for uncertainty in these values. Up to 8 ship speeds in knots can be entered as integer values.

6.3.1.2. *Loading*

Multiple loading conditions can be defined for each ship in CADET. The ship loading parameters that affect the three vertical motions of heave, pitch, and roll include (a) longitudinal center of gravity L_{CG}, (b) vertical center of gravity V_{CG}, (c) roll damping factor, (d) roll mass radius of gyration k_4, and (e) pitch and yaw mass radius of gyration k_6. Figure 6.4 is an example of the built-in calculated hydrostatics that provides some insight into the values for the loading section.

Figure 6.5 illustrates ship stability for static equilibrium and free unresisted rolling. Static equilibrium is based on Archimedes principle where the weight W of the ship and cargo is balanced by the weight B of the water displaced by the ship. The longitudinal center of gravity L_{CG} is located approximately midway along the longitudinal axis of the ship. The center of buoyancy is the center of gravity of the fluid displaced by the ship (i.e., midships). The longitudinal center of buoyancy (as shown in the calculated

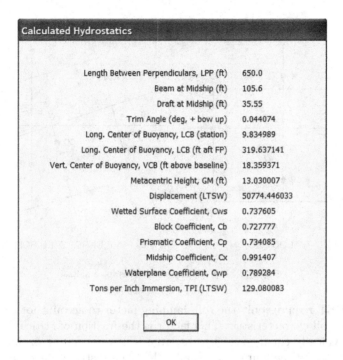

Fig. 6.4. Example CADET hydrostatics.

hydrostatics in Fig. 6.4) can be used to estimate of the L_{CG} for a prescribed ship loading condition, where both are measured positive, aft of the forward perpendicular. The vertical center of gravity V_{CG} is located along the vertical axis of the ship, approximately midway in the cargo as measured from the keel. In CADET, the V_{CG} is measured positive up from the waterline and varies with the type of ship. Tankers typically have negative values around 1.5 to 6.1 m (-5 to -20 ft) since their cargo is lower in the ship. Containerships, however, have positive values of approximately 1.5 to 4.6 m (5 to 15 ft) since their cargo is stacked on top of the deck as well as in the holds.

For stability the metacenter M needs to be above the CG. The metacentric height $\overline{GM}$ is the distance from the CG to M. When the ship rolls, the CB moves to a new position that is no longer in line with the CG and the vertical axis of the ship. The intersection of the vertical from the new CB' with the vertical axis defines the location of M. The righting moment of the ship is thus a function of the angle of roll, the weight of the ship and cargo, and the $\overline{GM}$.

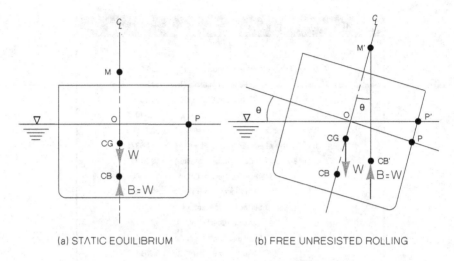

(a) STATIC EQUILIBRIUM   (b) FREE UNRESISTED ROLLING

Fig. 6.5.   Equilibrium conditions for a ship.

CADET requires only the roll damping factor to account for the ship's dynamic roll characteristics. This factor is the fraction of critical damping and is typically equal to 0.08 to as large as 0.4 depending on hull shape and appendage configuration. Finally, the mass distribution properties of the ship are defined by the roll and pitch gyradii k_4 and k_6 approximations given by

$$k_4 = 0.41B$$
$$k_6 = 0.25L_{pp} \tag{6.5}$$

6.3.1.3. *Motion risk parameter α*

The motion risk parameter α typically has a value of 0.01 for most design applications. However, for validation exercises where the motions are assumed to be known with certainty, a value of $\alpha = 1$ can be used.

6.3.1.4. *Sinkage and trim*

Underway sinkage and trim may be provided externally by calculations or by model test data and imported into CADET. Alternatively, it can be calculated within CADET using the BNT (Beck, Newman, and Tuck) potential flow program by Beck *et al.*[16] Although included in CADET, BNT is completely independent and standalone. Since channel geometry can vary from reach to reach, CADET supports the ability to define multiple sets of sinkage and trim data sets for the same ship and loading condition.

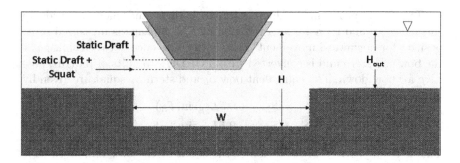

Fig. 6.6. BNT channel geometry variables.

The BNT sinkage and trim prediction program is based on early work by Tuck investigating the dynamics of a slender ship in shallow water at various speeds for an infinitely wide channel and for a finite width channel such as a canal.[17,18] This work was expanded to include a typically dredged channel with a finite-width inner channel of a certain depth and an infinitely wide outside channel of shallower depth.[16]

Figure 6.6 is a schematic of the simplified channel cross-section used in BNT. In addition to the automatically specified inside channel depth H, the user has the option to include the channel width W and outside channel depth H_{out} (i.e., similar to PIANC h_T trench height for restricted channels, but measured from the water surface to the top of the trench). The value of H_{out} remains the same for all H values. For unrestricted channel applications, the user can input "-1" in the H_{out} input space to automatically insure that the outer depths are equivalent to the inner channel depths, regardless of depth increment.

In his early work, Tuck calculated the dynamic pressure of slender ships in finite-water depth and infinite and finite-water width by modeling the underwater area of the hull.[17] This underwater area was defined by the 21 equally spaced stations along the ship's length. Therefore, the ship's geometry file, draft, speeds, and water depths are used in the BNT squat calculations. Within this analysis, the fluid is assumed to be inviscid and irrotational, and the hull long and slender. Input hull definition is provided in terms of the waterline beam and sectional area at 20 stations along the hull. The dynamic pressure is obtained for each depth Froude number (F_{nh}) by differentiating the velocity potential along the length of the hull. The sinkage and trim predictions are obtained from the dynamic pressure by calculating the vertical force and pitching moment that are translated to vertical sinkage and trim angle. The proper use of this BNT program requires that channel depths be of the same order as the draft of the ship, therefore satisfying the shallow-water approximations assumed in Tuck.[17]

The BNT program produces tabular listings and plots of midship sinkage S_{Mid} and trim T_R as a function of F_{nh}. Sinkage is measured in ft, positive for downward movement. Trim is the difference between sinkage at the bow and stern and is converted within CADET to units of radians, positive for bow down. The equivalent bow S_b and stern S_s squat are given by

$$S_b = S_{Mid} + 0.5L_{pp}\sin(T_R)$$
$$S_s = S_{Mid} - 0.5L_{pp}\sin(T_R)$$

(6.6)

This is a simplistic representation of the squat at the bow and stern as it assumes they are equally distant from the midpoint of the ship. In CADET, the squat is calculated for the actual distances to individual control points.

Finally, sinkage and trim errors can be specified to account for uncertainty in the predicted values. Typical values might be 3 to 6 cm (0.1 to 0.2 ft) in sinkage and 0.01 to 0.05 deg in trim.

6.3.1.5. *Critical point locations*

The red dots on Fig. 6.2 correspond with the critical point locations j previously discussed. The five primary control points are located on the centerline at the T_{FP} and T_{AP} to examine the effects of pitch at the bow and rudder(s), and along the port and starboard bilge to include the effects of roll. Four alternative (optional) control points can be added anywhere along the hull and they may include a minimum required vertical standoff distance from the channel bottom.

6.3.1.6. *Wave frequencies*

A total of 30 wave frequencies are input to define the range of frequencies containing significant wave energy that will be used to calculate the ship motion transfer functions or response amplitude operators (RAOs) for heave, pitch, and roll. The user inputs initial, final, and increment values of frequency in rad/s. Although the RAOs at a particular frequency are internally interpolated to match the specific wave frequencies in the project module, it is important that the RAO frequency range covers the range for which there is a nominally nonzero response amplitude over the range of nonzero wave spectral amplitudes, as they will be used in Eq. (6.3) to determine the CADET predictions.

6.3.1.7. *Ship motion transfer functions*

Finally, ship motion heave, pitch, and roll RAOs are calculated using a frequency domain, shallow water, strip-theory program (SCORES).[14,15] In a

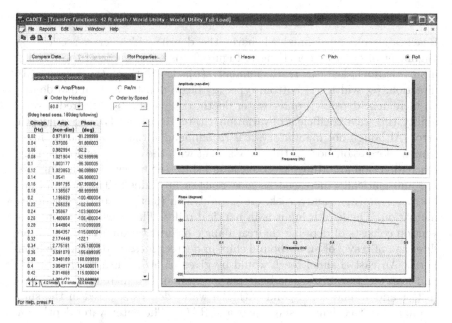

Fig. 6.7.   Example CADET roll ship motion transfer function.

manner similar to that used to calculate sinkage and trim, CADET generates SCORES input files from the defined hull geometry, draft and trim, ship speeds, water depths, roll damping coefficient, and wave frequencies. SCORES is run in the background by CADET, and the motion transfer functions are extracted from the SCORES output files. The extracted transfer functions are written to compressed binary files for later use in determining the A_j from Eq. (6.4). Plotting of the transfer functions can be performed with different representations as needed (real/imaginary or amplitude/phase versus frequency, frequency of encounter, or nondimensional wave length). Figure 6.7 is an example of a roll RAO showing amplitude and phase for the range of ship speeds and ship headings as a function of frequency.

6.3.2. *Project Parameters*

A project in CADET includes channel reaches, waves, ships, tides, and comments. While CADET keeps track of all of these direct and logical associations between projects, channel reaches, wave spectra, ships, loading conditions, and sinkage and trim data; the user is responsible for ensuring that these associations are coherent.

6.3.2.1. *Reaches*

Reaches should be defined whenever the depth, width, cross-section, or alignment of the channel changes. Figure 6.8 is an example of the reaches input. It includes (a) reach number, (b) description, (c) length, (d) direction, (e) width, (f) bottom type, (g) begin depth, (h) terminal depth, (i) increment depth, (j) outer water depth, (k) overdredge, (l) dredge variability, (m) wave coefficient of variation, (n) reach risk level β, and (o) depth error. Although not mandatory, it is recommended that reaches be defined from inshore to offshore as this is consistent with the convention for defining reach direction according to an outbound ship (see below). Reach numbers are automatically increased as new reaches are added. The description, bottom type (i.e., sandy, rock, etc.), and width (feet) are purely for documentation. The length input is in nautical miles. The reach direction convention consists of defining the reach angle or orientation according to the direction of travel of an outbound ship, in degrees measured clockwise from north. An east channel reach would be given as 90°, south reach as 180°, and west reach as 270°. The beginning, terminal, and increment depth are in feet and should correspond with the water depths selected in the ship module used to calculate ship squat and RAOs. The outer water depth can be defined as a fixed or variable value for the range of water depths. Again, a "−1" in this parameter will insure that an unrestricted channel cross-section is used for all depths. The next two inputs account for bottom variability. Overdredge is the amount of additional clearance assumed due to advance maintenance or dredging tolerance. Typical values are 61 cm (2 ft). The dredge variability is a tolerance for dredging execution tolerance to account for unevenness (i.e., nonhorizontal level) of the bottom. A typical value is

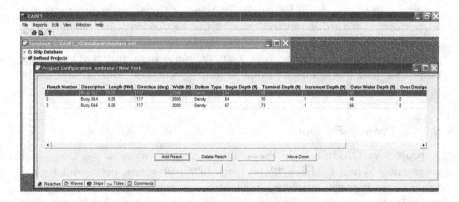

Fig. 6.8.   CADET reach example.

26 cm (0.85 ft). The wave coefficient of variation is an indication of the reliability of the waves with a typical value of 0.4 for estimated waves and 0.17 for measured waves. The reach risk level β is the risk of one of the five critical points of the ship touching the project depth in a particular reach. It can have values similar to the motions risk α, for example $\beta = 0.01$ would allow a touching of 1 out of every 100 transits. Finally, the depth error accounts for uncertainty in the depth measurements. Typical values are 3 cm (0.1 ft).

6.3.2.2. *Waves*

CADET requires directional wave spectra to predict vertical ship motions. The user can input one or more files as necessary for the project goals. Only one or two files might be necessary for validating CADET predictions with measured field or laboratory data for a particular ship, time, and transit direction. More than one file may be required to properly bracket transit times for validation. Design life predictions of channel accessibility, however, would require many spectra to properly represent the statistical variation in wave conditions over a 20-year design life.

6.3.2.2.1. Validation example wave input

For validation examples, measured laboratory or field data are usually available. This data may be from local unidirectional or directional wave gages or buoys. In some cases, a directional buoy is not on site near the channel location. In this case, a neighboring directional gage can be used if the distance is not too far away. Typically, a distance less than 5 nm may be considered "close enough." Otherwise, the waves should be transformed to the site using programs such as STWAVE to account for transformation mechanisms such as refraction, diffraction, and shoaling. If only a unidirectional wave gage is available, the directional spreading function $D(f, \theta)$ information can be provided using empirical equations. Depending on the wave gage distance to the site, travel time may also need to be considered.

6.3.2.2.2. Design life wave input

In the more typical application of CADET for design life predictions, it is customary to use something like a 20-year hindcast. The wave information study (WIS) is a good source of data for coasts around the United States (http://chl.erdc.usace.army.mil/wis). The user selects the WIS station that is closest to the project site and sorts it into joint distribution tables of wave height and period for fixed wave directions. Since WIS outputs data in 22.5 deg bins, it must be interpolated to a finer 15-deg resolution to accommodate CADET (SCORES) requirements. As before, if the WIS station is greater than about 5 nm from the project site, the data should be transformed to the site. Also, if the channel is long, waves may transform from one end to the other, so the STWAVE type of program can be used to predict ratios relating incident wave conditions to output stations along the reaches of the channel.[19,20] Except for reflections, waves do not travel off-shore from land. Therefore, the user can reduce the number of waves in this database by eliminating waves that are not possible due to blockage from land features. Wave directions should cover the full directional exposure of the channel.

Once the 20-year hindcast database has been transformed to the project site, a final postprocessing step is performed to compute statistical infor-mation. In this case, the user will want to minimize the number of individual cases according to combinations of wave height, period, and direction that are representative of the site and would significantly influence ship motions. Since deep draft ships are relatively large, one might want to limit the number of waves to those with longer wave periods and larger wave heights that would actually affect the vertical ship motions. One might think that since the largest vertical ship motions occur for wave periods that coincide with the natural oscillation periods in heave, pitch, and roll that are typi-cally of the order of 8 s or larger; it is reasonable to ignore wave spectra with peak wave periods below 5 or 6 s. Similarly, one might think that it is rea-sonable to ignore the insignificant ship motions due to waves with heights less than 0.5 to 1.0 m (1.6 to 3.3 ft). Of course, this would be dependant on the size of the ship(s) in the study.

However, a better procedure is to retain all of the data, but set up "bins" for the sorting that tend to isolate the "tails" data on the low and high ends of wave period and height. The coastal engineering design and analysis system (CEDAS) has a nearshore evolution modeling system (NEMOS) program that does sorting for joint distributions of wave period and height for fixed wave directions.[21] For instance, since the bins do not have to be evenly spaced, one can set up the lower and upper wave period and wave height bins to include relatively extreme or rare events in period and height. For instance, the lower wave period bin could include all wave periods from 0 to 5 s. The upper wave period bin might include all periods between 17 and 23 s, or whatever high period limit is contained in the dataset. Similarly for wave heights, bin size can be 0.5 m (2 ft) for the smaller waves with an upper bin to include all waves between 6 to 9 m (20 and 30 ft). Again, the number and increments for the bins should be based on the minimum and maximum values for the entire dataset. The NEMOS reports the distribu-tions in percent and number of occurrences. The program has the option to report the mean values for each bin, so these should be used in building the wave parameter statistics for generating the empirical directional wave spectra. The number of occurrences relative to the total provides the wave probabilities for CADET. A good rule of thumb is to ignore bins that have less than 0.05% of the total number of occurrences, as these represent very rare events on both low and high ends of the dataset. As mentioned previously, wave direction can be limited by the land features to include lower and upper directions that are possible. A fixed increment like 22.5 deg is a reasonable value although other values are also acceptable.

One of the main features in CADET is its risk-based predictions of UKC. The wave climatology for each reach is composed of the set of directional wave spectra and their associated probability of occurrence. This probability is converted into the number of days per year that each of the individual wave components contributes to the total wave environment. The total of all wave probabilities should equal 1.0 or 365 days. However, the total can be less than these values since missing values are assumed to represent wave conditions that are either (a) small and not a concern for safe navigation or (b) conditions that are very rare and do not represent more than 0.05% of the total number of observations. The small waves, or calm water, could represent a substantial part of the year, i.e., 103 calm water days at Pensacola, Florida.

6.3.2.2.3. CADET wave format

Figure 6.9 is an example of the wave record in CADET. Waves are listed for each reach and include (a) filename, (b) significant wave height $H_{1/3}$, (c) calculated significant wave height $H_{1/3}$, (d) wave period T_p, (e) mean wave direction θ_m, (f) wave probability of occurrence, (g) days per year for each wave, and (h) wave file location path. The significant wave height, modal period, and peak direction of each wave record in the wave climatology is then used to generate directional wave spectra. The calculated $H_{1/3}$ is a "check" on the input wave height that is calculated from the zero-moment wave height of the directional spectrum ordinates. The wave probability of occurrence is input with the individual directional spectra or can be manually entered after importing the wave file(s). The days/yr field is automatically populated based on these wave probabilities. All of the other parameters are input on the header line of the individual directional wave spectra files. The user can import these files individually or in batch import mode using a text file listing the filenames.

Figure 6.10 is an example of a directional spectrum in CADET. Individual spectral ordinates are listed in ft^2/Hz-rad as a function of wave frequency and wave direction. CADET allows up to 400 individual wave frequencies in units of Hz to define the directional wave spectrum. It requires a total of 24 wave directions in 15 deg increments from 0 to 345 deg, however.

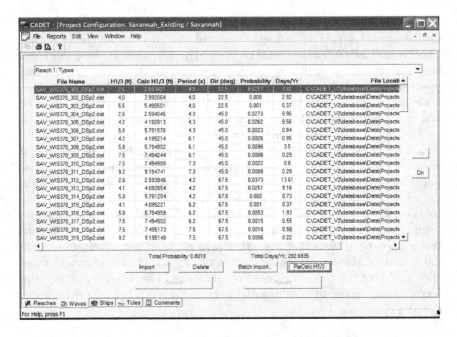

Fig. 6.9.   Example CADET wave record.

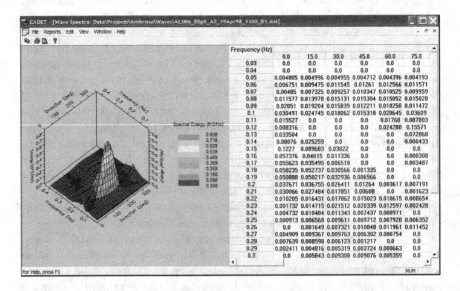

Fig. 6.10.   Example CADET directional spectrum.

6.3.2.2.4. Empirical directional wave spectrum

The directional wave spectrum $S(f,\theta)$ is typically created using empirical formulas for the frequency spectrum $S(f)$ and the directional spreading function $D(f,\theta)$ given by

$$S(f,\theta) = S(f)D(f,\theta) \tag{6.7}$$

It must satisfy the constraints that

$$S(f) = \sum_0^{2\pi} S(f,\theta)\Delta\theta \quad \sum_0^{2\pi} D(f,\theta)\Delta\theta = 1 \tag{6.8}$$

The TMA (Texel, MARSEN, and ARSLOE) is a shallow-water spectral form for $S(f)$ that characterizes waves generated in deepwater that have propagated into shallow water.[22] It is defined as

$$S(f) = \frac{\alpha g^2}{(2\pi)^4 f^5}\Phi(2\pi f, h)e^a\gamma^b \tag{6.9}$$

where α = Phillip's constant (defined below), g = gravitational acceleration, γ = peak enhancement factor, and the functions $\phi(2\pi f, h)$, a, and b are described below. The function $\phi(2\pi f, h)$ may be approximated as

$$\Phi(2\pi f, h) = \begin{cases} 0.5\omega^2 & \omega \leq 1 \\ 1 - 0.5(2-\omega)^2 & \omega > 1 \end{cases} \tag{6.10}$$

where $\omega = 2\pi f\sqrt{h/g}$. The functions a and b are given by

$$a = -1.25\left(\frac{f}{f_p}\right)^{-4} \quad b = \exp\left[\frac{-1}{2\sigma^2}\left(\frac{f}{f_p} - 1\right)^2\right] \tag{6.11}$$

where f_p = peak spectral frequency and σ is given by

$$\sigma = \begin{cases} 0.07 & f \leq f_p \\ 0.09 & f > f_p \end{cases} \tag{6.12}$$

Procedures for estimating α and γ are discussed by Hughes, Briggs *et al.*, Briggs, and U.S. Army Corps of Engineers.[23-26] The value for the Phillip's constant α is calculated using an iterative procedure that compares the target wave height to the calculated value for a required tolerance. The γ controls the width of the frequency spectrum (small values give broad frequency peaks and large values give narrow peaks). Considering sea and swell wave components, spectra with short (long) wave periods are generally

Table 6.2. TMA spectral peakedness γ and directional spreading n parameters.

T_p (s)	γ	n
≤ 10	3.3	4
11	4	8
12	4	10
13	5	12
14	5	16
15	6	18
16	6	20
17	7	22
18	7	26
19	8	28
20	8	30

broader (narrower) in frequency space. Table 6.2 provides some guidance in the selection of γ based on wave period. [27]

The TMA spectral parameters are the same as those in the more widely known JONSWAP (Joint North Sea Wave Program) spectrum, with the addition of $\varphi(2\pi f, h)$. The TMA spectrum reduces to the JONSWAP spectrum in the deep water limit.

The directional spreading function $D(f, \theta)$ can be approximated using several different empirical formulas. One of the simplest is a $\cos^n \theta$ directional distribution.[28] It is given by

$$D(f, \theta) = \frac{1}{C} \cos^n \left(\frac{\theta - \theta_m}{2} \right) \qquad (6.13)$$

where C = conversion constant to insure that the constraint in Eq. (6.8) is satisfied, θ_m = mean wave direction in deg, and θ = direction of the spectral component. The SCORES module requires that θ be in twenty-four 15-deg increments from 0 to 345 deg. The n is the multiplier that determines the width of the directional spreading. As for frequency spreading, a small n gives broader directional spreading and a large n gives narrower spreading. Guidance as a function of wave period is provided in Table 6.2.

6.3.2.3. *Ships*

Multiple ships and loading conditions can be associated with each reach in a project in CADET. Thus, several types of "design" ships can be included in the overall evaluation of UKC and channel accessibility. Since ship squat (i.e., sinkage and trim) is influenced by the channel cross-section, the user can accommodate changes in reach bathymetry by specifying different BNT squat files for different reaches for the same ship.

6.3.2.4. *Tides*

A recent addition to CADET allows for the use of an optional program that calculates the tidal duration for a project and includes the effect of the extra water depth in the calculated accessibility. This program can be used if the selected dredge depth from CADET does not allow for enough accessibility for a specific deep draft ship through the year. From the astronomic tide constants for a local port, the increased water level can provide additional accessibility by calculating a tidal window.

To generate the tide window table, two input data files and a tide prediction program are required. The first input file contains the 37 tidal constituents for orbital computations that are relevant for any port and required to calculate the tide level. The second input file contains the location-specific tidal constituents. The first line in this file contains the formal geographic name of the location. The second line is the name of the tide datum that is used in the tide level calculation. This can be mean lower low water (MLLW), as used in the United States, or mean low water springs (MLWS), as used in Europe, or any other datum. The third line contains three values. The first value is the difference between the water depth at mean sea level (MSL) at the port and the datum. The second value is the number of lines of tide constants that follow the header information. The third value is the difference, in number of hours, between the local standard time and Greenwich Mean Time (GMT) since the tide is calculated based on local time. The final value on this line is a unique four-digit number identifying the tide station for these tide constants. The succeeding lines in the table include three values: the index number for the tidal constituent, the amplitude of the tide constant in ft, and the phase of the tide constant for the specific port location.

These tables are then input to the tide prediction program that calculates the tides and the average tide range for 20 years between 1986 and 2006. Using this information, the number of hours the depth of the water in the channel is 0.3 to 3 m (1 to 10 ft) in 0.3 m (1 ft) increments above the datum is calculated. The tide prediction program then determines the number of days per year the water level is 0.3 to 3 m (1 to 10 ft) above the user-specified tide datum for 1 to 12 hours duration in 1 h increments.

The output from the tide prediction program information can then be applied to the channel accessibility plots by the user. If the deep draft ship requires a channel depth that is too deep to warrant the cost of dredging, then the use of tide levels can increase the accessibility. For example,

Table 6.3.   Number of days tide level is predicted to be 1, 2, 3, and 4 ft above MLLW for 2, 3, and 4 hours at a specific entrance channel.

Tidal window (h)	Water level above MLLW (ft)			
	1	2	3	4
2	360	357	356	324
3	360	357	344	290
4	360	357	309	235

suppose there is a deep draft ship that requires 15.2 m (50 ft) of water in the entrance channel to safely transit at all times and this channel is 10 miles long. The budget for dredging can only afford to dredge the channel so that the water in the channel is 14.3 m (47 ft) MLLW. That means the ship would need an additional 0.9 m (3 ft) to safely transit at all times. If the ship generally would be traveling at a speed of 5 kt in the channel, then it would need the extra 0.9 m (3 ft) of water for at least 2 h to safely transit the channel. The tide prediction program calculates that the tide would be 0.9 m (3 ft) higher than MLLW for a period of 2 h for 356 days per year. If the ship used the daily astronomic predictions, the channel and port would be accessible for almost every day during any given year. Table 6.3 is an example of the output from the tidal prediction option in CADET.

6.3.3. *Analyses*

Three analysis modules run in both foreground and background modes. As previously discussed, CADET utilizes external programs for computing underway sinkage and trim and ship motions due to waves. These programs are run in the background with all necessary input data files being created as needed from the ship and loading condition information managed by the CADET interface. Output from these programs is also parsed and manipulated as needed, creating the appropriate data files in the analysis of a project.

6.3.3.1. *Motion transfer analysis module 1*

The first analysis module uses the previously computed motion transfer functions for each selected ship and loading condition and the wave spectra data files defined for each project reach. From these input data files, the vertical motion variances (displacement σ_j and velocity σ_{vj} according to Eq. (6.3)) are computed for each ship speed, for both inbound and outbound transit directions at the motion control points defined for each ship. There is one motion variance output file created for each ship/loading condition, channel reach, and project depth combination. These motion variance output files are considered to be intermediate data and are not viewable directly within CADET.

6.3.3.2. *Risk analysis module 2*

The second analysis module performs the risk analysis and generates three output files for each combination of ship/loading condition (one with minimum clearance), channel reach, and water depth. The first output file is a summary file and includes limiting primary and alternative control point and corresponding clearance (includes sinkage and trim, motions allowance, etc.), and risk of touching a flat bottom and a bottom with random variation for each reach, water depth, wave condition, ship heading, and ship speed. The second output file contains the ship motion allowances (according to Eq. (6.4)) at all nine control points for each reach, water depth, wave condition, ship heading, and ship speed combination. The third and final output file is similar to the second output file. It contains the clearance values at all nine control points for each reach, water depth, wave condition, ship heading, and ship speed combination.

6.3.3.3. *Accessibility analysis module 3*

Finally, the third analysis program determines the accessibility for each channel reach and each ship/loading condition combination. Output files of days of accessibility are generated based on the reach risk level β of grounding for each particular reach and water depth. Note that β is not the same as the ship motions risk α, although they can have the same values (i.e., 0.01 for a 1 in 100 occurrences). The reach risk level β is the risk of a critical point of the ship touching the project depth in a particular reach, whereas the motions allowance α is the risk that the motions allowance would represent the largest excursion during 1 out of 100 transits (if $\alpha = 0.01$). Accessibility data is provided for primary and alternative control points for each ship loading condition, ship speed, inbound or outbound ship heading, and each channel reach.

6.3.4. *Results*

Multiple sets of analysis results may be logically associated with each project. This allows for different studies to be performed where project parameters are varied. This is useful for determining the sensitivity of the results to the varied parameters or for investigating design alternatives. When viewing the results of a study, CADET allows the user to change the data view between accessibility and risk results. Inbound and/or outbound transit directions may be selected and the view toggled between tabular or graphical presentations. Additionally, the data views may be filtered by the different channel reaches and by ship/loading conditions (or a composite of all ships).

Accessibility results are given in terms of the number of days of accessibility versus the range of project water depths. Risk results are somewhat more complicated in that they can be given in terms of multiple parameters including control points on a ship, ship speed, wave height, water depth, etc. Figure 6.11 illustrates two separate analysis results, one plotted in terms of the accessibility and the other in terms of the risk (as a 3D scatter plot of three parameters with the points color coded according to risk value).

While the viewing capabilities provided by CADET are useful for the analyst performing a study, they are not necessarily exhaustive. Therefore, CADET is capable of transferring tabular data from any data view to the Windows clipboard so that it may be pasted into other programs. Of course, output files can always be imported into spreadsheet programs like Excel. Images of all plots and graphical presentations of data may also be copied to the clipboard, allowing the user to paste report quality graphics into

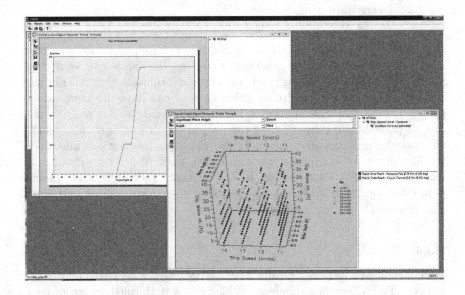

Fig. 6.11.   CADET analysis results plots.

Word or PowerPoint documents. CADET can also be used to create printed summary reports or listings of the ship and project data and results.

6.4. Barbers Point Harbor Validation Example

6.4.1. *Project Site*

Barbers Point Harbor (BPH) is located on the southwest coastline of Oahu (Fig. 6.12). It consists of a deep-draft harbor, barge basin, resort marina, and entrance channel. The deep-draft harbor has a depth of 11.6 m (38 ft) and an area of 0.364 km² (90 acres). A harbor extension was added in 1998. The entrance channel is approximately 1,160 m (3800 ft) long, 137.2 m (450 ft) wide, and 12.8 m (42 ft) deep. It is aligned in a northeast/southwest (harbor to offshore) direction. It has a shallow "trench" cross-section that can be approximated as an unrestricted or open channel. Stations are located every 30.5 m (100 ft) along the channel for ship positioning.

Fig. 6.12.　Barbers Point Harbor, HI.

"Station 0" is located at the offshore end of the channel, approximately 243.8 m (800 ft) from the 30-m (100-ft) depth contour. The 1-way channel is designed to accommodate containerships, tankers, and bulk carriers. Bulk carriers make approximately 120 transits per year.[29]

6.4.2. *Field Measurements*

In May 1999, six degrees-of-freedom (DOF) vessel motions were recorded using a differential global positioning system (DGPS) on two bulk carriers transiting the entrance channel at BPH.[30] The *World Utility* (WU) is typical of the vessel size and shape ($L_{OA} = 196.1$ m (643 ft), $B = 32.2$ m (106 ft)) of the Panamax bulk carriers (Fig. 6.13). The WU was light loaded on the inbound run and fully loaded on the outbound run. Table 6.4 summarizes the transit parameters including date, time, loading, and drafts at the bow and stern for inbound and outbound runs.

Fig. 6.13. *World Utility* bulk carrier during Barbers Point transit (model inset).

Table 6.4. *World Utility* transit parameters, Barbers Point, HI.

				Draft, m (ft)	
Transit	Loading	Date	Time	Bow	Stern
Inbound	Light	May 20, 1999	6:30	4.29 (14.1)	6.43 (21.1)
Outbound	Full	May 30, 1999	10:00	10.75 (35.3)	10.92 (35.8)

6.4.2.1. *DGPS sensors*

Three DGPS sensors were positioned on each vessel, one at the bow and two on either side of the bridge. The bow sensor was located about 1 m (3.3 ft) from the bow on the longitudinal centerline and the port and starboard sensors were located on both sides of the bridge wings, approximately 30.5 m (100 ft) forward of the stern. A land-based GPS receiver was used in this dual-frequency, DGPS system to insure "cm- level" accuracy of the measurements. All GPS data were referenced to a static ship survey. Engine speed and rudder commands were also recorded.

The vertical motions from the three GPS sensors were averaged to obtain the net vertical excursion for each ship transit. The location of the "average" of the three sensors is approximately equal to the center of gravity (CG). A digital filter was applied to the data with a cutoff of 30 s to separate low- and high-frequency components. The low-frequency portion corresponds to the vessel squat (sinkage and trim) and the high frequency to the wave-induced vertical motions from heave, pitch, and roll. Measured squat was less than 0.30 m (1 ft), and very transient due to the relatively short channel and varying ship speeds.

6.4.2.2. *Prototype wave conditions*

Environmental data were collected at 1 Hz for 20 min every 3 h using a PUV (pressure, u-, and v-velocity) gage located in the entrance channel and a NOAA deepwater buoy located approximately 363 km (196 nm) from the site. The pressure gage was positioned approximately 0.9 m (3 ft) above the bottom and 1.5 m (5 ft) from the south channel wall near station 2200.

Offshore wave heights varied from 1.2 to 2.7 m (4 to 9 ft), while inshore values were between 0.3 to 0.6 m (1 to 2 ft). Large offshore wave heights at the buoy do not necessarily correspond to large wave heights in the channel because of directional characteristics of the waves and nearshore processes such as diffraction, reflection, and wave breaking that strongly affect wave energy inside the channel. Visual observations during ship transits indicated that waves were primarily from the south with some directional spreading.

6.4.3. *Three-Dimensional Physical Model*

An undistorted, three-dimensional model of BPH (Fig. 6.14) was constructed at a model to prototype scale L_r = 1:75.[31] It included offshore contours to 30 m (100 ft) and wave absorption along the perimeter. Waves were generated with a unidirectional wavemaker. The total area of the model was over 1,022 m^2 (11,000 ft^2).

Fig. 6.14.   Physical model of Barbers Point Harbor and entrance channel.

6.4.3.1. *Model ship*

A 1:75 scale model of the WU was constructed for the laboratory study (see inset in Fig. 6.13). It had a draft of 10.9 m (35.8 ft) at a full-load displacement, and a draft of 6.5 m (21.3 ft) at lightship displacement. Remote-controlled features included forward and reverse speeds and rudder angle. The vessel was statically balanced with measured weight distribution and a digital level. Dynamic balance was performed with checks of roll and pitch natural frequencies.

6.4.3.2. *MOTAN inertial navigation system*

The wave-induced vertical motions of the model WU ship were measured using a motion analysis system (MOTAN) that contained accelerometers and angular rate sensors to measure the 6 DOF motions.[32] The MOTAN was installed on the scaled WU vessel near the CG of the model ship. The accuracy of MOTAN is 0.8 mm for data over a minimum duration of 10 (regular waves) up to 15 (irregular waves) times the encounter period.[33] The data acquisition system uses a data logger to record the analog voltage signals from the six sensors. A sampling rate of 50 Hz was used and synchronized with the wave measurements.

6.4.3.3. *Model waves*

Two wave gages were used to calibrate the wave conditions in the physical model: one at the 30 m (100 ft) contour and one in the channel at the same relative location as the field PUV at Station 2200. The PUV field data was used to simulate the laboratory wave conditions. Most of the observed directions during the field measurements were for waves from the south (i.e., 205 deg). Most of the ship transits occurred between the 3-h measurement times of the channel gage. Therefore, to properly bracket all possible wave conditions during the ship transit, the PUV gage data at the beginning and end of this 3-h interval were both simulated. Thus, two wave conditions were created for each of the inbound and outbound transits.

Table 6.5 lists target and measured peak wave period T_p and zero-moment wave height H_{m0} for the four waves for the WU wave conditions at gage 4. Figure 6.15 shows target and measured laboratory spectra for the sea-dominant DDU422 (wave ID) wave case, corresponding to the outbound WU transit. Most of the target wave conditions were multimodal because of wave transformation in the entrance channel due to wave shoaling, breaking, refraction, and diffraction. Since the field measurements were cut off at 0.23 Hz, the differences in measured H_{m0} are probably due to the inclusion of energy in the high-frequency range in the laboratory spectrum between 0.23 to 0.35 Hz. Because ship motions are not significantly affected by high-frequency energy, it was not felt to be necessary to perform any additional corrections to the control signals.

6.4.3.4. *Testing procedure*

For each wave case, two runs (i.e., one repeat) were made for each inbound and outbound transit at two different ship speeds and drafts. A slow and fast speed were selected to bracket the range of possible ship speeds. These speeds were calibrated with a series of speed trials. Typically, eight runs (i.e., 2 runs × 2 speeds × 2 transits) were made for each wave combination and ship draft.

Table 6.5. Laboratory wave conditions.

Direction	Loading	Case	Target, Gage 4		Measured, Gage 4	
			H_{m0}, m	T_p, sec	H_{m0}, m	T_p, sec
Inbound	Light	DDU512	0.41 (1.4 ft)	6.0	0.45 (1.5 ft)	6.0
		DDU522	0.43 (1.4 ft)	5.7	0.45 (1.5 ft)	5.7
Outbound	Full	DDU412	0.58 (1.9 ft)	6.6	0.75 (2.5 ft)	5.4
		DDU422	0.58 (1.9 ft)	6.7	0.65 (2.1 ft)	6.8

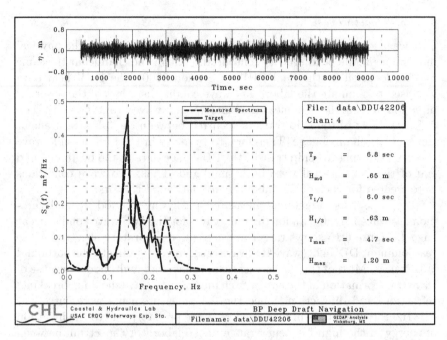

Fig. 6.15. Target and laboratory wave case DDU422 for $T_p = 6.7$ sec and $H_{m0} = 0.58$ m.

Inshore and offshore start/stop lines were located across the entrance channel for measuring ship motions. The inshore start/stop line was located near the shoreline at station 3430 and the offshore line near station 610. The beginning and ending times for each transit were recorded as the model ship crossed these lines. For inbound (outbound) transits, the ships stern (bow) was used as the reference to insure that the ship was completely within the entrance channel. The ship speed was calculated by dividing the distance between these start/stop lines of 858.4 m (2816 ft) prototype units by the travel time.

6.4.4. *Wave-Induced Vertical Motion Comparisons*

6.4.4.1. *Field and laboratory motion time series*

Time series of the wave-induced vertical motions at the ship's CG as a function of channel location x ($h_{Wa,CG}(x)$) were measured for the field transits and calculated for the laboratory runs. Figure 6.16 shows field and laboratory $h_{Wa,CG}(x)$ for the outbound run of the fully loaded WU from the harbor station 3800 to the offshore station 0. The match between the

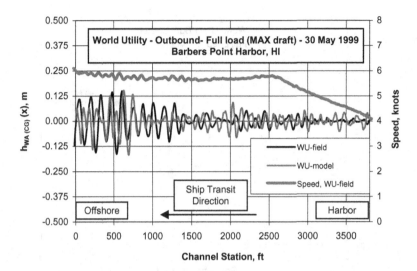

Fig. 6.16.   Field and laboratory vertical excursion and field ship speed, outbound *World Utility*.

field (blue) measurements and the laboratory (red) data is very good. The difference in phasing between the two time series is insignificant as each is only one realization of an infinite ensemble of possible outcomes because of the randomness of ocean waves. The important consideration is that both field and laboratory values are the same order of magnitude and show the same trends within the entrance channel.

The measured ship speed V_k for the field data is also shown on the right-hand side of this plot. It increases from a little over 4 to 6 kt as the ship leaves the entrance channel. The V_k for the model run was constant at about 4.2 kt. The F_{nh} for this range of ship speeds are in the subcritical range. Values are between $0.18 \leq F_{nh} \leq 0.28$ for $V_k = 4$ and 6 kt, respectively.

6.4.4.2. *Wave-induced vertical motion allowance*

Equation (6.4) defines the vertical motion allowance A_j used in the CADET program. An equivalent A_{CG} was calculated for field and laboratory measurements at the ship's CG location for comparison with the CADET predictions. Figures 6.17(a) and 6.17(b) show these comparisons as a function of ship speed for the outbound and inbound WU transits, respectively. Laboratory and CADET A_{CG} values are included for both wave conditions that bracket the outbound (i.e., DDU412 and DDU422) and inbound

M. J. Briggs, A. L. Silver and P. J. Kopp

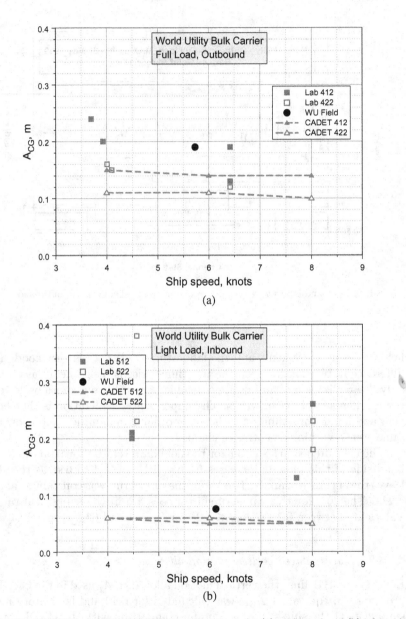

Fig. 6.17.   CADET vs. field and laboratory allowances for (a) outbound and (b) inbound *World Utility*.

(i.e., DDU512 and DDU522) transits. The field measurement (i.e., the large closed circle) is shown for reference.

In general, the slow-speed cases are affected by the waves more than the high-speed cases since their A_{CG} values are larger. For both outbound and inbound transits, CADET slightly underpredicts the field value of A_{CG}. For the fully loaded outbound transit, measured and laboratory values are in relatively good agreement, but the CADET predictions are 6 cm (2.4 in) smaller on average. For the light-loaded inbound transit, CADET and measured values are in excellent agreement, only 2 cm (0.8 in) smaller on average, but both are smaller than the laboratory values. Inbound (light-loaded) laboratory cases might be affected more by inaccuracies in the dynamic loading of the model ship relative to the field loading. Also, directional spreading in the waves affecting the field results is not necessarily present (other than reflections and transformations due to the channel) in the laboratory measurements since a unidirectional wavemaker was used. Overall, the agreement between laboratory, CADET, and measured values is very good considering the relatively small wave heights. Therefore, CADET is a very reasonable predictor of vertical ship motions.

6.5. Ambrose Channel Validation Example

6.5.1. *Channel Description*

Ambrose Channel is the main entrance channel to New York Harbor. It extends 16.4 km (8.8 nm) from the Sea Buoy at the seaward limit of the defined channel to the Narrows in the Lower Bay. Figure 6.18(a) is an overview map of the New York channel area and Figure 6.18(b) is a schematic of the Ambrose Channel sections in this study. Two bends in the channel at mile markers 7.5 (27.6 deg) and 8.5 (23.4 deg) align the channel from its approach course of approximately 296 deg into the Narrows. In 1998, it had a minimum depth of 13.7 m (45 ft) MLLW and a width of 610 m (2000 ft). However, due to commercial sand mining, depths up to 20.4 m (67 ft) MLLW are present. Ambrose Channel is flanked by East Bank Shoal on the east and West Bank and Romer Shoals on the west, all of which are approximately 3 to 4.5 m (10 to 15 ft) deep. Tides provide up to an additional 1.5 m (5 ft) of depth.

Traffic is primarily two-way for deep-draft vessels, with occasional overtaking of one vessel by another in the same direction. Inbound vessels reduce speed to 12–14 kt as they reach Ambrose Light Station (ALSN6). Pilots board within the Pilot Area between ALSN6 and the Sea Buoy. Navigation buoys 1 & 2, 3 & 4, and 5 & 6 delineate the channel boundaries over the

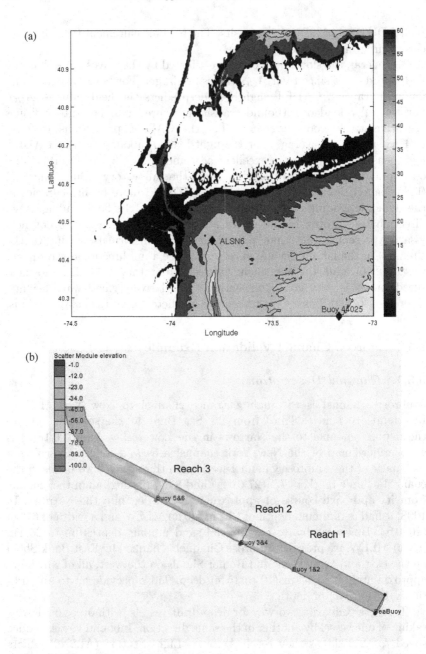

Fig. 6.18. Map and bathymetry of Ambrose Channel. Note that channel extends to ALSN6. Scale for bathymetry is in units of feet.

Table 6.6. Distances along Ambrose Channel.

From	To	Distance, km (ft)	Cumulative distance, nm
ALSN6	Sea Buoy	5.730 (18,800)	3.1
Sea Buoy	Buoys 1 & 2	3.658 (12,000)	5.1
Buoys 1 & 2	Buoys 3 & 4	2.256 (7,400)	6.3
Buoys 3 & 4	Buoys 5 & 6	1.768 (5,800)	7.2

portion of the channel of interest for this study. Table 6.6 lists the distances and cumulative distances in nautical miles (nm) between these buoys and stations.

6.5.2. *Field Measurements*

In April 1998, the U.S. Army Corps of Engineers conducted a DGPS survey in Ambrose Channel to measure vertical ship motions.[34] Three DGPS receivers were positioned on the bow and port and starboard bridge wings of 12 containerships and tankers for inbound and outbound transits. A reference GPS station was located on Coney Island. The *Bear G* (BG) tanker and *Frankfort Express* (FE) containership are typical of the ships in this study and are used as examples (Figs. 6.19(a) and 6.19(b), respectively). Table 6.7 lists ship parameters during inbound transits (full load conditions) that include date and time they passed the ALSN6, V_k, L_{pp}, B, T, and tide E_t. Note that times are Eastern Daylight Time (EDT).

Data were recorded at 1-Hz for 488-m-lengths (1,600-ft) centered at each of the three gated Buoy pairs 1 & 2, 3 & 4, and 5 & 6 (Fig. 6.18). DGPS accuracies of $\pm 3\,\mathrm{mm}$ were obtained for horizontal and vertical motions. Table 6.8 lists total depth including tide E_t (Eq. (6.1)) for both ships at each of the three reaches. An interesting note is that the inshore reaches are deeper than the offshore reaches due to sand mining.

6.5.3. *Waves*

Waves were not measured specifically for this study. However, unidirectional spectra were measured at ALSN6 (40.46 N 73.83 W), a distance of 5.1 nm from the first reach at Buoys 1 & 2 (Table 6.6). Initially, it was planned to pair these ALSN6 frequency spectra with measured directional spreading functions from the NDBC directional buoy Station 44025 (40.25 N 73.17 W) that is located 33 nm southeast of ALSN6 in deeper water (Fig. 6.18). The ALSN6 data were sampled from 0.03 to 0.40 Hz over the last 17 min of each hour and recorded at the end of that hour. The 44025 data were sampled

(a)

(b)

Fig. 6.19. Ambrose Channel ships (a) *Bear G* (BG) tanker and (b) *Frankfort Express* (FE) containership (Photos courtesy of Michael Schindler).

Table 6.7.   Ship parameters.

Ship ID	Type	ALSN6 Date	ALSN6 Time	V_k kt	L_{pp} m (ft)	B m (ft)	T m (ft)	E_t m (ft)
BG	T	19	1600	10.3	236 (774)	32.2 (106)	12.7 (41.6)	0.9 (2.9)
FE	C	19	700	13.3	264 (865)	32.2 (106)	11.3 (37.0)	0.3 (0.9)

Table 6.8.   Channel depths with tidal advantage.

		Depth including tide, m (ft)	
Reach	Buoys	*Bear G* (BG)	*Frankfort Express* (FE)
1	1 & 2	17.4 (57)	16.8 (55)
2	3 & 4	20.4 (67)	19.8 (65)
3	5 & 6	21.3 (70)	20.7 (68)

Table 6.9.   Wave parameters at Light Station ALSN6 and Buoy 44025.

Ship ID	Date	ALSN6 Time (GMT)	T_p (sec)	H_{m0}, m (ft)	Buoy 44025 Θ (deg)
BG	4/19	2000	7.69	0.62 (2.0)	153
FE	4/19	1100	8.33	0.72 (2.4)	147

from 0.03 to 0.35 Hz in the middle third of each hour (i.e., minute 20 to minute 40) and also recorded at the end of the hour. Therefore, the data from the 44025 is approximately 20 min earlier than that from the ALSN6. Both were recorded according to Greenwich Mean Time (GMT), which is 4 h ahead of EDT (i.e., GMT = EDT + 4).

After preliminary analysis, it was decided to use empirical spreading functions instead of the 44025 measured spreading due to (a) missing 44025 data, (b) differences in measurement times, (c) uncertainties in travel times, and (d) possible changes in spreading due to bathymetry and winds between the two locations. The use of the ALSN6 wave spectra ensured that the wave height and period experienced during each ship transit were accurately modeled. The empirical spreading function was calculated from 0.03 to 0.40 Hz using the $\cos^n\theta$ spreading function (Eq. (6.13)). The exponent n was selected according to Table 6.2, with a value of $n = 4$ since the wave period was less than 10 s. The value of θ_m was selected according to the closest record to the desired time from the 44025 analysis. Table 6.9 lists the date and GMT time pairings and the wave parameters for the waves.

Table 6.10.   Variation in wave height along Ambrose Channel reaches.

Reach	Buoy	Reduction Factor	Wave Heights, m (ft)	
			BG	FE
1	1 & 2	0.62	0.39 (1.3)	0.45 (1.5)
2	3 & 4	0.58	0.36 (1.2)	0.42 (1.4)
3	5 & 6	0.55	0.34 (1.1)	0.39 (1.3)

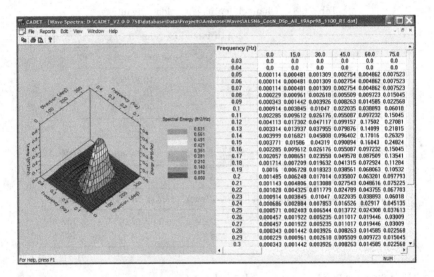

Fig. 6.20.   Transformed directional spectrum at Reach 1 for *Frankfort Express* (FE).

In an earlier study, the numerical model STWAVE was used to predict wave transformation from Buoy 44025 to the different navigation buoys along Ambrose Channel. Table 6.10 summarizes the reduction factors from ALSN6 to the three different reaches and the corresponding reductions in wave height for the two ships.[35] These reduction factors and reduced wave heights were used in this study. Finally, Fig. 6.20 shows the transformed directional spectrum in Reach 1 (Buoys 1 & 2) during the transit of the *Frankfort Express*.

6.5.4. *DGPS Calculations*

6.5.4.1. *DGPS vertical motions*

As mentioned previously, DGPS vertical motions z_i were recorded at the bow and stern every $\Delta t = 1\,$s during transits of the three gated reaches

Table 6.11. Transit duration and average speed in Ambrose Channel reaches.

Reach	Buoy	*Bear G* (BG)		*Frankfort Express* (FE)	
		Duration (s)	V_k (kt)	Duration (s)	V_k (kt)
1	1&2	94	10.1	64	14.8
2	3&4	90	10.5	64	14.8
3	5&6	90	10.5	65	14.6

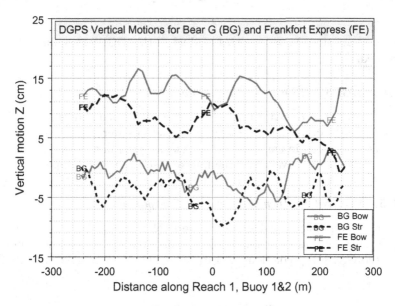

Fig. 6.21. DGPS Vertical motions in Reach 1 for *Bear G* (BG) and *Frankfort Express* (FE).

from $-244\,\mathrm{m}$ $(-800\,\mathrm{ft})$ to $244\,\mathrm{m}$ $(800\,\mathrm{ft})$. Table 6.11 lists durations T_d and average V_k at each reach from the vertical motion data. Figure 6.21 shows the vertical motions for the two ships. One way to estimate a total vertical motion is to take the difference between the largest crest and trough in the time series. However, this has no corresponding value in CADET to facilitate comparisons. Therefore, it was determined that the best "apples to apples" comparisons would be to calculate a vertical motions allowance A_j for the measured DGPS data using Eq. (6.4). The α risk parameter was set to 1.0 for both DGPS and CADET calculations since the data were actually measured during the transits. In reality, we are mainly comparing and validating the SCORES portion of CADET rather than the entire program.

6.5.4.2. *DGPS vertical velocities*

The first step in the calculation of the A_j for the DGPS data is to calculate the corresponding vertical velocities. These were calculated as

$$v_i = \frac{\Delta z}{\Delta t} = \frac{z_{i+1} - z_i}{1} \tag{6.14}$$

where a time step $\Delta t = 1\,\text{s}$ corresponding to this DGPS sampling interval was used.

6.5.4.3. *DGPS RMS vertical motion and velocity*

The next step was to estimate the RMS vertical motion and velocity values for the DGPS data. Sample standard deviations were calculated for the vertical motions σ_j in cm and vertical velocities σ_{vj} in cm/s at each reach. Table 6.12 lists these values at the bow and stern for both ships.

6.5.5. *Vertical Motion Allowance Comparisons*

The final step was to calculate the vertical motion allowance A_j using Eq. (6.4) (note that the variances were converted to meters) and compare it to the CADET predictions. A ratio R_j of CADET to measured DGPS

Table 6.12. Standard deviations for vertical motions (cm) and velocities (cm/s).

Reach	Buoy	Bear G (BG)		Frankfort Express (FE)	
		σ_j	σ_{vj}	σ_j	σ_{vj}
		Bow			
1	1 & 2	2.33	0.88	2.63	1.03
2	3 & 4	1.96	0.96	4.94	1.03
3	5 & 6	3.66	1.74	3.98	2.07
		Stern			
1	1 & 2	2.33	0.91	3.03	0.85
2	3 & 4	1.69	0.74	2.39	0.88
3	5 & 6	2.94	1.47	4.19	1.70

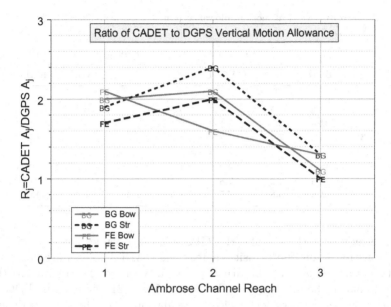

Fig. 6.22. Ratio of CADET to measured vertical allowance for *Bear G* (BG) and *Frankfort Express* (FE).

A_j was then calculated as

$$R_j = \frac{A_j^{\text{CADET}}}{A_j^{\text{DGPS}}} \tag{6.15}$$

Figure 6.22 is a plot of R_j for the bow and stern for both ships as a function of reach. Table 6.13 compares the measured DGPS values with the CADET predictions for A_j in cm. A value of $R_j = 1.0$ represents perfect correlation. Ratio values larger than 1.0 indicate overprediction by CADET.

In general, CADET overpredicted the DGPS measured values up to 2.4 times. The best agreement occurred in Reach 3 for both ships. The BG predictions varied from 1.1 for the bow to 1.3 for the stern. The FE predictions ranged from 1.0 (near exact match) for the stern and 1.3 for the bow. The largest overprediction of 2.4 occurred for the stern of the BG in Reach 2. The largest FE overprediction of 4.9 cm or 2.1 times occurred at the bow in Reach 1.

Although these ratios seem somewhat large, one must remember that we are talking about relatively small waves and correspondingly small allowances. The differences in UKC corresponding with these A_j allowances would be very small for these cases with the maximum actual variation of less than 4.9 cm (2 inches) for the bow of the FE in Reach 1. Another

Table 6.13. Vertical motion allowance (cm) comparisons for DGPS and CADET.

Reach	Buoy	Bear G (BG)			Frankfort Express (FE)		
		DGPS	CADET	Ratio	DGPS	CADET	Ratio
		Bow					
1	1&2	4.3	8.8	2.0	4.4	9.3	2.1
2	3&4	3.9	8.1	2.1	6.1	9.5	1.6
3	5&6	7.2	7.8	1.1	7.3	9.3	1.3
		Stern					
1	1&2	4.4	8.2	1.9	4.4	7.4	1.7
2	3&4	3.2	7.8	2.4	3.9	7.7	2.0
3	5&6	5.8	7.5	1.3	7.1	7.4	1.0

difference is that the calculated DGPS allowance values are discrete calculations based on one transit whereas the CADET values are based on a statistical representation with a probability of exceedance due to the uncertainty in the waves. As can be seen in Fig. 6.21, the measured DGPS values represented very few cycles of oscillation (typically 3 to 9) during the relatively short transits (typically 90 s) in each reach. The wave input in CADET consisted of the full 20-min records over the frequency range from 0.03 to 0.40 Hz. In the short 90-s transits it is impossible to know what portion of this frequency range the ships actually experienced. The ships might have been in relatively calm portions of the wave spectra relative to wave energy and corresponding wave heights.

6.6. Summary and Conclusions

This chapter summarized the CADET channel design tool for predicting the accessibility of deep-draft ships for a range of entrance channel depths. It is based on probabilistic risk analysis techniques to evaluate the accessibility of a series of channel reaches for multiple vessel geometries and loading conditions. It allows the planner to determine the "optimum" channel depth, given acceptance of a user-determined level of risk of grounding and reduced accessibility. The CADET model can be used while the channel is being designed to optimize its usefulness and cost–benefit ratio.

Two validation examples were presented comparing CADET vertical ship motion allowances with measured field and laboratory data for a range of ship and wave conditions. The allowances are the main component of the UKC predictions as they represent the ship motions due to wave-induced heave, pitch, and roll.

In the first validation example, field and laboratory measurements of wave-induced vertical motions at Barbers Point Harbor, HI, were compared. Prototype ship motions and environmental data were obtained in May 1999. These field measurements were reproduced in a controlled laboratory study with a model of the *World Utility* bulk carrier. In general, CADET predictions matched the field and laboratory data reasonably well, although slightly underpredicted.

In the second validation example, field measurements of wave-induced vertical motions at Ambrose Channel, New York, were presented and compared. Prototype ship motions were measured with DGPS in April 1998 along three reaches of 488-m (1,600 ft) length. Comparisons were made with CADET for the *Bear G* tanker and the *Frankfort Express* containership during inbound transits. In general, the CADET predictions overpredicted the field measurements as much as 2.4 times, but were still considered within reasonable limits. Explanations for these differences include (a) the wave heights were very small during the measurement period, (b) the reaches were not long enough to obtain statistically meaningful data, (c) the wave frequency content experienced by the ships in these short reaches is unknown, and (d) the calculated DGPS values are discrete calculations based on one transit whereas the CADET values are based on a statistical representation with a probability of exceedance due to the uncertainty in the waves. Additional validation examples are being investigated and will be reported in the future. Meanwhile, CADET predictions can be used with confidence that they have demonstrated reasonable agreement with measured laboratory and field data.

Acknowledgments

The authors wish to acknowledge the Headquarters, U.S. Army Corps of Engineers, and the Naval Surface Warfare Center, Carderock Division for authorizing publication of this book chapter. It was prepared as part of the Improved Ship Simulation work unit in the Navigation Systems Research Program. We would also like to acknowledge the assistance of Frank Santangelo, New York District, and Milton Yoshimoto and Tom Smith, Honolulu District, and Stan Boc, CHL.

References

1. B. McCartney, Report on ship channel design, ASCE Manual and Report No. 80 (1993).

2. USACE, Hydraulic design guidance for deep-draft navigation projects, Engr. Man. EM 1110-2-1613, Washington, DC (1995).

3. PIANC (Permanent International Association of Navigation Congresses). Approach channels: A guide for design, Supplement No. 95 (1997).

4. M. W. McBride, J. V. Smallman and S. W. Huntington, Guidelines for design of approach channels, presented at Ports 98, Long Beach, CA (1998), pp. 1315–1324.

5. ROM, Recommendations for maritime works — ROM 3.1-99, Puertos del Estado, Spain (1999).

6. A. L. Silver, Environmental monitoring and operator guidance system (EMOGS) for shallow water ports, presented at Ports '92, ASCE, Seattle, WA, July (1992), pp. 535–547.

7. M. J. Briggs, L. E. Borgman and E. Bratteland, Probability assessment for deep draft navigation channel design, *Coastal Engng J.* **48**, 29–50 (2003).

8. M. J. Briggs, A. L. Silver and L. E. Borgman, *Risk-Based Predictions For Ship Underkeel Clearance* (ICCE, San Diego, CA, 2006).

9. M. J. Briggs, I. Melito, Z. Demirbilek, F. Sargent, Deep-Draft entrance channels: preliminary comparisons between field and laboratory measurements, ERDC/CHETN-IX-7, U.S. Army Engineer Waterways Experiment Station, Vicksburg, MS, December (2001).

10. M. J. Briggs, A. L. Silver and P. J. Kopp, *CADET: A tool for predicting underkeel clearance in deep-draft entrance channels* (CEO6, Baltimore, MD, October 2004).

11. M. J. Briggs, P. J. Kopp, F. A. Santangelo and A. L. Silver, Comparison of CADET Vertical Ship Motions with DGPS in Ambrose Channel, presented at PORTS 2010, Jacksonville, FL, April (2010).

12. P. J. Kopp and A. L. Silver, Program documentation for the channel analysis and design evaluation tool (CADET), David Taylor Model Basin, Carderock Division, Naval Surface Warfare Center NSWCCD-50-TR-2005/004, May (2005).

13. M. K. Ochi, On prediction of extreme values, *J. Ship Res.*, 17 (1973).

14. P. Kaplan, Technical manual for SCORES II program — finite depth version, Hydrodynamics, Inc., Report No. 96-101A, June (1996).

15. P. Kaplan, Sample calculations and verification of SCORES II — finite depth program, Hydrodynamics, Inc., Report No. 96-101B, June (1996).

16. R. F. Beck, J. N. Newman and E. O. Tuck, Hydrodynamic forces on ships in dredged channels, *J. Ship Res.* **9**(3), September (1975).

17. E. O. Tuck, Shallow-water flows past slender bodies, *J. Fluid Mechanics* **26**(1), 81–95 (1966).

18. E. O. Tuck, Sinkage and trim in shallow water of finite width, Schiffstechnik, **14**(73), 92–94 (1967).

19. D. K. Stauble, J. E. Davis, J. Z. Gailani, L. Lin, E. F. Thompson, H. Benson, T. C. Pratt and M. P. Rollings, Construction, Monitoring and Data Analysis of a Nearshore Mixed-sediment Mound, Mobile Bay Entrance, Alabama.